Electronic Circuits by System and Computer Analysis

Electronic Circuits by System and Computer Analysis

Wendell H. Cornetet, Jr.
Professor of Electrical Engineering
The Ohio State University

Frank E. Battocletti
Associate Professor of Electrical Engineering
The Ohio State University

McGRAW-HILL BOOK COMPANY

*New York   St. Louis   San Francisco   Auckland   Düsseldorf   Johannesburg
Kuala Lumpur   London   Mexico   Montreal   New Delhi   Panama   Paris   São Paulo
Singapore   Sydney   Tokyo   Toronto*

**Electronic
Circuits by
System and
Computer
Analysis**

1 2 3 4 5 6 7 8 9 0 K P K P 7 9 8 7 6 5

This book was set in Modern 8A by Maryland Composition Incorporated.
The editors were Kenneth J. Bowman and Michael Gardner; the cover was designed by Joseph Gillians; the production supervisor was Sam Ratkewitch. The drawings were done by Vantage Art, Inc.
Kingsport Press, Inc., was printer and binder.

Library of Congress Cataloging in Publication Data

Cornetet, Wendell Hillis, date.
    Electronic circuits by system and computer analysis.

    1. Integrated circuits. 2. Electronic circuits. 3. Electronic data processing—Electronic circuit design. I. Battocletti, Frank E., joint author.
II. Title.
TK7874.C69   621.3815'3   74-13986
ISBN 0-07-013174-0

To Our Wives
June
and
Martha

Contents

Preface

Most electronics textbooks introduce basic electronic circuits separately as individual elements. A complete system that interconnects these elements is seldom introduced. The advent of the integrated circuit has changed this approach because complete systems may be obtained in one small package. Our fundamental method of presentation here is to analyze some integrated circuit systems and break them down into basic-component circuits. In this manner the reader can see how the building blocks are applied, learn the reasons for their use, and understand the operation of each block, as well as that of the system. The interface problems of interconnection of the basic circuits are automatically included. The importance of integrated circuits in present and future applications is continually implied.

The basis for computer analysis and design programs is included. However, the emphasis is placed on the fact that such programs exist. We show the usefulness and strength of their application to electronic circuits.

The introductory chapter treats an electronic amplifier as a "black box." The terminal characteristics are emphasized, and the application of two-port network expressions is included. The fact that voltage, current, and power gain can be obtained is explained. The frequency-response characteristics and the Bode plot of these responses are introduced. The general purpose of this chapter is to arouse an interest in the way these black boxes work and what is inside them.

The second chapter introduces the operational amplifier. The black-box approach is continued, but the basic operational amplifier applications and terms are defined. The importance of frequency response and stability is emphasized. After this discussion, the internal parts of operational amplifiers are ready to be explained.

Chapter 3 introduces the pn-diode equation and practical diode characteristics and applications. No attempt is made to include physical electronics in this or succeeding chapters. It is easy to describe diode operation qualitatively and to verify the diode expression experimentally. Thus, the diode equation is accepted as a proven postulate. The authors believe that physical electronics should be presented in a separate text. Once the student understands electronic-circuit applications from the terminal characteristics, he has a better reason to study the internal physics of devices. The physics is also more useful to device designers than to the majority of engineers and technicians who apply the devices in circuits.

Chapter 4 presents a description of the operation of bipolar and unipolar transistors. The small- and large-signal models of transistors have been included. The Ebers-Moll model is discussed because of its applicability to computer programs. The regions of operation are described, and practical limitations on voltage and current values are presented.

The basic circuit models for diodes and transistors having been developed, Chap. 5 applies these models to several linear integrated circuits. The complete IC is shown and is then broken down into component circuits. Upon completion of the analysis, the following circuits have been analyzed: the differential amplifier, emitter follower, common-emitter and common-base amplifiers, Darlington transistor pairs, phase splitter, push-pull amplifiers, and biasing circuits. Miller's theorem is developed and applied. The frequency response of an internally compensated IC is calculated. Finally, linear MOSFET and CMOS ICs are presented by analyses of gain and transfer functions.

Chapter 6 presents the types of computer programs for design or analysis that are available. Examples applying LINCAD and OSUCAD developed at The Ohio State University indicate the simplicity of using existing computer programs for the analysis of linear and nonlinear circuits. The methods employed in using the computer programs are developed in this chapter, and the complete programs are given in Appendices C and D for entering into a computer. Both passive and active elements can be included. The programs are intended for "on-line" use, and they are conversational in nature.

Chapters 7 and 8 introduce a simplified voltmeter system. A circuit has been chosen that employs a unijunction transistor and digital ICs. A description of UJT characteristics and negative-resistance operation follows. A voltage-to-frequency converter is developed, along with Miller and bootstrap sweep circuits. The basic type of digital or logic ICs are presented by transfer characteristics and circuit diagrams. The application of these ICs is presented by examining a pulse generator and astable, bistable, and monostable multivibrators. Counting and decoding circuitry are described, and the completed digital voltmeter is presented near the end of Chap. 8.

The final chapter applies the computer programs to circuit design. The advantages of computer use is emphasized here, and a few topics not previously presented in the text are included by example; e.g., the optimization of a cascode video amplifier by computer is illustrated.

The text is oriented toward junior-level college students and possibly for students of a technical college. A background including calculus and electrical circuits is assumed.

An explanation of the notations regarding capital- and lowercase symbols and subscripts as used in the circuit theory of this text is appropriate. Voltages and currents are written as follows: capital letters with capital letter subscripts denote dc values; capital letters with lowercase letter subscripts denote rms values; lowercase letters with capital letter subscripts represent total instantaneous values; and lowercase letters with lowercase letter subscripts represent instantaneous values referenced to the average of the waveform.

By selecting topics, this basic text can be applied for one or two quarters (or one semester) depending upon the pace desired by the user and the number of problems to be solved by the student. The text is also applicable to nonelectrical engineering students who have a similar basic background. We suggest the deletion of the following sections for short-term use or nonelectrical students: 3-9, 4-1.9, 4-1.10, 4-1.11, 4-1.12, 5-4 and all its subsections, 5-5, 5-6, and 5-7, 7-5.1, 7-5.2, and 8-8. The text should also be usable in any English-reading country.

The authors wish to thank the many people who helped in the discussions and preparation of this book. In particular, we wish to thank our department chairman, Professor M. O. Thurston for his encouragement, our department secretaries for their help, and especially our wives, June and Martha, for their patience and encouragement.

Wendell H. Cornetet, Jr.
Frank E. Battocletti

1

Linear Amplifiers

INTRODUCTION

Most electronics textbooks introduce basic electronic circuits separately as individual elements. A complete system which interconnects these elements is seldom introduced. The advent of the integrated circuit has changed this approach because complete systems may be obtained in one small package. Our fundamental method of presentation here is to analyze some integrated circuit systems and break them down into basic component circuits. In this manner we see how the building blocks are applied, the reasons for their use, and understand the operation of each block as well as that of the system. The interface problems of interconnection of the basic circuits in present and future applications will be continually implied.

This chapter treats an electronic amplifier as a "black box." The terminal characteristics are emphasized, and the application of two-port network expressions is included to show that voltage, current, and power gain can be obtained. The frequency-response characteristics and the Bode plot of these responses are introduced. The general purpose of this chapter is to arouse an interest in the way these black boxes work and in what is inside them. The chapter also introduces fundamental ideas which will be applied in later chapters.

Linear amplifiers consist of active circuits which include transistors connected in such a way that the power level of a signal can be increased without distorting the signal waveform. We shall define parameters of interest concerning linear amplifiers and analyze a few systems using linear amplifiers before we worry about what makes them operate.

1-1   THE BASIC CIRCUIT

A transformer can be applied to give a time-varying signal a voltage gain or a current gain, but it cannot produce a power gain. Active circuits employing devices, such as transistors, can convert dc power into increased signal power. A time-varying input signal controls the power-conversion process so that the output waveform corresponds to the input.

Consider an integrated-circuit linear amplifier as indicated by a triangle in Fig. 1-1. An integrated circuit (IC) contains active and passive components in a small package, and many designs are available. Two dc power sources are shown as $V+$ and $V-$ since most linear ICs must operate in this manner. Two input terminals are available, and these are indicated by the "$+$" and "$-$" signs. The signs denote that the input and output signals are either in phase ($+$ input) or 180° out of phase ($-$ input) when we consider sinusoidal waveforms. The signal source is connected to the inverting input in Fig. 1-1, and the noninverting input is grounded. We need to know the terminal characteristics of the IC before we can predict the amplification properties of the circuit shown. It is possible to replace the symbol for the amplifier by an equivalent circuit composed of linear

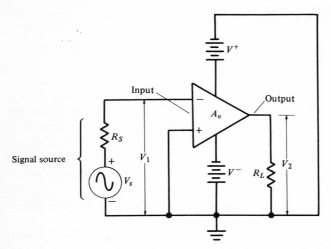

FIG. 1-1 An integrated-circuit amplifier with a voltage gain $\mathbf{A_v}$ connected between a signal source and a resistive load.

components R, L, C and voltage and current sources. The combination of these elements must yield the same characteristics at the network terminals as the actual circuit for the voltage and current levels of interest. The problem of determining equivalent circuits for amplifiers and devices is encountered frequently in the study of electronics.

1-2   CIRCUIT PARAMETERS

To find an equivalent circuit for the IC of Fig. 1-1, we shall utilize two-port parameters from circuit theory. Figure 1-2 defines the terminal voltage polarities and assumed current directions for two-port analysis. Since there are two voltages and currents and any two of the four parameters may be selected for independent variables, there are six possible sets of two-port parameters. Any one of the six sets of parameters could be used here to yield correct answers. Which set is applied is a matter of convenience, and we emphasize that these parameters can only be used for linear circuits—not including transformers—with the voltages and currents of Fig. 1-2 representing rms phasor values.

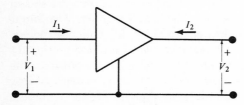

FIG. 1-2   The circuit defining two-port terminal voltages and currents.

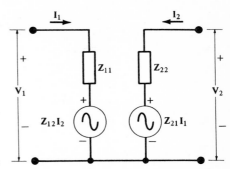

FIG. 1-3 The **Z**-parameter equivalent circuit.

For our analysis, assume that I_1 and I_2 are independent variables yielding a set of **Z** parameters defined as:

$$V_1 = Z_{11}I_1 + Z_{12}I_2 \qquad\qquad\qquad (1\text{-}1)$$

$$V_2 = Z_{21}I_1 + Z_{22}I_2 \qquad\qquad\qquad (1\text{-}2)$$

The **Z** parameters can be measured by applying a signal voltage at V_1 or V_2 and allowing I_1 or I_2 to be zero. With the output unloaded, a sinusoidal signal applied at V_1 will determine Z_{11} and Z_{21} if I_1 and V_2 are measured. Thus,

$$Z_{11} = \left(\frac{V_1}{I_1}\right)_{I_2=0} \qquad \text{and} \qquad Z_{21} = \left(\frac{V_2}{I_1}\right)_{I_2=0}$$

Similarly, with a signal source applied at V_2, measurement of V_1 and I_2 yield $Z_{12} = (V_1/I_2)_{I_1=0}$ and $Z_{22} = (V_2/I_2)_{I_1=0}$. Having determined the **Z** parameters, we obtain the equivalent circuit of Fig. 1-3 by inspection of Eqs. 1-1 and 1-2. There are two current-controlled voltage generators in this circuit in addition to two impedances.

We shall restrict our analysis to large-gain IC amplifiers operating at low frequencies. These ICs are so constructed that $Z_{12}I_2$ of Fig. 1-3 can be neglected, and we represent the circuit of Fig. 1-1, as shown in Fig. 1-4. The polarity of generator $Z_{21}I_1$ has been inverted to account for the phase inversion between the input and output voltages as required by the input connecting to the inverting terminal. Terminology commonly applied to linear amplifiers will be defined and expressions for the defined terms will be developed with the aid of Fig. 1-4.

The representation of Fig. 1-4 does not show how the $V+$ and $V-$ supplies of Fig. 1-1 are connected internally, but their effect is manifested in the **Z** parameters. Since $I_1 = V_1/Z_{11}$, the current-controlled voltage generator of Fig. 1-4 can be expressed in terms of the input voltage as $(-Z_{21}/Z_{11})V_1 = -A_{vo}V_1$ where A_{vo} is defined as the open-loop gain or gain with the output open circuit.

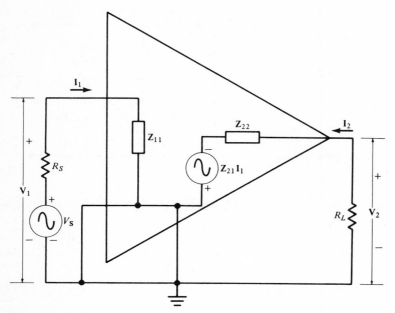

FIG. 1-4 The equivalent circuit for the amplifier of Fig. 1-1.

1-3 AMPLIFIER CHARACTERISTICS

The important characteristics of an amplifier are the voltage and current gains and the input and output impedances. These characteristics are defined for Fig. 1-4 as follows:

1. *Voltage gain* $\mathbf{A}_v$ is a ratio of the load voltage to the voltage at the input terminals or

$$\mathbf{A}_v = \frac{\mathbf{V}_2}{\mathbf{V}_1} \tag{1-3}$$

2. *Current gain* is a ratio of the output current to the input current or

$$\mathbf{A}_i = \frac{\mathbf{I}_2}{\mathbf{I}_1} \tag{1-4}$$

3. *Input impedance* is the ratio of the input terminal voltage to input current when the output is terminated in a load or

$$\mathbf{Z}_i = \frac{\mathbf{V}_1}{\mathbf{I}_1} \tag{1-5}$$

4. *Output impedance* is the ratio of an applied voltage at the output terminals to the current entering the output when the input is terminated with the

driving generator impedance. If we remove R_L and apply an output voltage $\mathbf{V}_2$, measure $\mathbf{I}_2$ with $\mathbf{V}_s = 0$, the output impedance is

$$\mathbf{Z}_o = \frac{\mathbf{V}_2}{\mathbf{I}_2} \tag{1-6}$$

It is important to note the method of measurement of $\mathbf{Z}_o$. If it is not convenient to remove R_L for this measurement, a voltage generator can be placed in series with R_L and the output impedance determined by measuring $\mathbf{V}_2$ and $\mathbf{I}_2$ and applying Eq. (1-6).

5. *Power gain* is the ratio of the power developed in the load P_o to the power into the input terminals P_i or

$$G_p = \frac{P_o}{P_i} \tag{1-7}$$

We are now ready to determine expressions for the terms defined above for the circuit of Fig. 1-4.

1-3.1 VOLTAGE GAIN

The total-circuit voltage gain differs from $\mathbf{A}_{vo}$, the open-circuit voltage gain of the amplifier. For our circuit,

$$\mathbf{A}_v = \frac{\mathbf{V}_2}{\mathbf{V}_1} = \frac{-\mathbf{I}_2 R_L}{\mathbf{V}_1} = \frac{-\mathbf{Z}_{21} R_L}{\mathbf{Z}_{11}(\mathbf{Z}_{22} + R_L)}$$

$$= \frac{-\mathbf{A}_{vo} R_L}{\mathbf{Z}_{22} + R_L} \tag{1-3a}$$

Unless $\mathbf{Z}_{22}$ is much less than R_L, we see that it has a pronounced effect on $\mathbf{A}_v$. We conclude that $\mathbf{Z}_{22}$ should be small for maximum voltage gain.

1-3.2 CURRENT GAIN

For our example circuit,

$$\mathbf{A}_i = \frac{\mathbf{I}_2}{\mathbf{I}_1} = \frac{\mathbf{Z}_{21}\mathbf{I}_1}{\mathbf{I}_1(\mathbf{Z}_{22} + R_L)} = \frac{\mathbf{Z}_{21}}{\mathbf{Z}_{22} + R_L} \tag{1-4a}$$

As R_L increases, $\mathbf{A}_i$ decreases in magnitude.

1-3.3 INPUT IMPEDANCE

By inspection of the circuit,

$$\mathbf{Z}_i = \frac{\mathbf{V}_1}{\mathbf{I}_1} = \mathbf{Z}_{11} \tag{1-5a}$$

If $\mathbf{Z}_{12}$ were not zero, then $\mathbf{Z}_i$ would depend on the output termination in addition to $\mathbf{Z}_{11}$. The input impedance is important since it acts as a load on the signal source.

1-3.4 OUTPUT IMPEDANCE

Assuming $\mathbf{V}_2$ to be a driving voltage and $\mathbf{V}_s = \mathbf{V}_1 = 0$, at the output $\mathbf{V}_2 = \mathbf{Z}_{22}\mathbf{I}_2 - \mathbf{Z}_{21}\mathbf{I}_1 = \mathbf{Z}_{22}\mathbf{I}_2 - \mathbf{A}_{vo}\mathbf{V}_1 = \mathbf{Z}_{22}\mathbf{I}_2$ and

$$\mathbf{Z}_o = \frac{\mathbf{V}_2}{\mathbf{I}_2} = \mathbf{Z}_{22} \tag{1-6a}$$

If $\mathbf{Z}_{12}$ were not zero, the R_S input termination would affect this expression for $\mathbf{Z}_o$.

1-3.5 POWER GAIN

Since the voltages and currents for our signals are rms values, the output power is given by $P_o = I_2{}^2 R_L$, and the input power is expressed as $P_i = I_1{}^2 \operatorname{Re}(\mathbf{Z}_{11})$ where $\operatorname{Re}(\mathbf{Z}_{11})$ means the real part of $\mathbf{Z}_{11}$. Applying Eq. (1-4a),

$$G_p = \frac{P_o}{P_i} = \frac{R_L I_2{}^2}{I_1{}^2 \operatorname{Re}(\mathbf{Z}_{11})} = \left| \frac{R_L \mathbf{Z}_{21}{}^2}{\operatorname{Re}(\mathbf{Z}_{11})(\mathbf{Z}_{22} + R_L)^2} \right| \tag{1-7a}$$

It is also convenient to show that

$$G_p = \frac{\mathbf{V}_2\mathbf{I}_2^*}{\mathbf{V}_1\mathbf{I}_1^*} = |\mathbf{A}_v| \cdot |\mathbf{A}_i| \tag{1-7b}$$

Unfortunately for us, $\mathbf{Z}$ parameters are not published in data sheets for IC amplifiers. Remembering that $\mathbf{Z}_o = \mathbf{Z}_{22}$, $\mathbf{Z}_i = \mathbf{Z}_{11}$, and $\mathbf{A}_{vo} = \mathbf{Z}_{21}/\mathbf{Z}_{11}$, we find that the parameters $|\mathbf{Z}_o| = R_{\text{out}}$, $|\mathbf{Z}_i| = R_{\text{in}}$, and $\mathbf{A}_{vo}$, the open-loop gain, are specified in the data sheets (see Chap. 3). The $\mathbf{Z}$ parameters were introduced here only as a mechanism for explaining the equivalent circuit.

1-4 DECIBEL MEASURE OF GAIN

The term *decibel* (abbreviated as dB) is applied to the gains we have defined. The fundamental definition of decibel involves the logarithm of a power ratio as,

$$dB = 10 \log_{10} \frac{P_2}{P_1} = 10 \log_{10} \frac{P_{\text{out}}}{P_{\text{in}}} \tag{1-8}$$

The term evolved from the fact that the human ear can detect a minimum power-level change of 1 dB. Its use in electronic circuits is a matter of convenience for representing large values of gain or gains of cascaded amplifiers. It has other uses as well, such as for attenuation in transmission lines.

A reference level should be stated when using the decibel for measuring a power-level change. If P_1 is 1 W (watt), then the unit should be abbreviated dBW; and if P_1 is 1 mW (milliwatt), the abbreviation should be dBm. Confusion sometimes arises when voltage gain of an amplifier is expressed in decibels. Assuming that the power levels in Eq. (1-8) are being delivered to resistors R_1

and R_2 by rms (root mean square) voltages V_1 and V_2, respectively, then we have

$$dB = 10 \log_{10} \frac{V_2{}^2/R_2}{V_1{}^2/R_1}$$

$$= 10 \log_{10} \frac{V_{out}{}^2/R_L}{V_{in}{}^2/R_{in}}$$

(1-9)

If R_1 and R_2 are equal, then

$$dB = 20 \log_{10} \frac{V_2}{V_1}$$

(1-10)

The form of Eq. (1-10) is often used for the voltage gain of an amplifier even though the input and load resistances may not be the same. If power gain is specified in decibels, Eq. (1-8) is implied.

Example 1-1 We are now ready to check some numerical values for the terms defined. For simplicity, we select an RCA-CA3000 IC amplifier. This amplifier can amplify direct as well as time-varying voltages, but we shall assume that sinusoidal signal frequencies of less than 10,000 Hz (hertz) are of interest. The CA-3000 is adaptable to several applications, but we shall consider its use as a voltage amplifier. The IC package has 10 leads for connections, not all of which will be used, and Fig. 1-5 shows the connections

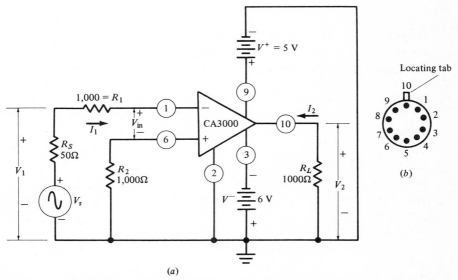

(a)

FIG. 1-5 (a) The amplifier circuit for Example 1-1 where the pin connections are given inside the circles; (b) the terminal connections looking at the bottom or pin side of the IC package.

we shall consider. The $V+$ and $V-$ voltages have been chosen so that the output dc voltage is zero when the circuit is connected as shown. Complete data sheets are available from the manufacturer or in data books which they publish (see Chap. 2). We shall not present all the data given for this IC since certain terms will be defined later. Typical specifications of interest are: (1) $\mathbf{Z}_{11}$, the input impedance, is resistive and equal to 195 kΩ (kilohms); (2) $\mathbf{Z}_{22}$, the output impedance, is also resistive and equal to 8 kΩ; (3) A_{vo}, the open-loop voltage gain, is 30 dB. The limits on peak output voltage for linear amplification are ±1.5 V (volts) which places a limit on the amplitude of input signal that may be applied.

The 1-kΩ resistors connected to pins 1 and 6 in Fig. 1-5 are necessary so that the IC will operate correctly. Since $\mathbf{Z}_{11} = 195$ kΩ, we see that $V_{\text{in}} = V_s\mathbf{Z}_{11}/(R_S + R_1 + R_2 + \mathbf{Z}_{11}) = V_s(195)(10^3)/197{,}050$ or V_{in} and $\mathbf{V}_1$ are approximately equal to V_s, and the high input resistance does not load the signal source.

The voltage gain of the circuit in Fig. 1-5 can now be obtained. Since the voltage gain is expressed in decibels, we apply Eq. (1-10) to obtain the magnitude of the open-loop gain or

$$30 = 20 \log_{10}\mathbf{A}_{vo} \qquad \text{from which } \mathbf{A}_{vo} = 31.7$$

From Eq. (1-3a),

$$\mathbf{A}_v = \frac{\mathbf{V}_2}{\mathbf{V}_1} = -\frac{\mathbf{A}_{vo}R_L}{\mathbf{Z}_{22} + R_L} = -\frac{(31.7)(1{,}000)}{8{,}000 + 1{,}000}$$

$$= -3.52$$

Because of the large value of the output resistance $\mathbf{Z}_{22}$, the gain is much lower for our value of load resistance than the amplifier is capable of delivering. If R_L is greater than 100 kΩ, then $\mathbf{A}_v$ is approximately equal to $\mathbf{A}_{vo}$.

The peak magnitude of a usable input signal follows from the specification that $V_2(\text{max}) = \pm 1.5$ V, or

$$V_{\text{in}}(\text{max}) = \frac{V_2(\text{max})}{\mathbf{A}_v} = \frac{\pm 1.5}{-3.52} = \mp 0.426 \text{ V}$$

and this value will be the approximate maximum value for V_s and V_1. If the input sinusoidal signal has a peak amplitude greater than this value, the output signal will be grossly distorted. It is always important to check the signal amplitude for linear amplifiers to be certain that the circuit is operating linearly.

From Eq. (1-4a), the current gain is

$$\mathbf{A}_i = \frac{\mathbf{Z}_{21}}{\mathbf{Z}_{22} + R_L} = \frac{\mathbf{Z}_{21}/\mathbf{Z}_{11}}{(\mathbf{Z}_{22} + R_L)/\mathbf{Z}_{11}}$$

$$= \frac{\mathbf{A}_{vo}}{(\mathbf{Z}_{22} + R_L)/\mathbf{Z}_{11}} = \frac{31.7}{(8{,}000 + 1{,}000)/195{,}000} = 687$$

For $V_2(\text{max}) = 1.5$ V, $I_2(\text{max}) = V_2(\text{max})/R_L = 1.5/1{,}000 = 1.5$ mA (milliamperes) or the maximum input current for linear operation is

$$I_1(\text{max}) = \frac{I_2(\text{max})}{\mathbf{A}_i} = \frac{1.5(10^{-3})}{687} = 2.18(10^{-6}) \text{ A (amperes)}$$

For power gain, we can use Eq. (1.7b) to obtain

$$G_p = |\mathbf{A}_v| \cdot |\mathbf{A}_i| = (3.52)(687) = 2{,}418 \text{ or } 33.8 \text{ dB}$$

This example illustrates that active ICs can develop voltage, current, and power gain.

1-5 AMPLIFIER FREQUENCY RESPONSE

Measurements on amplifiers indicate that the gain decreases at high frequencies, and for many amplifiers, the gain decreases at low frequencies as well. An amplifier may have a voltage-gain response as plotted in Fig. 1-6 in which there is a region where the gain is constant and the gain decreases from this constant value at both low and high frequencies.

Amplifier bandwidth is important since it indicates the range of signal frequencies in which gain is within specified limits. Bandwidth is defined as the difference between the two frequencies where the gain drops by 3 dB from the constant-gain value. From Fig. 1-6, the bandwidth is $(f_2 - f_1)$ Hz and the gain transfer function in the frequency domain can be represented as

$$\mathbf{A}_v(j\omega) = -\frac{k}{(1 - j\omega_1/\omega)(1 + j\omega/\omega_2)} \tag{1-11}$$

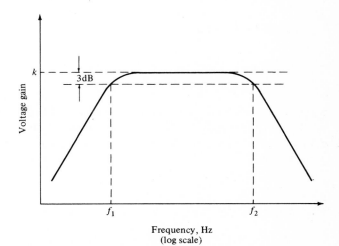

FIG. 1-6 The voltage gain response of an amplifier.

where $\omega_1 = 2\pi f_1$

$\omega_2 = 2\pi f_2$

$k = $ gain at midrange

The negative sign represents a 180° phase shift between input and output as in our earlier discussion. Usually, ω_1 and ω_2 are separated by several decades, and we can look at each end of the spectrum separately. For example, if ω_1 is much less than ω, but ω is in the range of ω_2, then

$$\mathbf{A}_v(j\omega) = -\frac{k}{(1 + j\omega/\omega_2)} \tag{1-12}$$

A similar expression is obtained at the low-frequency end of the spectrum (see Prob. 1-8).

1-6   CORNER PLOTS (BODE PLOTS)

The corner or Bode plotting technique is a fairly simple method for approximating the frequency response of a network or amplifier. Assuming a steady-state sinusoidal input signal, the technique involves the logarithm of the gain function and what happens as frequency changes. The current or voltage gain can be expressed in terms of decibels as in Eq. (1-10).

The high-frequency gain function as given by Eq. (1-12) may be written in polar form as

$$\mathbf{A}(\omega) = \frac{k}{(1 + \omega^2/\omega_2{}^2)^{1/2}} e^{j(\theta - \pi)} \tag{1-13}$$

Using decibel gain expressions, the magnitude function of Eq. (1-13) becomes,

$$A_{\mathrm{dB}}(\omega) = 20 \log \frac{k}{(1 + \omega^2/\omega_2{}^2)^{1/2}}$$

$$= K_{\mathrm{dB}} - 20 \log \left[1 + \frac{\omega^2}{\omega_2{}^2}\right]^{1/2} \tag{1-14}$$

where K_{dB}, defined by $20 \log k$, is the decibels value of the gain in the constant-gain region. Applying Eq. (1-14) and allowing the frequency to vary, for ω much greater than ω_2, ω/ω_2 is much larger than unity, and

$$A_{\mathrm{dB}}(\omega) \simeq K_{\mathrm{dB}} - 20 \log \frac{\omega}{\omega_2}. \tag{1-15}$$

For values of ω much less than ω_2, ω/ω_2 is much less than unity, and

$$A_{\mathrm{dB}}(\omega) \simeq K_{\mathrm{dB}} - 20 \log(1) = K_{\mathrm{dB}} \tag{1-16}$$

Equations (1-15) and (1-16) represent asymptotes of the actual frequency response and plot, as shown in Fig. 1-7, where the frequency has a logarithmic scale. These asymptotes intersect at the frequency where Eq. (1-15) is equal to Eq. (1-16) or where

$$K_{dB} = K_{dB} - 20 \log \frac{\omega}{\omega_2} \tag{1-17}$$

and we see that the intersection occurs at $\omega = \omega_2$. This frequency ω_2 is also called the half-power point and occurs where the voltage or current gain is 0.707 of the constant-gain value. Frequencies that are in the ratio of 2:1 are said to be separated by one octave, and if $\omega = 2\omega_2$ in Eq. (1-15), $A_{dB} = K_{dB} - 20 \log(2) = K_{dB} - 6$, or the slope of the high-frequency asymptote is approximately -6 dB/octave. If $\omega = 10\omega_2$ is substituted into Eq. (1-15), the gain is found to decrease by 20 dB, or the slope may also be plotted as -20 dB/decade.

The transition of the actual amplitude characteristic from one asymptote to the other is very simple in form. For the case where $\omega = \omega_2$ in Eq. (1-14), $A_{dB}(\omega_2) = K_{dB} - 20 \log(2)^{1/2} = K_{dB} - 3$ dB. At the break frequency or corner, the amplitude characteristic is 3 dB below the constant-gain asymptote. When $\omega = 2\omega_2$, the actual response from Eq. (1-14) is $A_{dB}(2\omega_2) = K_{dB} - 20 \log(1+4)^{1/2} = K_{dB} - 10 \log(5) = K_{dB} - 7$ dB. Therefore, one octave above the break frequency, the actual amplitude characteristic lies 7 dB below the constant-gain asymptote, and since the high-frequency asymptote has a slope of -6 dB/octave, the actual characteristic lies 1 dB below the -6 dB/octave asymptote at $\omega = 2\omega_2$. Finally, when $\omega = \omega_2/2$, the actual response from Eq. (1-14) is $A_{dB}(\omega_2/2) = K_{dB} - 20 \log(1 + \frac{1}{4})^{1/2} = K_{dB} - 10 \log(\frac{5}{4}) = K_{dB} - 1$ dB. Thus, one octave below the break frequency, the actual amplitude characteristic lies 1 dB below the constant-gain asymptote. These results are summarized in Fig. 1-8 where we see that it is only necessary to plot the asymptotes when we know K_{dB} and ω_2. Knowing the actual response at $\omega_2/2$, ω_2, and $2\omega_2$, as shown, makes sketching easy.

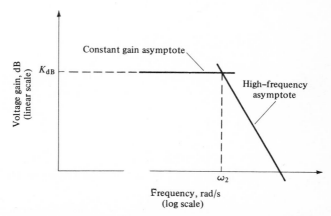

FIG. 1-7   The asymptotic voltage gain plot at high frequencies.

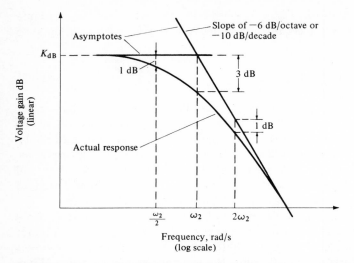

FIG. 1-8   Construction of the actual high-frequency response from the asymptotic plot.

The phase-shift characteristic has equally simple properties. From the $\mathbf{A}(\omega)$ expression in polar form, Eq. (1-13),

$$\theta(\omega) \;=\; -\tan^{-1}\frac{\omega}{\omega_2} \tag{1-18}$$

The phase angle from Eq. (1-18) is zero degrees when ω approaches zero and $-90°$ when ω approaches an infinite value, with the phase angle being $-45°$ when $\omega = \omega_2$. These asymptotes are joined at $\omega_2/10$ and $10\omega_2$ by a line as in Fig.

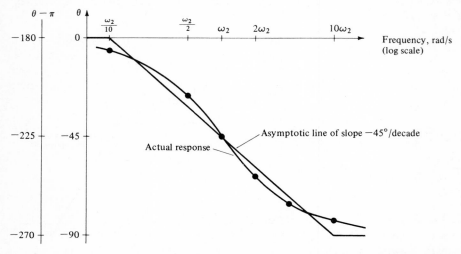

FIG. 1-9   Asymptotic and actual phase plot at high frequencies for an amplifier with gain transfer function given by Eq. (1-12).

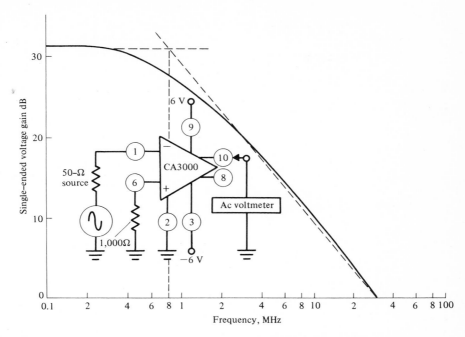

FIG. 1-10 Single-ended voltage gain of CA3000 as a function of frequency in test circuit shown. (By permission from RCA Corporation, Somerville, N.J.)

1-9, and we see that the slope is $-45°$/decade. The actual phase response is also sketched to indicate the rather small deviation away from the straight-line plot.

Example 1-2 Figure 1-10 shows the response of voltage gain versus frequency for the RCA CA3000 IC. What is the gain transfer function?

The corner-plot asymptotes have been sketched by extension of the response slopes as dashed lines in Fig. 1-10. We see that these asymptotes intersect at approximately $f_2 = 0.8$ MHz (megahertz). The constant gain at $f = 0.1$ MHz is 31 dB, or a magnitude of 35.5, so that a transfer function similar to Eq. (1-12) is

$$\mathbf{A}_{vo}(j\omega) = -\frac{35.5}{1 + j\omega/2\pi(0.8)(10^6)}$$

$$= -\frac{35.5}{1 + j\omega/5(10^6)}$$

The phase response would be as in Fig. 1-9 with $\omega_2 = 5(10^6)$ rad/s (radians per second). Since the CA3000 is a dc-voltage amplifier, the plot of Fig. 1-10 extends with constant gain to zero frequency.

Example 1-3 If the response function of an amplifer is given by

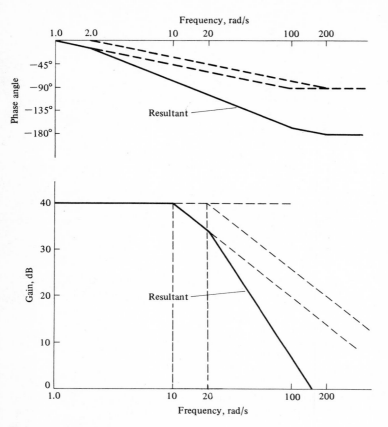

FIG. 1-11   The solution to Example 1-3.

$$\mathbf{A}_v(s) = \frac{100}{(1 + s/10)(1 + s/20)}$$

sketch the asymptotic gain and phase function for $s = j\omega$.

By inspection of $\mathbf{A}_v(s)$, we know that the break frequencies or poles occur at $s_1 = \omega_1 = 10$ and $s_2 = \omega_2 = 20$ rad/s. To solve this problem in asymptotic form, write $\mathbf{A}_v$ in decibels as

$$A_{v\mathrm{dB}}(\omega) = 20\log(100) - 10\log\left(1 + \left[\frac{\omega}{10}\right]^2\right) - 10\log\left(1 + \left[\frac{\omega}{20}\right]^2\right)$$

By comparison with Eq. (1-17), the response can be sketched in the same manner as Fig. 1-7 except that there is a second term to subtract from K_{dB}. We locate each breakpoint, and draw a line decreasing at -20 dB/decade and add the results algebraically.

The phase angle is given by

$$\theta(\omega) = -\tan^{-1}\left[\frac{\omega}{10}\right] - \tan^{-1}\left[\frac{\omega}{20}\right]$$

and we again see that a plot similar to Fig. 1-9 can be drawn for each break frequency and the two lines added algebraically. The results are shown in Fig. 1-11 where the dashed lines are the individual break-point asymptotes, and the solid line is the resultant.

REFERENCES

Angelo, E. J., Jr., "Electronic Circuits," McGraw-Hill Book Company, New York, 1964.
Radio Corporation of America, *RCA Integrated Circuits Application Note* ICAN-5030, New York, 1967.

PROBLEMS

1-1. Two other sets of small-signal parameters are commonly used to represent electronic amplifiers. These are the *h* or *hybrid* and *y* or *admittance* parameters. (Refer to Fig. 1-2.) These parameters are defined as:

$$\left.\begin{aligned}\mathbf{V}_1 &= \mathbf{h}_{11}\mathbf{I}_1 + \mathbf{h}_{12}\mathbf{V}_2 \\ \mathbf{I}_2 &= \mathbf{h}_{21}\mathbf{I}_1 + \mathbf{h}_{22}\mathbf{V}_2\end{aligned}\right\}$$

$$\left.\begin{aligned}\mathbf{I}_1 &= \mathbf{y}_{11}\mathbf{V}_1 + \mathbf{y}_{12}\mathbf{V}_2 \\ \mathbf{I}_2 &= \mathbf{y}_{21}\mathbf{V}_1 + \mathbf{y}_{22}\mathbf{V}_2\end{aligned}\right\}$$

Find the units associated with each parameter (ohms, mhos, etc.). Outline a procedure for measuring the parameters of the amplifier only in Fig. 1-1. Use a 1,000-Hz signal source, and consider that dc voltages may appear at the input and output terminals of the amplifier.

1-2. Apply the *h* parameters defined in Prob. 1-1 to the circuit of Fig. 1-1, and determine expressions for signal voltage gain $\mathbf{V}_2/\mathbf{V}_1$ and input impedance $\mathbf{V}_1/\mathbf{I}_1$. Do not assume that any *h* parameters are negligible. Draw an equivalent circuit derived from the defining equations of Prob. 1-1. Note that $\mathbf{Z}_{in}$ could become negative which is undesirable.

1-3. For the *h* and *y* parameters defined in Prob. 1-1, sketch equivalent circuits which represent these equations. Derive expressions for the *y* parameters in terms of the *h* parameters so that these two circuits are equivalent. *Hint*: Note that $y_{11} = \mathbf{I}_1/\mathbf{V}_1|_{V_2=0}$. Make $\mathbf{V}_2 = 0$ for the *h*-parameter circuit, and solve for the ratio of $\mathbf{I}_1/\mathbf{V}_1$. Follow a similar procedure to evaluate y_{12}, y_{21}, and y_{22}. This method may be applied to find the equivalence between any two sets of parameters.

1-4. Refer to Fig. P 1-4, and determine the **Z** parameters of Eq. (1-1) from the measured rms values indicated. The capacitors block direct current from the signal source and can be considered to have zero reactance to signal.

1-5. For the circuit of Fig. 1-4, $\mathbf{Z}_{11} = R_{in} = 195\text{k}\Omega$, $R_L = R_S = 5\Omega$, $\mathbf{Z}_{22} = R_{out} = 8\text{ k}\Omega$, $A_{vo} = 30\text{ d}B$. Determine:

    (*a*) $\mathbf{Z}_{21}$

    (*b*) $\mathbf{A}_v = \mathbf{V}_2/\mathbf{V}_1$

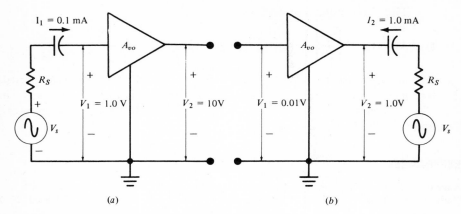

FIG. P1-4 Circuits for determining **Z** parameters in Prob. 1-4.

(c) $A_i = I_2/I_1$
(d) $G_p = P_o/P_{in}$ in dB
(e) V_{smax} if $V_{2max} = 1$ V

1-6. Express the following in decibels:

(a) Power gain = 100 for an amplifier
(b) Voltage gain = 100 for an amplifier
(c) Current gain = 100 for an amplifier
(d) 10 W in dBW
(e) 10 W in dBm

1-7. Repeat Example 1-1 in the text if

$$\mathbf{Z}_{11} = R_{in} = 1 \text{ k}\Omega, \mathbf{Z}_{22} = R_{out} = 100 \ \Omega, \text{ and } A_{vo} = 60 \text{ dB}$$

1-8. The low-frequency response of an amplifier is given by

$$\mathbf{A}(j\omega) = \frac{-K}{1 - j(\omega_1/\omega)} \qquad \omega \leq 10 \ \omega_1$$

Following the procedure of the text, plot the asymptotic and accurate gain and phase response.

1-9. Consider the amplifier circuit of Fig. 1-5 and the gain response of the CA3000 given in Fig. 1-10. Plot the circuit voltage gain (dB) as a function of frequency (MHz) over a frequency range of 0.1 to 30 MHz. The **Z** parameters are given in the text. Assume $\mathbf{Z}_{11}$ and $\mathbf{Z}_{22}$ remain constant. What maximum value does V_1 have at $f = 3$ MHz if the output is 1 V? What is the value of $\mathbf{Z}_{21}$ at this same frequency?

1-10. The measured gain- vs.-frequency response curve for an amplifier is shown in Fig. P1-10. (a) Write the gain response function. (b) Sketch the asymptotic gain and phase response plots. Assume that the amplifier is an inverting type having $-180°$ phase shift in the constant-gain region.

1-11. Plot the asymptotic and actual phase response for the amplifier of Example 1-2 in the text.

1-12. An amplifier response function is found to be

$$\mathbf{A}_v(f) = 1,000 / \left(1 + j\frac{f}{0.2}\right)\left(1 + j\frac{f}{2}\right)\left(1 + j\frac{f}{20}\right)$$

where f is in MHz. Sketch the asymptotic gain and phase plots for this function.

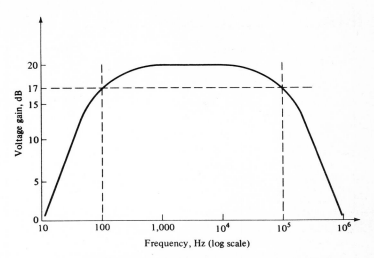

FIG. P1-10   Amplifier response curve for Prob. 1-10.

1-13. An amplifier is designed to have a double pole in its response function so that

$$\mathbf{A}_v(s) = \frac{10^5}{(2 + s)^2 \, (20 + s)}$$

Plot the asymptotic magnitude and phase functions for $s = j\omega$ rad/s.

2

Introduction to Operational Amplifiers

INTRODUCTION

The operational amplifier is a versatile linear amplifier which has numerous applications. The integrated circuit form of the operational amplifier has become very common. It appears that most new amplifier circuit designs will involve the use of linear ICs because of their low cost, small size, low power consumption, and the limited number of external components required.

We continue the black-box approach in this chapter so that basic operational amplifier (op-amp) terms and applications may be defined. The importance of frequency response, slew rate, and stability will be emphasized. After this discussion, we shall be ready to investigate the internal components of op-amps.

The equivalent circuit for the ideal operational amplifier is shown in Fig. 2-1 (without dc supply voltages) along with its specifications. Actual op-amps meet the specifications with varying degrees of success, but most of them are found to be satisfactory. It should be remembered that other connections are made to the amplifier as shown in Fig. 1-5. The "$-$" and "$+$" input terminals have the same connotation as discussed previously; i.e., when the $-$ terminal has a positive going input signal with respect to the $+$ terminal, v_{out} is negative going, etc. If only one terminal is indicated in a diagram, measurement is made with respect to a common terminal which is usually ground. The output of Fig. 2-1 is said to be "single ended."

The voltage at the $-$ terminal is measured with respect to the voltage at the $+$ terminal and both voltages may be referenced to ground rather than as shown in Fig. 2-1. In this case, the amplifier responds to the difference between the voltages at the $-$ and $+$ terminals. The amplifier is said to have "differential input."

If the amplifier has two output terminals as does the CA3000 shown by pins 8 and 10 in Fig. 1-10, the voltage on pin 8 is approximately equal and opposite in polarity to the voltage on pin 10. When the output is between pins 8 and 10 with no ground reference, "differential output" occurs by the same reasoning applied to the "differential input."

2-1 INVERTING AMPLIFIER

Consider the operational amplifier circuit in Fig. 2-1 which is connected as an "open-loop" amplifier. Since $R_{\text{in}} = Z_{11} \rightarrow \infty$, no current flows into the $-$ terminal which is defined as the summing point, and $I_1 = 0$ is referred to as the summing-point restraint. When $V_s = 0$, $V_o = 0$, and if V_s has any value, the output voltage would be infinite since $A_{vo} \rightarrow \infty$, and $A_{vo}V_s \rightarrow \infty$. In a practical op-amp, an internal limiting process occurs so that the output saturates at about the value of the supply voltage V^+ or V^-. We can see that high gain op-amps are difficult to use in open-loop configurations.

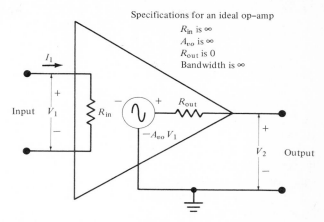

FIG. 2-1 An ideal op-amp equivalent circuit and its specifications.

In order to limit the gain, negative feedback is obtained by introducing a portion of the output signal voltage or current into the input circuit. For the feedback to be negative, the signal fed back must be out of phase with the input. If the two signals are in phase, oscillations can occur. The result is an output voltage with no input signal, an undesirable situation for amplifiers.

Figure 2-2 indicates a "closed-loop" circuit. It can be seen that some of the output voltage is returned to the inverting input through the feedback loop involving R_F. Since $R_{in} \rightarrow \infty$, no current flows into the summing point, and

$$I = \frac{V_1 - V_i}{R_1} = \frac{V_i - V_2}{R_F} \tag{2-1}$$

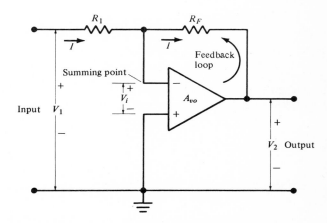

FIG. 2-2 The closed-loop inverting amplifier where R_1 is the input resistance and R_F is the feedback resistance.

By definition,

$$V_2 = -A_{vo}V_i \quad \text{or} \quad V_i = -\frac{V_2}{A_{vo}} \tag{2-2}$$

Letting $A_{vo} \to \infty$, V_i becomes negligible, and from Eq. (2-1),

$$\frac{V_1}{R_1} = -\frac{V_2}{R_F} \quad \text{or} \quad \frac{V_2}{V_1} = -\frac{R_F}{R_1} \tag{2-3}$$

In fact, it is only necessary that A_{vo} be large for Eq. (2-3) to be accurate. We now see that the gain is set by the ratio of R_F to R_1 and can be set precisely and independently of the exact value of A_{vo}. The importance of negative feedback in this case is evident, and there are many useful variations of this circuit.

Since V_i is very small, the summing point *appears* to be at ground potential. Thus the driving source is loaded by R_1, and R_1 is the input resistance of the closed-loop inverting amplifier meaning that we can set both the input resistance and gain of the circuit with R_1 and R_F.

Example 2-1 An op-amp has an open-loop gain A_{vo} which may be assumed infinite. Design a circuit having an input resistance of 1,000 Ω (ohms) and a voltage gain of (-100).

Since R_1 is the input resistance, R_1 of Fig. 2-2 is 1,000 Ω. The gain is

$$A_v = \frac{V_2}{V_1} = -\frac{R_F}{R_1} \quad \text{or} \quad -100 = -R_F/1{,}000 \quad \text{and} \quad R_F = 100{,}000 \ \Omega$$

In summary, for an ideal op-amp and as a good first approximation to a practical op-amp, the current into the summing point ("− terminal") is negligible and the differential input voltage is nearly zero when negative feedback is employed. We know that the differential input voltage cannot actually be zero, but it is so small compared to other circuit voltages that it is negligible.

2-2 NONINVERTING AMPLIFIER

It is sometimes necessary to maintain the same phase relations between input and output signals in control and instrumentation circuits. A noninverting amplifier may be easily analyzed using the results of the above summary.

Referring to Fig. 2-3, the open-loop gain A_{vo} will be considered infinite so that $V_i \to 0$ due to negative feedback introduced at the − terminal. Since V_i is approximately zero, the voltage at the − terminal with respect to ground must be equal to V_1, the applied voltage. No current flows into either amplifier terminal so that, by voltage division,

$$V_1 + V_i \simeq V_1 = \frac{V_2 R_1}{R_1 + R_F}$$

so that the gain is

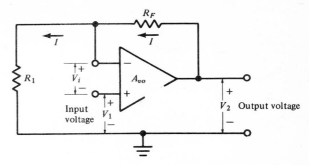

FIG. 2-3 A noninverting amplifier circuit.

$$\frac{V_2}{V_1} = \frac{R_1 + R_F}{R_1} = 1 + \frac{R_F}{R_1} \tag{2-4}$$

and the input impedance is infinite, the output impedance is zero, and the input and output signals are in phase.

Example 2-2   Determine the voltage gain of Fig. 2-3 if $R_1 = R_F = 10$ kΩ. What is the resulting gain if R_1 is removed ($R \rightarrow \infty$)?

From Eq. (2-4), $A_v = 1 + R_F/R_1 = 1 + 10$ kΩ/10 kΩ $= 2$. If $R_1 \rightarrow \infty$, then $A_v = 1$ and the circuit is said to be a "voltage follower." In the latter case, R_F is usually made to be zero ohms. It can be seen that the voltage follower will decouple a load from the driving source due to its high input impedance.

2-3   DIFFERENTIAL INPUT-DIFFERENTIAL OUTPUT AMPLIFIER

Servo motors, push-pull amplifier stages (see Chap. 5), and symmetrical transmission lines can be driven by the differential output of an amplifier. The circuit model for a differential input–differential output operational amplifier and its use as a differential balanced output amplifier are displayed in Fig. 2-4. Applying the usual input restraints that the same voltage V_i appear at the − and + input terminals and that there is no input current to these terminals, a voltage gain expression may be derived for Fig. 2-4b. At the input terminals,

$$I_1 = I_2 \quad \text{or} \quad \frac{V_1 - V_i}{R_1} = \frac{V_i - (V_o + V_p)}{R_F} \tag{2-5}$$

or

$$I_3 = I_4 \quad \text{or} \quad \frac{V_2 - V_i}{R_1} = \frac{V_i - V_p}{R_F} \tag{2-6}$$

Subtracting Eq. (2-6) from Eq. (2-5),

$$\frac{V_1 - V_2}{R_1} = \frac{-V_o}{R_F} \tag{2-7}$$

or

$$A_v = \frac{V_o}{V_2 - V_1} = \frac{R_F}{R_1} \tag{2-8}$$

We can see that the ground reference is not critical and that the input can be floating. The input can also be single ended by allowing either V_1 or V_2 to be zero. The values of V_i and V_p are determined by the internal circuitry of the op-amp and have not been determined in the above analysis.

For single-ended output with double-ended input, the circuit of Fig. 2-5 can be used. The reader can verify that Eq. (2-8) applies for this circuit when V_o is referenced to ground (see Prob. 2-5).

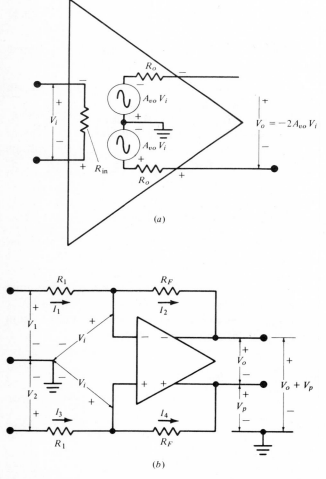

FIG. 2-4 (a) The equivalent circuit of the DIDO amplifier; (b) the differential balanced-output amplifier.

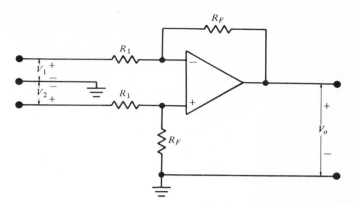

FIG. 2-5   The gain for this double-ended-input single-ended-output amplifier is given in Eq. (2-8).

2-4   ACTUAL OPERATIONAL AMPLIFIER

The manufacturer's data sheets for a $\mu A741$ operational amplifier are presented in Fig. 2-6. A study of these sheets can be very useful and informative. Some of the information presented will be discussed in this section, and the remainder will be used for reference in later sections.

The equivalent circuit shown on the first data page will be analyzed in Chap. 5. Most of the data is self-explanatory, and with the definitions given on the sixth data page, the electrical characteristics given on the second data page can be understood. It is suggested that the reader study these pages carefully to become familiar with the terms.

Several terms in the description have not been defined. These are:

Monolithic   all circuit elements are made on a single silicon wafer.

"Latch-up"   a saturation or nonfunctioning of an input circuit due to the input voltage exceeding the input-voltage-range specification.

Integrator   an application in which the output signal is the integral of the input signal (see sixth data page for the circuit).

Summing amplifier   with two or more input signals introduced through resistances to the — input of an inverting amplifier, the output is proportional to the sum of the input signals (see Prob. 2-4).

Short-circuit protected   the output can be short-circuited without damaging the IC.

Notice that the terms "input voltage range" and "input common-mode voltage range" have the same meaning. These terms specify the maximum voltage that can be applied to either input terminal or both terminals (with respect to ground) without causing the circuit to cease functioning as an amplifier.

ABSOLUTE MAXIMUM RATINGS

Supply Voltage	
Military (312 Grade)	±22 V
Commercial (393 Grade)	±18 V
Internal Power Dissipation (Note 1)	
Metal Can	500 mW
Ceramic DIP	670 mW
Silicone DIP	340 mW
Mini DIP	310 mW
Flatpak	570 mW
Differential Input Voltage	±30 V
Input Voltage (Note 2)	±15 V
Storage Temperature Range	
Metal Can, Ceramic DIP, and Flatpak	$-65°$C to $+150°$C
Mini DIP and Silicon DIP	$-55°$C to $+125°$C
Operating Temperature Range	
Military (312 Grade)	$-55°$C to $+125°$C
Commercial (393 Grade)	$0°$C to $+ 70°$C
Lead Temperature (Soldering)	
Metal Can, Ceramic DIP and Flatpak (60 seconds)	$300°$C
Mini DIP and Silicone DIP (10 seconds)	$260°$C
Output Short Circuit Duration (Note 3)	Indefinite

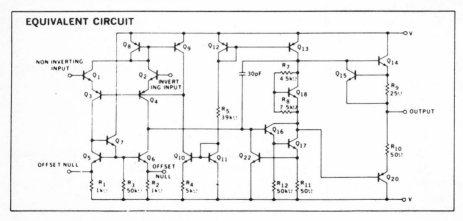

FIG. 2-6 (*By permission Fairchild Semiconductor Corporation, Mountain View, Calif.*)

ELECTRICAL CHARACTERISTICS ($V_S = \pm 15$ V, $T_A = 25°C$ unless otherwise specified)

PARAMETERS (see definitions)	CONDITIONS	MIN.	TYP.	MAX.	UNITS
Input Offset Voltage	$R_S \leq 10$ kΩ		1.0	5.0	mV
Input Offset Current			20	200	nA
Input Bias Current			80	500	nA
Input Resistance		0.3	2.0		MΩ
Input Capacitance			1.4		pF
Offset Voltage Adjustment Range			± 15		mV
Large-Signal Voltage Gain	$R_L \geq 2$ kΩ, $V_{out} = \pm 10$ V	50,000	200,000		
Output Resistance			75		Ω
Output Short-Circuit Current			25		mA
Supply Current			1.7	2.8	mA
Power Consumption	—		50	85	mW
Transient Response (unity gain)	$V_{in} = 20$ mV, $R_L = 2$ kΩ, $C_L \leq 100$ pF				
Risetime			0.3		μs
Overshoot			5.0		%
Slew Rate	$R_L \geq 2$ kΩ		0.5		V/μs
The following specifications apply for $-55°C \leq T_A \leq +125°C$:					
Input Offset Voltage	$R_S \leq 10$ kΩ		1.0	6.0	mV
Input Offset Current	$T_A = +125°C$		7.0	200	nA
	$T_A = -55°C$		85	500	nA
Input Bias Current	$T_A = +125°C$		0.03	0.5	μA
	$T_A = -55°C$		0.3	1.5	μA
Input Voltage Range		± 12	± 13		V
Common Mode Rejection Ratio	$R_S \leq 10$ kΩ	70	90		dB
Supply Voltage Rejection Ratio	$R_S \leq 10$ kΩ		30	150	μV/V
Large-Signal Voltage Gain	$R_L \geq 2$ kΩ, $V_{out} = \pm 10$ V	25,000			
Output Voltage Swing	$R_L \geq 10$ kΩ	± 12	± 14		V
	$R_L \geq 2$ kΩ	± 10	± 13		V

TYPICAL PERFORMANCE CURVES
312 GRADE

OPEN LOOP VOLTAGE GAIN AS A FUNCTION OF SUPPLY VOLTAGE

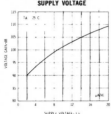

OUTPUT VOLTAGE SWING AS A FUNCTION OF SUPPLY VOLTAGE

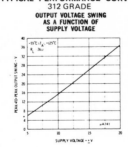

INPUT COMMON MODE VOLTAGE RANGE AS A FUNCTION OF SUPPLY VOLTAGE

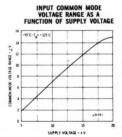

FIG. 2-6 Continued

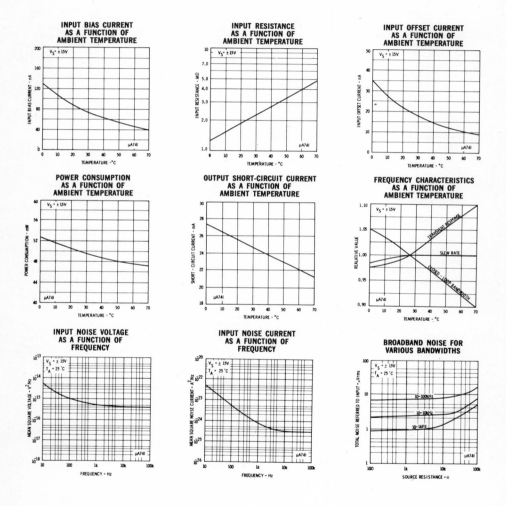

FIG. 2-6 Continued

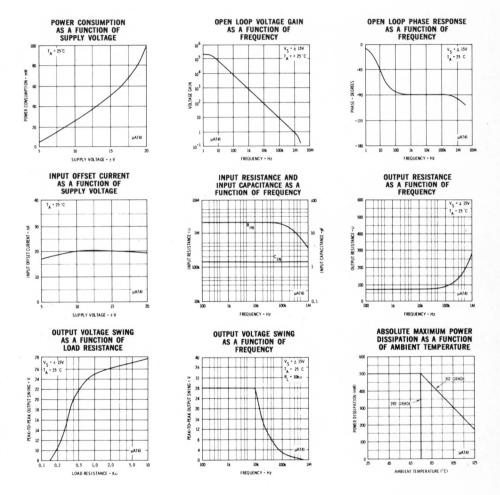

FIG. 2-6 Continued

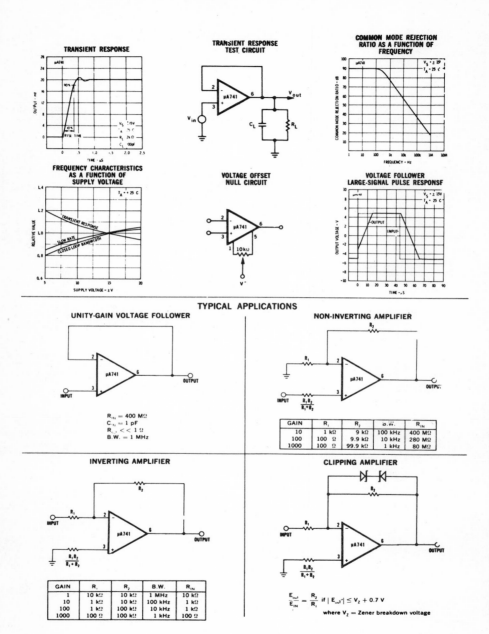

FIG. 2-6 Continued

TYPICAL APPLICATIONS

SIMPLE INTEGRATOR

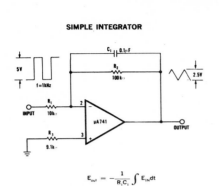

$$E_{out} = -\frac{1}{R_1 C_1} \int E_{IN} dt$$

SIMPLE DIFFERENTIATOR

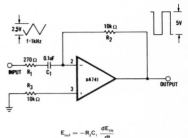

$$E_{out} = -R_2 C_1 \frac{dE_{IN}}{dt}$$

LOW DRIFT LOW NOISE AMPLIFIER

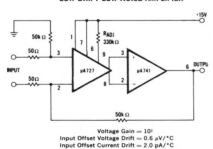

Voltage Gain = 10³
Input Offset Voltage Drift = 0.6 μV/°C
Input Offset Current Drift = 2.0 pA/°C

HIGH SLEW RATE POWER AMPLIFIER

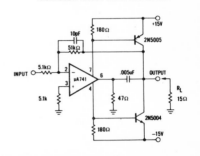

FAIRCHILD LINEAR INTEGRATED CIRCUITS μA741

DEFINITION OF TERMS

INPUT OFFSET VOLTAGE — That voltage which must be applied between the input terminals to obtain zero output voltage. The input offset voltage may also be defined for the case where two equal resistances are inserted in series with the input leads.

INPUT OFFSET CURRENT — The difference in the currents into the two input terminals with the output at zero volts.

INPUT BIAS CURRENT — The average of the two input currents.

INPUT RESISTANCE — The resistance looking into either input terminal with the other grounded.

INPUT CAPACITANCE — The capacitance looking into either input terminal with the other grounded.

LARGE-SIGNAL VOLTAGE GAIN — The ratio of the maximum output voltage swing with load to the change in input voltage required to drive the output from zero to this voltage.

OUTPUT RESISTANCE — The resistance seen looking into the output terminal with the output at null. This parameter is defined only under small signal conditions at frequencies above a few hundred cycles to eliminate the influence of drift and thermal feedback.

OUTPUT SHORT-CIRCUIT CURRENT — The maximum output current available from the amplifier with the output shorted to ground or to either supply.

SUPPLY CURRENT — The DC current from the supplies required to operate the amplifier with the output at zero and with no load current.

POWER CONSUMPTION — The DC power required to operate the amplifier with the output at zero and with no load current.

TRANSIENT RESPONSE — The closed-loop step-function response of the amplifier under small-signal conditions.

INPUT VOLTAGE RANGE — The range of voltage which, if exceeded on either input terminal, could cause the amplifier to cease functioning properly.

INPUT COMMON MODE REJECTION RATIO — The ratio of the input voltage range to the maximum change in input offset voltage over this range.

SUPPLY VOLTAGE REJECTION RATIO — The ratio of the change in input offset voltage to the change in supply voltage producing it.

OUTPUT VOLTAGE SWING — The peak output swing, referred to zero, that can be obtained without clipping.

FIG. 2-6 Continued

2-5 EFFECTS OF OFFSET

Two offset terms are defined to op-amps. The input offset voltage is the dc voltage which must be applied between the input terminals to obtain zero output voltage with no other input signal. The input offset current is the difference of the currents into the two input terminals with the output at zero volts, again with no input signal.

Direct-current amplifiers and integrators are circuits which are affected by input offset and offset changes with time, temperature, supply voltage, and common-mode voltage. An ideal op-amp has zero offset voltage and current, but a practical op-amp has both factors. The offset is referred to the input of the amplifier, and since it is independent of the gain of the amplifier, it can be compared with the input signal. Note on data page 5, Fig. 2-6, that an external circuit can be connected to null or zero the effect of offset voltage at fixed temperatures.

Figure 2-7 shows an equivalent circuit of an op-amp with the offset represented by a battery V_{os} in the input. The dc voltage V_{os} has the same magnitude as the input voltage which must be applied to obtain zero output voltage. Input bias current when the output is zero is defined as

$$I_{\text{bias}} = \frac{I_{D1} + I_{D2}}{2} \tag{2-9}$$

where I_{D1} and I_{D2} are the input currents, and the input offset current

$$I_{os} = I_{D1} - I_{D2} \tag{2-10}$$

The offsets V_{os} and I_{os} are of random and independent polarities and must be considered.

The circuit of Fig. 2-7 will have an output voltage because of V_{os} and I_{os}. The output voltage is given by

$$V_o = I_1 R_F + (I_1 - I_{D1})R_1 = I_1(R_1 + R_F) - I_{D1}R_1 \tag{2-11a}$$

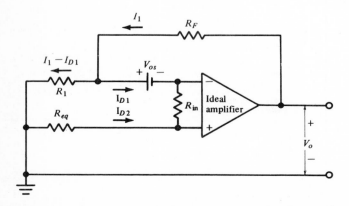

FIG. 2-7   The equivalent circuit of an op-amp with offset.

We also can write a Kirchhoff loop from the output through the input as,

$$V_o = [V_{os} + (I_{D1} - I_{D2})R_{in} - I_{D2}R_{eq}] + I_1 R_F \qquad (2\text{-}11b)$$

With the op-amp input voltage given by $(I_{D1} - I_{D2})R_{in}$, we see that

$$(I_{D1} - I_{D2})R_{in} = \frac{-V_o}{A_{v(ideal)}} \to 0$$

and Eq. (2-11b) becomes

$$V_o = V_{os} - I_{D2}R_{eq} + I_1 R_F$$

or

$$I_1 = \frac{V_o - V_{os} + I_{D2}R_{eq}}{R_F}$$

and Eq. (2-11a) becomes

$$V_o = \frac{(R_1 + R_F)}{R_F}(V_o - V_{os} + I_{D2}R_{eq}) - I_{D1}R_1 \qquad (2\text{-}11c)$$

Solving Eq. (2-11c) for V_o,

$$V_o\left[1 - \left(\frac{R_1}{R_F} + 1\right)\right] = (I_{D2}R_{eq} - V_{os})\left(1 + \frac{R_1}{R_F}\right) - I_{D1}R_1$$

$$V_o = -(I_{D2}R_{eq} - V_{os})\left(\frac{R_F}{R_1} + 1\right) + I_{D1}R_F$$

$$V_o = V_{os}\left(\frac{R_F}{R_1} + 1\right) + I_{D1}R_F - I_{D2}R_{eq}\left(\frac{R_F}{R_1} + 1\right)$$

$$= V_{os}\left(\frac{R_F}{R_1} + 1\right) + R_F\left[I_{D1} - I_{D2}R_{eq}\left(\frac{1}{R_1} + \frac{1}{R_F}\right)\right]$$

$$\qquad (2\text{-}11d)$$

We see that the effect of bias currents passing through external elements can be minimized by making $R_{eq} = 1/(1/R_1 + 1/R_F) = R_1 R_F/(R_1 + R_F)$. Then Eq. (2-11d) becomes

$$V_o = V_{os}\left(\frac{R_F}{R_1} + 1\right) + (I_{os})R_F \qquad (2\text{-}12)$$

It is common practice to make the impedance to ground at both amplifier input terminals the same. This was done above when we made $R_{eq} = R_1 \parallel R_F$, and the reason can be seen by comparing Eqs. (2-11d) and (2-12).

Note from Eq. (2-12) that the minimum-output offset voltage is obtained in a voltage follower, where $R_1 \to \infty$ (or removed) and $R_F = 0$. Then, $V_o = V_{os}$.

The effect of the static input offset voltage can be balanced out by nulling circuits which effectively introduce an input dc voltage to counteract the effects of V_{os} and I_{os}. However, drifts of V_{os} and I_{os} with temperature and other factors will appear as an input signal error.

2-6 SLEW RATE

The response of an op-amp to large change in input signal is not as fast as might be expected from the small-signal bandwidth due to the internal or external frequency compensation networks which must be employed. A large step change in signal forces the feedback to overdrive the input, and the clipped signal is integrated by the compensation capacitors resulting in an output voltage that rises at a fixed rate. The rate limit is the slew rate and determines the speed with which the amplifier can respond to large signals. The large-signal pulse response of the voltage follower on data page 5, Fig. 2-6, for the μA741 indicates the effect. From this figure, it can be seen that the slew rate for an increasing or decreasing signal is 0.5 V/μs (volts per microsecond).

An op-amp may be limited by its slew rate so that the maximum frequency for linear operation with sinusoidal signals is less than that predicted by the frequency response function. If the output is given by $v_2 = V_{2m} \sin(\omega t)$, the time rate of change of voltage is given by $dv_2/dt = V_{2m}\omega \cos(\omega t)$, and the maximum value is $(dv_2/dt)_{\max} = V_{2m}\omega_{\max}$. When dv_2/dt is the slew rate of an op-amp, the maximum operating frequency due to the slew rate is

$$f_{\max} = \frac{1}{2\pi V_{2m}} \frac{dv_2}{dt} \tag{2-13}$$

and it can be seen that $f_{\max}$ decreases as V_{2m} increases. Sinusoidal waves appear to be triangular when viewed on an oscilloscope if $f_{\max}$ is exceeded.

2-7 FREQUENCY RESPONSE AND STABILITY

Practical operational amplifiers have a gain which varies with the frequency of applied signals as shown by the open-loop voltage-gain curve of the μA741 in Fig. 2-6, data page 4. In application we find that some externally connected elements may cause an amplifier to become unstable. The cause of the instability can be investigated by considering the gain-frequency response function.

In order to discuss stability, it is necessary to derive an expression for the gain of the feedback circuit of Fig. 2-2 when A_v is finite while Z_{in} is very large and Z_o is small.

Starting with Eq. (2-1), we rewrite it as

$$\frac{V_1}{R_1} = \frac{V_i}{R_1} + \frac{V_i - V_o}{R_F} = V_o \left[\frac{V_i}{V_o R_1} + \frac{(V_i/V_o) - 1}{R_F} \right] \tag{2-14a}$$

Defining $(+A_{vo}(\omega)) = +(V_o/V_i) = -\,|\,A_{vo}(\omega)\,|$,

$$\frac{V_1}{R_1} = \frac{V_o}{-\,|\,A_{vo}(\omega)\,|}\left(\frac{1}{R_1} + \frac{1 + |\,A_{vo}(\omega)\,|}{R_F}\right) \tag{2-14b}$$

Solving for V_o/V_1,

$$\frac{V_o}{V_1} = \left(\frac{R_F}{R_F + R_1}\right)\left(\frac{-\,|\,A_{vo}(\omega)\,|}{1 + \dfrac{R_1\,|\,A_{vo}(\omega)\,|}{R_1 + R_F}}\right) \tag{2-14c}$$

The second denominator term of Eq. (2-14c), defined as the loop gain, is the term of interest regarding stability. The gain $A_{vo}(\omega)$ is complex in that it has a magnitude and phase angle associated with it. (As an example see Figs. 1-8 and 1-9.) If the term $R_1\,|\,A_{vo}(\omega)\,|/(R_1 + R_F)$ in Eq. (2-14c) is unity (or greater) in magnitude and becomes negative (or the phase angle of $A_{vo}(\omega)$ becomes 180° but not asymptotically), then the denominator of Eq. (2-14c) becomes zero indicating that $V_o/V_1 \to \infty$. This is the condition for oscillations which means we will have an output signal with no input, and the circuit ceases to be an amplifier. If $R_1\,|\,A_{vo}(\omega)\,|/(R_1 + R_F) < (-1)$, oscillations build up until limiting in the internal amplifier circuit occurs. The oscillating condition is to be avoided in amplifier applications.

In some applications, R_F and R_1 of Fig. 2-2 are replaced by more general impedances Z_F and Z_1, respectively (see data page 6 of Fig. 2-6 for an integrator and differentiator circuit). Then, for Fig. 2-2 with $V_{os} = 0$ and $R_{in} \to \infty$, Eq. (2-14) becomes

$$\frac{V_o}{V_1} = \left(\frac{-Z_F}{Z_F + Z_1}\right)\left[\frac{+\,|\,A_{vo}(\omega)\,|}{1 + \dfrac{|\,A_{vo}(\omega)\,|}{1 + (Z_F/Z_1)}}\right] \tag{2-15}$$

By sketching the Bode plot of the amplitude and phase of the closed-loop function $(|\,A_{vo}(\omega)\,|/[1 + Z_F/Z_1])$, we can quickly determine if the magnitude is greater than zero decibels when the phase angle passes through 180° so that the amplifier is unstable. We state without proof that the closed-loop function must have at least three closely spaced poles to be unstable.

Example 2-3 The loop gain function of an amplifier is found to be

$$\frac{|\,A_{vo}(f)\,|}{1 + Z_F/Z_1} = \frac{10^3}{[1 + j(\,f/0.2)\,]^2[1 + j(\,f/20)\,]}$$

where f is in MHz.

The open-loop gain of $A_{vo}(f)$ is given by

$$|\,A_{vo}(f)\,| = \left|\frac{1{,}000}{\left(1 + j\dfrac{f}{0.2}\right)\left(1 + j\dfrac{f}{2.0}\right)\left(1 + j\dfrac{f}{20}\right)}\right|$$

and we see that

$$1 + \frac{Z_F}{Z_1} = \frac{1 + jf/0.2}{1 + jf/2}$$

Determine whether or not the system is stable.

The asymptotic magnitude and phase plots of the loop gain are drawn in Fig. 2-8. The double pole at $f = 0.2$ MHz causes the -40 dB/decade drop in the magnitude and $-90°$/decade decrease in phase. The final decrease in phase beyond 2 MHz is due to the pole of the amplifier response function at $f = 20$ MHz. We see that the gain is greater than zero decibels when the phase is 180°. Solving the loop gain equation for the exact frequency at which the imaginary part of the denominator is zero, we find that $f = 2.84$ MHz when the phase of the function is 180° and the system is unstable.

In the study of feedback control systems, there are many good methods, one of which is the root-locus plot, that we may apply to determine the stability of a feedback system. Our purpose is not to present these methods, but we wish to emphasize that the frequency characteristics of an op-amp are of concern. One reasonable method for a quick check of stability involves the asymptotic plot of the function $(1 + Z_F/Z_1)$ on the asymptotic open-loop function. If these two asymptotic functions intersect with a slope equal to or greater than 12 dB/octave, the system will be unstable. The plot of these asymptotes and their resultant for $(1 + Z_F/Z_1) = (1 + jf/0.2)^2(1 + jf/2)$ as introduced in Example 2-3 is sketched in Fig. 2-9. We see that the intersection is at a slope of 12 dB/octave, and the system is unstable as shown in the example.

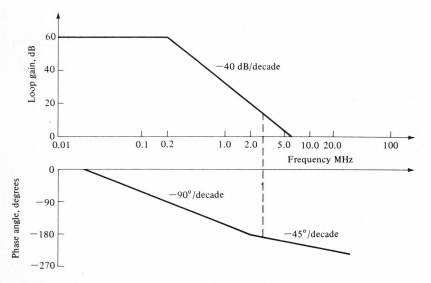

FIG. 2-8 Asymptotic magnitude and phase plots of the loop gain of Example 2-3.

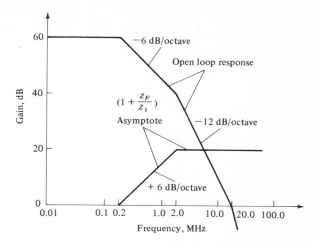

FIG. 2-9 The asymptotic plots of $(1 + Z_F/Z_1)$ and the open-loop response to determine stability.

The μA741 IC has been internally compensated (see Chap. 5) so that the gain is less than unity before the phase angle becomes 180°. For IC amplifiers which are not internally compensated, data sheets present gain frequency plots for external elements added to specified terminals, and the response may be "shaped" as desired.

We have introduced many factors which are of interest in amplifiers, and, hopefully, our curiosity is aroused as to how an amplifier operates. We start by qualitatively describing electron devices in the next chapter.

REFERENCES

Burr-Brown Research Corporation, "Handbook of Operational Amplifier Applications," 1963. Tucson, Ariz.

Giles, James N., "Fairchild Semiconductor Linear Integrated Circuits Applications Handbook," Fairchild Semiconductor, Mountain View, Calif., 1967.

Radio Corporation of America, *RC Integrated Circuits Application Note* ICAN-5290, Princeton, N. J., 1967.

PROBLEMS

2-1. An operational amplifier which can be represented as in Fig. 2-1 is to be used in Fig. P2-1. Resistor R_3 is adjusted to minimize the effects of the input bias current and has a value of 900 Ω. Determine R_1 and R_2 so that the circuit has a gain of (-10). What is the input resistance, V_1/I_1? Assume $R_{in} \rightarrow \infty$.

2-2. The noninverting amplifier of Fig. 2-3 is driven from a source having an internal resistance of 1,000 Ω. To minimize error due to input bias current, R_1 in parallel with R_F is made equal to the driving-source resistance. Design an amplifier to have a gain of 1.5.

2-3. A "unity-gain buffer" or voltage follower is shown in Fig. P2-3. Minimum error resulting from input bias current is obtained by making R equal to the driving-source resistance. Determine R, the resistance seen by the driving source, and the gain of the amplifier.

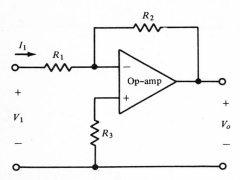

FIG. P2-1 Circuit for Prob. 2-1.

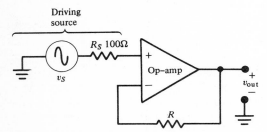

FIG. P2-3 Circuit for Prob. 2-3.

2-4. (*a*) A summing amplifier is shown in Fig. P2-4. Since no current flows into the summing point, no input current flows through R_5, and the summing point is at virtual ground potential. Show that the *inverted* output is proportional to the sum of V_1, V_2, and V_3, and determine the proportionality factors.

(*b*) To minimize bias current error, R_5 is made equal to the parallel combination of R_4 and R_1, R_2, and R_3 in parallel. If $R_1 = R_2 = R_3 = R_4 = 1,000\ \Omega$, determine V_{out} in terms of V_1, V_2, and V_3. Also determine the value for R_5.

2-5. Verify that Eq. (2-8) applies to Fig. 2-5.

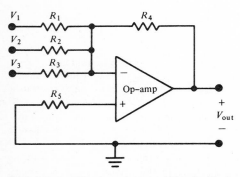

FIG. P2-4 The summing amplifier for Prob. 2-4.

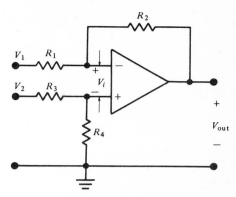

FIG. P2-6   The difference amplifier for Prob. 2-6.

2-6. A difference amplifier is indicated in Fig. P2-6. The circuit is useful as a computational amplifier in making a differential to single-ended conversion or in rejecting a common-mode signal.

(a) Noting that no current flows into the op-amp and that $V_i \cong 0$, show that

$$V_{\text{out}} = -\frac{R_2}{R_1} V_1 + \frac{V_2 R_4}{R_1} \frac{R_1 + R_2}{R_3 + R_4}$$

Note that if $R_1 = R_3$ and $R_2 = R_4$, $V_{\text{out}} = (R_2/R_1)(V_2 - V_1)$.

(b) Determine the input resistance at each input assuming the second terminal to be at ground. Note that these resistances are unequal. (For minimum error due to input bias current, R_1 and R_2 in parallel equals R_3 and R_4 in parallel.)

2-7. A practical differentiator is shown in Fig. P2-7. Stability and noise problems are improved by elements R_1 and C_2. Differentiation is caused by C_1 and R_2.

(a) Determine the voltage gain function, $A_v(\omega) = -Z_F/Z_1$, where Z_F is composed of R_2 and C_2.

(b) With $\omega_o = 1/R_1C_1 = 1/R_2C_2$ and $\omega_L = 1/R_2C_1 = \omega_o/10$, sketch the ideal asymptotic gain function $A_v(\text{dB})$ versus ω for $\omega_o/100 \leq \omega \leq 100\,\omega_o$. Note that ideally $A_v(\omega_o) = 20$ dB and the open-loop unity-gain frequency must be much greater than ω_o. Assume open-loop unity-gain radian frequency ω to be $100\,\omega_o$ and the high-frequency slope is -20 dB/decade.

(c) What is the actual gain in decibels at $\omega = \omega_o$?

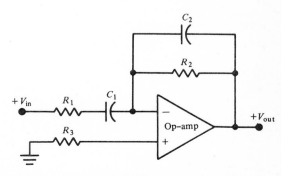

FIG. P2-7   A practical differentiator for Prob. 2-7.

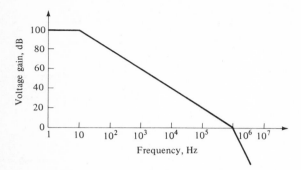

FIG. P2-8 The open-loop asymptotic response of the 741 for Prob. 2-8.

2-8. The 741 IC op-amp open-loop response (asymptotic) is shown in Fig. P2-8.

(a) Sketch the asymptotic closed-loop response function,

$$1 + Z_F/Z_1 = (1 + (jf/1,000))$$

on the above characteristic, and determine the stability. Design a differentiator circuit to yield this closed-loop response. (Use Fig. 2-6, (data page 6), and eliminate R_1.)

(b) From Prob. 2-7, $1 + Z_F/Z_1$ has the form,

$$1 + (jf/f_L)/(1 + (jf/f_1))^2 \qquad \text{when } f_1 = f_2$$

Sketch the asymptotic response of this function on the above characteristic and discuss stability for $f_L = 100$ and $f_1 = 1,000$ Hz.

2-9. (a) For an ideal op-amp as in Fig. P2-9, determine $A_v = V_o/V_1 = -Z_F/Z_1$.

(b) Assume the op-amp to have the characteristics of Fig. P2-8, and sketch the asymptotic response of part (a) if $C = 0.02$ μF (microfarads), $R = 5 \times 10^4$ Ω. Is the circuit stable?

Discussion. In an actual circuit, switches must be included as in Fig. P2-9b to establish initial conditions. With switches in position 1, the amplifier is connected for unity gain, and C is discharged setting an initial condition of zero volts. With the switch in position 2,

$$v_{\text{out}} = -\frac{1}{R_1C} \int_{t_1}^{t_2} v_{in} \, dt$$

For minimum error due to bias current, $R_1 = R_2$.

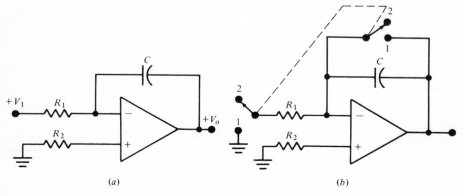

FIG. P2-9 Circuit for Prob. 2-9.

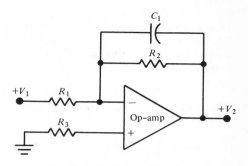

FIG. P2-10   The circuit for Prob. 2-10.

2-10. (*a*) Show that the circuit of Fig. P2-10 is a low-pass amplifier by determining the gain function $A_v = V_2/V_1 = -Z_F/Z_1$.

(*b*) If the gain at zero frequency is 20 dB, $R_2 = 50\,\text{k}\Omega$, $C_1 = 0.02\,\mu\text{F}$, sketch the asymptotic gain response, and determine R_1. *Note*: $R_1 = R_3$ for minimum bias error.

(*c*) For $R_1 = 5,000\,\Omega$, $R_2 = 50,000\,\Omega$, $C_1 = 0.2\,\mu\text{F}$, and if the amplifier has the response of Fig. P2-8, sketch the asymptotic gain response.

2-11. The asymptotic (solid lines) and actual (dashed curve) for the differentiator of Fig. P2-7 are shown in Fig. P2-11. The gain function is given by

$$A_v(f) = -\frac{Z_F}{Z_1} = -\frac{j(f/f_L)}{(1 + jf/f_1)(1 + jf/f_2)}$$

where $f_1 = 1/2\pi\,R_1C_1$
     $f_L = 1/2\pi\,R_2C_1$
     $f_2 = 1/2\pi\,R_2C_2$

Verify these expressions and design the circuit (determine the element values) for the response of Fig. P2-11. *Note*: $f_h <$ unity-gain-f of the op-amp. (*Hint*: $f_1 = f_2$ for the desired response.) Determine the actual gain magnitude at $f = 1,000$ Hz.

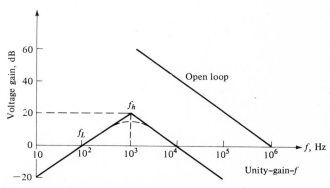

FIG. P2-11   A differentiator response for Prob. 2-11.

2-12. Show that the circuit of Fig. P2-12 operates as a current-to-voltage converter. This circuit may be used as a photoconductive cell or low-current amplifier.

2-13. Assume that the amplifier of Fig. 2-2 has the response curve of Fig. P2-8. Sketch the asymptotic closed-loop function $(1 + R_F/R_1)$ for $R_F/R_1 = 1,000$ and $R_F/R_1 = 100$. Determine the bandwidth for each case.

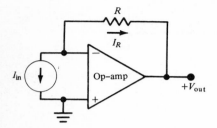

FIG. P2-12   A current-to-voltage converter for Prob. 2-12.

3

Diodes

INTRODUCTION

Chapter 3 introduces the *pn*-junction diode and practical diode characteristics and applications. No attempt is made to include physical electronics in this or succeeding chapters. It is easy to qualitatively describe diode operation and experimentally verify a diode equation. Thus, the diode equation is accepted as a proven postulate. Once we understand electronic-circuit applications from the terminal characteristics, we have a better reason to study the internal physics of devices. The physics is also more useful to device designers rather than to the majority of engineers and technicians who apply the devices in circuits.

3-1 DIODES

A diode is a two-element nonlinear electron device which acts as an electrically controlled switch. It can connect or disconnect sections of electrical circuits according to the polarity of the applied voltage. We find that numerous applications of diodes are possible.

There are three general types of diodes which may be classified according to their physical construction. These are: the semiconductor diode, the thermionic vacuum tube, and the gas tube. The latter two types have been replaced in most applications by solid-state diodes. The physical electronics explaining the operation of solid-state devices will not be discussed here since the principal purpose of this part of the text is to present the circuit properties of electron devices. A qualitative description of device operation will be given.

3-2 SEMICONDUCTOR DIODES

The semiconductor diode is fabricated from carefully processed germanium (Ge) or silicon (Si). Other materials such as selenium or copper oxides can be employed, but Ge and Si diodes are more prevalent. By adding carefully controlled amounts of impurities to Si or Ge, the material may have current conduction due to a majority of positive or negative charge carriers. The positive charge carriers are designated as "holes" while the negative charge carriers are the familiar electrons. Experimental measurements verify the polarity of these charge carriers. If conduction is predominantly due to holes, the material is designated *p* type (*p* for *positive*); and if conduction is predominantly due to electrons, the material is designated *n* type (*n* for *negative*). Even though the conduction may be due to majority holes or electrons, it will be necessary to understand that minority carriers are also present. For example, *p*-type material will have some minority-carrier electrons and vice versa. The density of minority carriers is temperature dependent, and this effect plays an important role in device characteristics.

In Fig. 3-1*a* a representation of a *pn*-junction diode is indicated. The plus

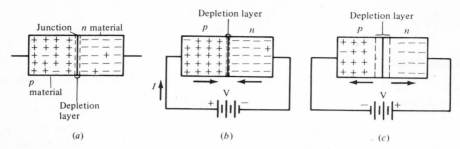

FIG. 3-1 (a) Representation of a *pn*-junction diode, (b) a forward-biased *pn* diode, (c) a reverse-biased *pn* diode.

signs represent movable holes and the negative signs represent electrons. Note that a few minority carriers are included in each region and metal leads are connected on each end to form ohmic contacts.

In Fig. 3-1b a voltage source has been connected so that the diode is said to be forward-biased. The term *bias* defines a fixed or dc operating condition for voltage or current. In this case, voltage V and current I are the bias voltage and the current for the diode. When forward-biased, it will be noted in Fig. 3-1b that the holes and electrons are forced to move in the direction of the arrows, i.e., toward the junction. The carriers flow by diffusion, and holes are injected into the *n* region from the *p* region, and electrons are injected into the *p* region from the *n* region. These carriers combine a very short distance from the junction with the opposite polarity particle which moves by diffusion toward the junction, and current flows easily. Thus, with the *p* end positive and the *n* end negative, current flows easily with little applied bias.

Consider what occurs when the source voltage polarity is reversed as in Fig. 3-1c. The majority carriers diffuse away from the junction, and minority carriers diffuse toward the junction. A barrier is formed, and very little current flows. A high field region exists in the vicinity of the junction, defined as the depletion layer, which sweeps the minority carriers across. The current flowing is due to these minority carriers. This small and essentially constant current is called the reverse saturation current.

It can be shown theoretically and verified experimentally that the current-voltage characteristics of the junction diode is given by

$$i = I_S(e^{qv/k\eta T} - 1) \qquad \text{Amperes} \tag{3-1}$$

The factor η accounts for carrier recombination in the junction region and has a value near unity for germanium and 2 for silicon. In Eq. (3-1), q is the electron charge $(1.602)(10^{-19})$ C (coulomb), v is the positive applied potential for forward bias, k is Boltzmann's constant $(1.38)(10^{-23})$ J/°K (joules per °K), T is the absolute temperature (°K), and I_S is the reverse saturation current. For future reference, we calculate kT/q to be 0.026 V at room temperature, $T = 300$°K.

In Fig. 3-2a the diode electrical-circuit symbol and its nomenclature are shown in a circuit from which i-v characteristics, as in Fig. 3-2b, may be obtained. The terms anode and cathode derive historically from the thermionic diode, and the diode symbol is an arrow which points in the direction of conventional current flow in the forward-bias direction.

A plot of Eq. (3-1) for germanium and silicon diodes is displayed in Fig. 3-2b. These static terminal characteristics are plotted for small values of current and voltage to show the reverse current I_S. Germanium normally has a larger value of I_S than silicon (100 to 1,000 times) and I_S is temperature sensitive, increasing in magnitude as temperature increases. In general, silicon diodes have better

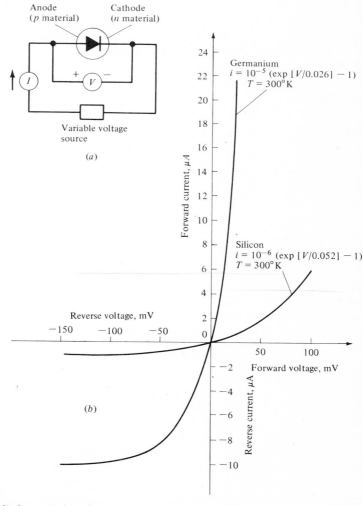

FIG. 3-2 (a) The diode symbol and a measurement system, (b) germanium and silicon pn-junction-diode characteristics; the saturation current for germanium assumed to be 10 times that of silicon.

high-temperature characteristics than germanium. A part of I_S due to the bulk properties of the material increases by about 7 %/°C for both Si and Ge, and the maximum usable temperatures are 75°C for Ge and 150°C for Si. In some diodes, surface leakage contributes in significant part to I_S. At constant current, we measure diode voltage changes with T, $\Delta v/\Delta T$, to be about -2 mV/°C (millivolts per degree Centigrade).

In practical applications, the pn-junction diodes are applied in circuits demanding larger currents than indicated in Fig. 3-2b. The i-v characteristics then appear as in Fig. 3-8c and d (to be discussed later) for the forward direction and in Fig. 3-8e for reverse bias. The values of Fig. 3-2 are no longer visible due to the large current and voltage scales. A reverse voltage breakdown is indicated in Fig. 3-8e. This is caused by a large multiplication of movable holes and electrons created by the high electric field in the vicinity of the junction. During device fabrication, this voltage can be controlled so as to occur at a few volts or a few hundred volts. Forward current values from milliamperes to many amperes can be obtained from available diodes. The current rating depends upon the heat dissipation properties of the device.

3-3 ANALYSIS OF DIODE CIRCUITS

We first consider an ideal diode which has characteristics as shown in Fig. 3-3. When the anode is positive with respect to the cathode, the diode acts as a short circuit. When the anode is negative with respect to the cathode, the diode is an open circuit. These two bias conditions indicate a general procedure for circuit analysis. Assume the diode to be open, and calculate the voltage appearing across the diode. If the anode is more positive than the cathode, the diode is on, and we modify the circuit by replacing the diode with a direct connection between its terminals. The validity of an on-diode equivalent circuit is determined by the direction of current which must be in the forward direction. The on state must be checked. If the anode is negative with respect to the cathode, the diode is open, and the assumption was correct. The calculation is completed.

Example 3-1   Consider Fig. 3-4 and sketch the i versus v characteristics of the circuit. Assume an ideal diode.

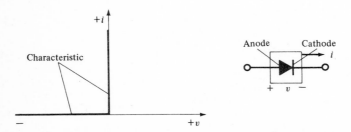

FIG. 3-3 Ideal diode characteristics.   The box around the diode symbol indicates that it is ideal.

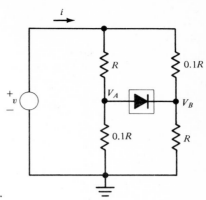

FIG. 3-4 Circuit for Example 3-1.

For $v \geq 0$, assume the diode to be open. With respect to ground reference, $V_A = [v(.1R)/1.1R] = v/11$ and $V_B = [v(R)/1.1R] = v/1.1$. Thus, $V_{AB} = V_A - V_B = v/11 - v/1.1 = 0.091v - 0.91v = -0.819$ V. The anode-to-cathode voltage is negative and the diode is open. The resulting circuit is shown in Fig. 3-5 for all $v \geq 0$.

For $v \leq 0$, replace v by $(-v)$ in the above calculation and $V_{AB} = 0.819$ V, or the anode-to-cathode voltage is positive. Thus we replace the diode with a short circuit and recalculate obtaining the circuit shown in Fig. 3-6.

We can now sketch the i-v characteristics from the circuits of Figs. 3-5 and 3-6. The result is shown in Fig. 3-7.

3-4 NONIDEAL DIODES

Typical diode data sheets are shown in Fig. 3-8. The FD 100 diode is constructed for high-frequency applications, but it can be operated at low frequencies. A study of Fig. 3-8 is useful because the parameters important in circuit design

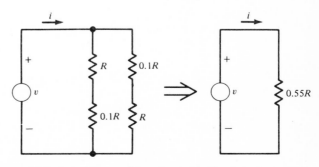

FIG. 3-5 The resulting circuit for Fig. 3-4 when $v \geq 0$.

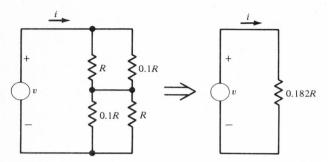

FIG. 3-6 The resulting circuit of Fig. 3-4 when $v \leq 0$.

are defined and typical values are given. We shall use the data to obtain equivalent circuits for use in analysis.

The equivalent circuit which is applicable in a particular case depends upon the current limits. Consider Fig. 3-8d which indicates operation over a current range of a few milliamperes to about 50 mA. We have approximated the actual characteristic by a straight line as indicated in the figure. We now devise a circuit which yields this straight-line approximation. No current flows until the voltage rises to 0.7 V, and thereafter the current-voltage relation is governed by a resistance. The resulting circuit is that of Fig. 3-9. The ideal diode is included so that reverse current cannot flow. The value of R is given by the reciprocal slope of the line sketched in Fig. 3-8d. Following a similar procedure, the circuit representing the diode with reverse bias can be obtained as in Fig. 3-10 which is obtained from Fig. 3-8e. Ideal diode D2 does not turn on until $v = -120$ V, and then the voltage remains constant at this value. It can be seen that the reverse resistance within the specified reverse-voltage rating of 50 V is large: 2.4 MΩ (megohms). If the input voltage reaches -120 V, the current is limited primarily by external circuit resistance, and the reverse current must be limited so that the power dissipation does not exceed the rating of 250 mW (milliwatts) or approximately, $P/BV = (+0.25/120) = +2$ mA.

Example 3-2 A half-wave rectifier or clipping circuit is shown in Fig. 3-11. Sketch the load waveform, and calculate the average load voltage.

For $v_i \geq 0$, we replace the diode by the equivalent circuit of Fig. 3-9 as shown in Fig. 3-12. The output voltage is

$$v_L = \frac{50 \sin 377t - 0.7}{1,010} \, 1,000 = (49.5 \sin 377t - 0.69) \text{ Volts for } v_i \geq 0$$

For $v_i \leq 0$, replace the diode by the circuit of Fig. 3-10 as drawn in Fig. 3-13. From this circuit, $v_L = (50 \sin 377t/2.4 \times 10^6)(10^3) = 0.021 \sin 377t$ volts is negligibly small for reverse bias. The resulting output-voltage waveform is given in Fig. 3-14.

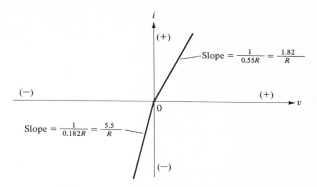

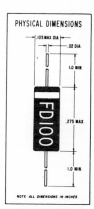

$$\text{Slope} = \frac{1}{0.55R} = \frac{1.82}{R}$$

$$\text{Slope} = \frac{1}{0.182R} = \frac{5.5}{R}$$

FIG. 3-7 The resultant $i\text{-}v$ characteristic of Fig. 3-4.

MAXIMUM RATINGS (25°C.) [Note 1]

WIV	–	Working Inverse Voltage	50 V
I_o	–	Average rectified current	75 mA
I_F	–	Forward current steady state d. c.	115 mA
i_f	–	Recurrent peak forward current	225 mA
i_f (surge)	–	Peak forward surge current pulse width of 1 second	500 mA
i_f (surge)	–	Peak forward surge current pulse width of 1 microsecond	2000 mA
P	–	Power dissipation	250 mW
$\frac{1}{\theta}$	–	Power derating factor	1.67 mW/°C
T_A	–	Operating temperature	-65° to + 175°C
T_{stg}	–	Storage temperature, ambient	-65° to + 200°C

PHYSICAL DIMENSIONS

FD100

NOTE: ALL DIMENSIONS IN INCHES

ELECTRICAL SPECIFICATIONS (25° C unless otherwise noted)

SYMBOL	CHARACTERISTIC	MIN.	TYPICAL	MAX.	TEST CONDITIONS
V_F	Forward Voltage			1.0 V	$I_F = 10$ mA
I_R	Reverse Current			0.1 μA	$V_R = 50$ V
I_R	Reverse Current (150° C)			100μA	$V_R = 50$ V
BV	Breakdown Voltage	75V			$I_R = 5 \mu$A
t_{rr} [Note 2]	Reverse Recovery Time			4.0 nsec	$I_f = 10$ mA
					$I_r = 10$ mA
					$R_L = 100\Omega$
t_{rr} [Note 2]	Reverse Recovery Time			2.0 nsec	$I_f = 10$ mA
					$V_r = 6.0$ V
					$R_L = 100\Omega$
C_o [Note 3]	Capacitance			2.0 pf	$V_R = 0$ V $f = 1$ mc
RE	Rectification Efficiency	45%			100 mc [Note 4]
ΔV_F/°C	Change of forward voltage per degree change in temperature		-1.8mV		

FIG. 3-8 Diode specifications. (*By permission Fairchild Semiconductor Corporation, Mountain View, Calif.*)

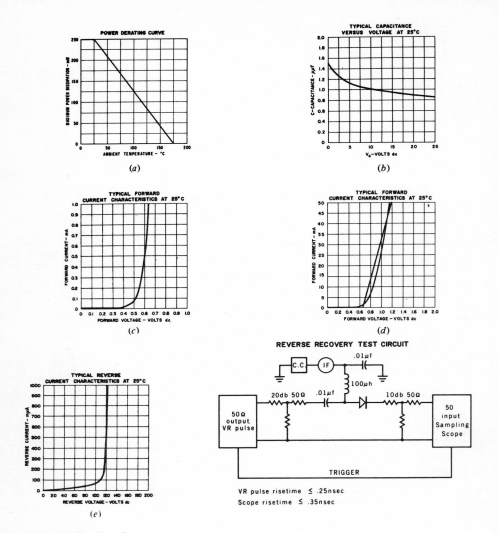

(a)

(b)

(c)

(d)

(e)

FIG. 3-8 Continued

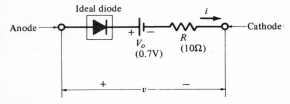

FIG. 3-9   The FD100-diode equivalent circuit for $v \geq 0$.

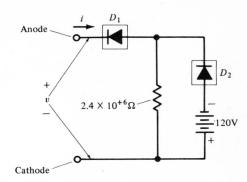

FIG. 3-10 The FD100-diode-circuit approximation for $v \leq 0$. The ideal diode D_1 blocks current when $v \geq 0$.

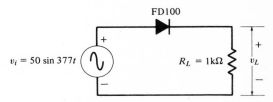

FIG. 3-11 A half-wave rectifier circuit.

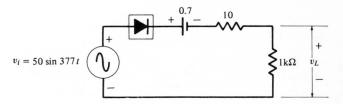

FIG. 3-12 The equivalent circuit of Fig. 3-11 for $v \geq 0$.

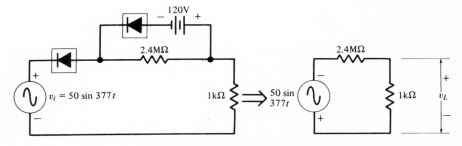

FIG. 3-13 The equivalent circuit for Fig. 3-11 when $v \leq 0$.

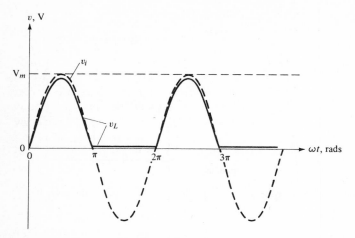

FIG. 3-14 The output voltage of Fig. 3-11.

In this example, the effect of the series voltage and resistance of the forward-biased diode is negligible. The average value of v_L is that of a half sine wave having a peak value V_p of approximately 50 V. Thus,

$$V_{L(\text{avg})} = \frac{1}{2\pi} \int_0^\pi 50 \sin \omega t \, d\omega t = \frac{25}{\pi} [-\cos \omega t]_0^\pi = 15.9 \text{ V}$$

A more precise calculation is not justified.

3-5 DYNAMIC DIODE RESISTANCE

If a time-varying signal is superimposed on a dc signal in a diode circuit, the dynamic resistance r to the time-varying signal differs from the static resistance. The static resistance R is equal to the reciprocal slope of a line joining the operating point on the i-v characteristic to the origin. The *dynamic, incremental,* or *slope resistance* r is defined as the reciprocal of the slope of the i-v characteristic at a particular value of static current I. Using Eq. (3-1),

$$g = \frac{di}{dv} = \frac{qI_S e^{qv/\eta kT}}{\eta kT} = \frac{q(I + I_S)}{\eta kT} \tag{3-2}$$

For reverse bias, $I = -I_S$ and g is zero (or $r = 1/g$ is infinite). For a reasonable forward-bias current, $I \gg I_S$, so that

$$r = \frac{\eta kT}{qI} = 0.026 \frac{\eta}{I} \qquad \text{at } T = 300°\text{K} \tag{3-3}$$

We see that r varies inversely with the dc current and, in Figs. 3-9 and 3-10 we have approximated r to be constant at 10 Ω and 2.4 MΩ, respectively.

3-6 DIODE CAPACITANCE AND CHARGE STORAGE

In addition to the static or low-frequency characteristics which have been discussed, pn-junction diodes may also be used in high-frequency or pulse applications. Since positive and negative charges are separated by the junction barrier, it is to be expected that a capacitance exists between the p and n sides (see Fig. 3-8b for reverse-bias values). This capacitance tends to shunt high-frequency signals around the diode; thus the capacitance acts as if it is in parallel with the device. The diode is made small physically when it is desired to reduce this capacitance. For very high frequency work (tens to thousands of MHz) the point-contact diode is employed. The point-contact device is made with a semiconductor and a pointed metal probe (tungsten) formed into its surface. While the frequency characteristics of this type of diode are satisfactory, reverse currents are normally larger than for pn-junction diodes. Schottky, pin and Back diodes, trade names for certain types of diodes, are also applied at high frequencies.

Varactor diodes actually make use of the capacitance existing when pn-junction diodes are reverse-biased. Capacitance values from 0.1 to 100 pF (picofarads) are available and can be varied electrically. This variable capacitor finds application in several high-frequency circuits.

A second effect occurs in semiconductor diodes which is important in switching networks. During the time that the diode is on, excess minority carriers are injected into the p and n regions, and these carriers do not recombine immediately. When the voltage is reversed so that the diode is switched to the *off* polarity, current continues to flow for a period of time since these excess minority carriers must be removed before steady-state conditions can be obtained. The result is as displayed in Fig. 3-15 where the circuit, applied voltage, and resulting current have been sketched. The forward- and reverse-storage currents depend on the magnitude of the forward-polarity voltage V_F and the reverse-polarity voltage V_R respectively. The storage time t_s also will be a function of these voltages with t_s decreasing when V_R is increased. Note that the fall time t_f is defined between the points marked 90 percent and 10 percent of total current excursion. This is a standard definition for decreasing pulse waveforms, and a similar definition applies for increasing waveforms. The storage time places a limit on the switching speed of the diode.

3-7 POWER DERATING

In Fig. 3-8a, a power-derating curve is given for the FD 100, and this curve is simply interpreted. As the ambient temperature rises, the diode can dissipate less electrical power safely. For example, if the surrounding temperature is 100°C (centigrade), the diode can work at a current and voltage level so that the total power dissipated is no more than 125 mW (from Fig. 3-8a).

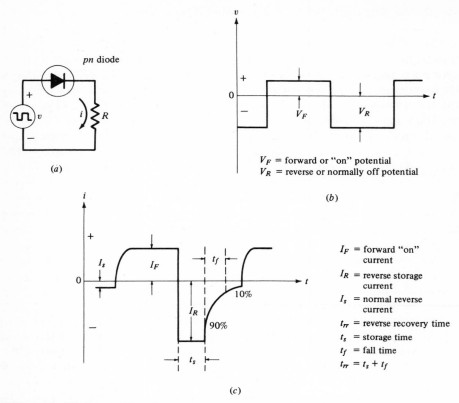

FIG. 3-15   Diode switching characteristics.

V_F = forward or "on" potential
V_R = reverse or normally off potential

(b)

I_F = forward "on" current
I_R = reverse storage current
I_s = normal reverse current
t_{rr} = reverse recovery time
t_s = storage time
t_f = fall time
$t_{rr} = t_s + t_f$

3-8   VOLTAGE REGULATOR DIODES

The reverse-breakdown characteristic of the pn diode is employed to maintain a constant dc voltage for regulator or voltage reference purposes. These pn diodes are called Zener, avalanche-breakdown, or regulator diodes. The regulator diode is formed by placing two avalanche diodes in series with the cathodes (or anodes) connected together. The diode names are derived from processes explaining the phenomena involved or the application, and typical characteristics are sketched in Fig. 3-8e. Zener breakdown occurs due to high electric fields developed in the diode at the junction when voltages of about two volts or less occur across the diode. Avalanche breakdown occurs for higher voltages, the process being similar to that explaining the operation of a gas tube (such as employed in fluorescent lamps). Due to an error in early experimentation, many diodes operating under avalanche-breakdown conditions were (and still are) designated Zener diodes.

Example 3-3   Voltage Regulator—An example will indicate the use of the avalanche diode. We will design a circuit to maintain a constant voltage of

24 V within ±5 percent across a 2,400-Ω load. The tolerance limit of ±5 percent is necessary since manufacturing techniques are not perfect and some variation in specified voltage must be expected. Standard Zener or breakdown diodes may be obtained with voltage tolerances specified between ±2½ percent and ±20 percent, power dissipation ratings from 0.25 to 50 W and voltages between 2.4 and 200 V. Not all values between these limits are possible so that it is necessary to consult technical data sheets or diode manuals. For this example, the input voltage will be assumed 50 V dc but varying ±10 percent due to fluctuations in the source of this voltage. Since the input is approximately 50 V, and the breakdown voltage is approximately 24 V, a series resistor will be necessary as shown in the circuit of Fig. 3-16, where the Zener-diode symbol and currents that flow are also defined. The cathode side of the diode is connected to the + side of the circuit since it is operating under reverse-polarity breakdown conditions. Zener diodes having a power dissipation rating of less than 50 W must have a minimum current of approximately $I_{Z\,min} = 1$ mA to operate in the breakdown region. Fifty-watt diodes have a minimum current of about 5 mA, and the usual design procedure allows the Zener current I_Z to be above these minima or 10 percent of the maximum load current, whichever is greater. For this example, load current will be very nearly $I_L = V_L/R_L = 24/2,400 = 0.01$ A so that the minimum device current $I_{Z\,min} = 1$ mA. To ensure this minimum current, the current through R_S must be $I_Z + I_L = 11$ mA when the input voltage is minimum or 45 V for this case. Therefore, $R_S = (45 - 24/0.011) = 21/0.011 = 1910$ Ω. This is not a standard-size resistor so that 1,800 Ω should be chosen to ensure $I_Z \geq I_{Z\,min}$. When the input voltage rises to 55 V, the increased current flows in the diode since the voltage across the diode and R_L remains essentially constant, and I_Z increases to approximately 7 mA with the power dissipated in the diode being $I_Z V_Z = 0.007(24) = 168$ mW. A diode rated at 0.25 W and ±5 percent voltage tolerance would be adequate as long as the surrounding temperature is approximately 25°C. Increased ambient temperature would require a derating of the diode-power-dissipation capability or a higher diode power rating would be necessary.

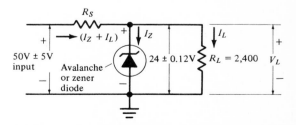

FIG. 3-16 A Zener-diode shunt-voltage regulator.

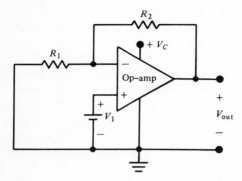

FIG. 3-17 Single supply voltage operation of an operational amplifier.

3-9 OP-AMP ZENER-DIODE VOLTAGE REGULATOR

Operational-amplifiers are normally utilized with V^+ and V^- supply voltages as in Fig. 1-1, so as to have zero output voltage for zero input voltage. However, it is also possible to operate these amplifiers with one supply voltage as in Fig. 3-17 where $V_C = |V^+| + |V^-|$ when it is not required to have zero-dc output voltage. For a finite value of R_1 in this circuit, $V_{\text{out}} = V_1$ if $R_2 = 0$, and $V_{\text{out}} = V_1(1 + R_2/R_1)$ if $R_2 > 0$. The voltage V_1 must be above 2 V but less than V_C or the amplifier will not operate correctly and latching will occur.

Single voltage operation of an op-amp leads to a useful voltage-regulator circuit as indicated in Fig. 3-18. Resistor R_4 is adjusted to maintain a minimum current in the Zener diode so that V_Z is approximately constant. If V_Z is constant, even though V_C is unregulated and varying in amplitude, the feedback loop constantly compares the output voltage with V_Z and maintains the output voltage constant. We note that V_Z of Fig. 3-18 acts in the same manner as V_1 in

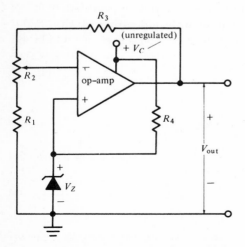

FIG. 3-18 An op-amp Zener-diode voltage regulator.

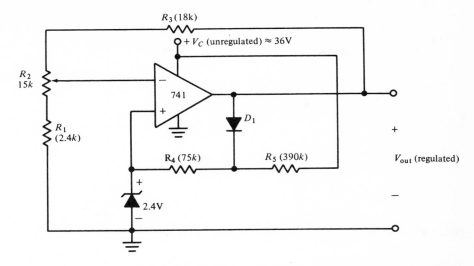

FIG. 3-19   The final version of the op-amp Zener-diode regulator.

Fig. 3-17. The output voltage is adjusted by varying R_2 and thus the closed-loop gain.

Since V_Z in Fig. 3-18 would vary slightly with the unregulated voltage, V_C, the output regulation may not be satisfactory. To correct this problem, the circuit of Fig. 3-19 is applied. The Zener-diode reference current is supplied from the amplifier output through D_1 and R_4 to improve the regulation and allow the regulator to operate from unregulated supplies. Resistor R_5 provides the necessary startup current for the Zener diode and diode D_1 decouples the reference string $R_4 - R_5$ from the amplifier output at startup. After startup, diode D_1 operates and the low output impedance of the op-amp reduces the effect of current variations due to the unregulated voltage and R_5. The output current is limited to that value which can be furnished by the op-amp and for the 741 IC, the maximum output current is about 15 mA. If $V_C = 36$ V, V_{out} can be adjusted between 5 and 35 V, and resistor R_2 should be a ten-turn potentiometer for accurate adjustment.

3-10   FURTHER EXAMPLES OF DIODE APPLICATIONS

The clipping or rectifying capabilities of the diode have been indicated. Other fundamental applications involve waveshaping, clamping, and switching; and examples of some of these circuits follow.

3-10.1   WAVESHAPING

It is possible to approximate many voltage waveforms by modifying an available source waveform. Triangular waveforms can be modified to represent sine waves;

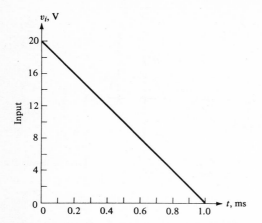

FIG. 3-20 The linearly decreasing input voltage for Example 3-4.

linear waveforms can be modified to represent exponential or logarithmic functions, etc.

Example 3-4 Consider the case where it is desired to modify a linearly decreasing potential across a 1,000-Ω load resistance. The available source waveform is represented in Fig. 3-20, and the desired exponential wave is represented by the points marked by x's in Fig. 3-21. These voltages could be repetitive waves of which only one cycle is shown. By comparing the two waveforms, we see that it is necessary to prevent the desired voltage from decreasing as rapidly as the source voltage, and this can be done by switching potential sources and resistances into the circuit at appropriate times. Straight-line segments intersecting at 8 and 6 V are also sketched in Fig. 3-21 and appear to represent the desired exponential waveform satisfactorily

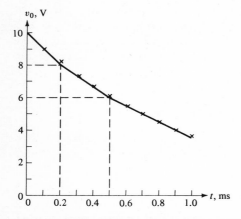

FIG. 3-21 The desired output voltage and its linear approximation for Example 3-4. The x's represent points on an exponential curve.

(piecewise-linear approximation). It will be necessary to cause a change to a different circuit at these voltages and their corresponding times. Between 0 and 0.2 ms (milliseconds), the first line segment decreases at the same rate as the input, but it is only half the magnitude of the input or source voltage. Therefore, the voltage divider circuit of Fig. 3-22 will perform this function. To introduce breakpoints at 6 and 8 V, the remaining components of Fig. 3-23 are added, and we must find appropriate values for R_2 and R_3. The diodes are biased by 8- and 6-V batteries so that they will not begin to conduct until the output voltage falls to these values. As a first approximation, assume the diodes to be ideal so that they begin to conduct as soon as the anode is positive with respect to the cathode. This is always a satisfactory starting point, and corrections can be made later to account for actual diode characteristics.

Values for the resistances of Fig. 3-23 can be deduced by noting the relative changes in voltage of the input and output for the distinct regions. For the region between $V_O = 8$ V and $V_O = 6$ V, the equivalent circuit of Fig. 3-24 applies since the output voltage has not decreased enough for D_2 to operate, and D_1 has zero resistance. The input is changing from 16 to 10 V during this time (0.2 ms $\leq t \leq$ 0.5 ms), and the relative changes are $\Delta v_i = 6$ V at the input and $\Delta v_O = 2$ V at the output or $(\Delta v_i / \Delta v_O)$ has a ratio of 3:1. The resistors again form a voltage divider and must be in this ratio. Thus, R_2 and R_L are in parallel, and by making $R_2 = 1$ kΩ, R_2 in parallel with R_L is 500 Ω, and the voltage division of the input is 0.5 k/(1 k + 0.5 k) = $\frac{1}{3}$, the desired ratio. To obtain the absolute voltage v_O for any v_i between 16 and 10 V, apply superposition. For example, with $v_i = 10$ V, we find that $i_2 = 6$ mA or $v_O = 6$ V. For v_O less than 6 V, diode D_2 now operates, and the equivalent circuit becomes that of Fig. 3-25, v_I changes from 10 to zero or $\Delta v_i = 10$ V while v_O changes from 6 to 3.5 or $\Delta v_O = 2.5$ V, and the ratio is (2.5/10) = $\frac{1}{4}$. A few calculations indicate that the parallel combination of R_2, R_3, and R_L must be $\frac{1}{3}$ kΩ and the value of R_3 is 1 kΩ to obtain this result. For absolute values of voltage at a particular time, apply superposition at $v_I = 0$ to find that $i_3 = 3.5$ mA or $v_O = 3.5$ V. Another method for determining R_3 involves converting each battery-resistance combination to its Norton equivalent, set $v_i = 0$ and solve for R_3 to see that $v_O = 3.5$ V. We see that the circuit of Fig. 3-23 will approximately produce the desired exponential waveform.

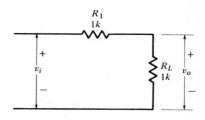

FIG. 3-22 The voltage divider circuit for Example 3-4 when $16 \leq v_i \leq 20$ V.

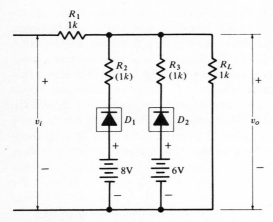

FIG. 3-23 The waveshaping circuit for Example 3-4.

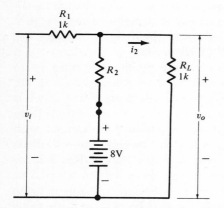

FIG. 3-24 The equivalent circuit of Fig. 3-23 for $10 \leq v_i \leq 16$ V.

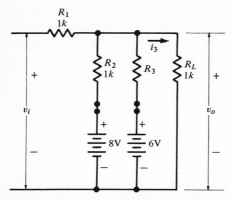

FIG. 3-25 The equivalent circuit of Fig. 3-23 when $10 \leq v_i \leq 0$ V.

If the diode of Fig. 3-9 is employed in Fig. 3-23, correction of the above solution involves replacement of the batteries by sources that are 0.7 V less than that indicated, and by modifying the resistors accordingly. Finally, voltages across the diodes when not conducting and the current passing through them when conducting should be checked to see if the values are within specifications.

The above example finds application in forming the necessary voltage waveform to obtain linear frequency output with time in some microwave backward-wave-oscillators (BWOs). It is necessary to apply a nonlinear voltage to BWO tubes for a linear sweep of output frequency with time. If more exact shaping is needed, more diodes, resistors, and batteries can be inserted with one diode for each breakpoint. In many cases, we may wish to remove the batteries and add Zener diodes having the desired breakdown voltages. Zener diodes cannot be applied to Fig. 3-23 because the batteries must furnish power when v_i drops below the output voltage.

If the resistor R_3 of Fig. 3-23 is reduced to zero ohms, the output voltage will be the same as in Fig. 3-21 until $t = 0.5$ ms, after which v_o will remain at 6 V until v_i again switches above 10 V. Diode D_2 is then called a "pickoff" diode, and the pickoff diode is applied in some switching circuits.

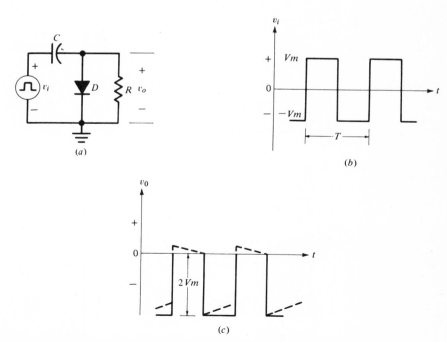

FIG. 3-26 A clamping circuit and its operation. (a) Top "clamping circuit"; (b) input waveform; (c) output waveform.

3-10.2 CLAMPING CIRCUITS

It is necessary to establish a new reference level to certain waveforms after they have been transmitted through electronic circuits. Figure 3-26*a* indicates a method of depressing the input waveform so that its positive peak value drops to ground potential. For the square-wave input of Fig. 3-26*b*, the output represented by the solid line of Fig. 3-26*c* would result if the diode were ideal and the *RC* time constant were long compared to the period of the input wave. The circuit operation is straightforward. Whenever the anode is positive, the output is connected to ground potential, and capacitor *C* is charged to peak input voltage. During the next half cycle of input voltage, the diode is open or nonconducting, and the source, load, and capacitor are in series so that the output is below ground at a value of twice the peak input voltage.

  Analysis of the circuit of Fig. 3-26*a* can be performed by applying equivalent circuits for the period of time representing the diode action for each half cycle of the input voltage. By the use of two equivalent circuits such as Figs. 3-9 and 3-10 to represent the actual diode, the output is modified as shown in Fig. 3-26*c*. Analysis for a particular case is left as an exercise for the student (see Prob. 3-15).

REFERENCES

1. Chen, Chi Ho, "Predicting Reverse Recovery Time of High Speed Semiconductor Junction Diodes," General Electric Application Note 90.36, Syracuse, N. Y., 1962.
2. Eimbinder, J., "Application Considerations for Linear Integrated Circuits," John Wiley & Sons, Inc., N. Y., 1970
3. Radio Corporation of America, "RCA Transistor Manual," Harrison, N. J., 1964.
4. Motorola, Inc., "Silicon Zener Diode and Rectifier Handbook," Phoenix, Ariz., 1961.

PROBLEMS

3-1. Sketch and label the *i-v* characteristics for the circuit of Fig. P3-1. Note that the diode is ideal.

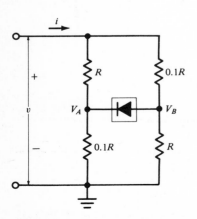

FIG. P3-1 The circuit for Prob. 3-1.

3-2. Sketch the i-v and i_2-v characteristics for the circuit of Fig. P3-2. Label the slopes and breakpoints.

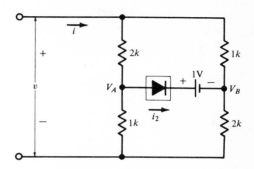

FIG. P3-2   The circuit for Prob. 3-2.

3-3. Repeat Example 3-2 of the text if $v_i = 2\sin(377t)$. What is the average voltage across the 1,000-Ω load resistor?

3-4. Evaluate the dynamic resistance of a germanium ($\eta = 1$) and a silicon ($\eta = 2$) diode at a forward-bias current of 5.0 mA. Neglect the leakage current I_S. What is the reverse dynamic resistance if $I_S = 10^{-8}$ A? Consider $T = 300°$K.

3-5. Sketch the i versus v and v_2 versus v characteristics of Fig. P3-5. Note that this is a case where the direction of current in diode D_1 must be checked to verify whether it is off or on.

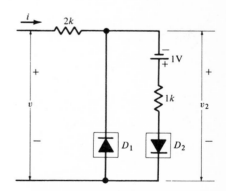

FIG. P3-5   The circuit for Prob. 3-5

3-6. A full-wave rectifier is drawn in Fig. P3-6a, and its equivalent circuit is given in Fig. P3-6b. The transformer is center-tapped so that $v_1 = -v_2$. For $v_1 = V_m\sin\omega t$, sketch the load voltage waveform as a function of time. Determine an expression for the average or dc voltage across R_L. Compare the results with Example 3-1 in the text.

3-7. A full-wave rectifier for changing alternating current to direct current is drawn in Fig. P3-7a. The capacitor helps to filter or smooth the dc output voltage.

   (a) With the capacitor removed, and noting that only two diodes are forward-biased at any instant, sketch v_L versus ωt, for $0 \leq \omega t \leq 2\pi$.

   (b) With the resistor removed, sketch v_L versus ωt. Assume $v = 0$ at $\omega t = 0$.

   (c) With the circuit as shown in Fig. P3-7, the output voltage is as in Fig. P3-7b.

Between θ_1 and θ_2, the output approximately follows the sine wave input with two diodes on,

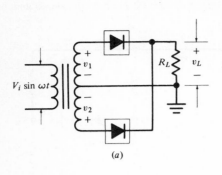

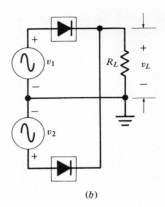

(a)

(b)

FIG. P3-6   A full-wave rectifier for Prob. 3-6.

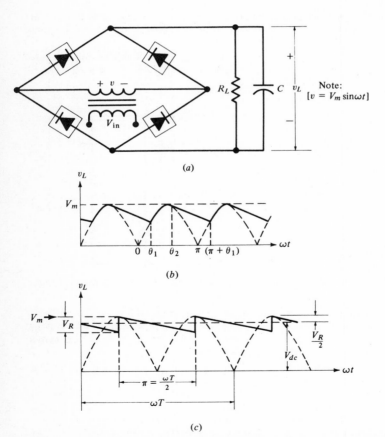

(a)

(b)

(c)

FIG. P3-7   The approximate analysis of a full-wave rectifier with filter for Prob. 3-7.

and between θ_2 and $\pi + \theta_1$, the capacitor discharges through R_L with all diodes open. We see that this is an intermediate case between what we found in parts (a) and (b). The exact analysis is tedious, and for engineering applications, we find that we can represent the output as in the approximation of Fig. P3-7c. Straight-line approximations have been made, assuming the capacitor to charge instantaneously and discharge at a constant rate of I_{dc} or the charge lost is $Q = I_{dc}T/2$. Note that $I_{dc} = V_{dc}/R_L$. Since the charge added to the capacitor is $V_RC = Q$, show that

$$V_{dc} = V_m - \frac{I_{dc}}{4fC}$$

where $f = $ input frequency

(d) Since the open-circuit output voltage is V_m, draw a Thévenin equivalent dc circuit from the results of part (c).

3-8. Repeat Prob. 3-7 for a half-wave rectifier as shown in Fig. 3-11, assuming an ideal diode and adding a capacitor across R_L.

3-9. The output of the filtered rectifier of Prob. 3-7 is assumed to be as in Fig. P3-7c. The peak-to-peak ripple voltage is defined as V_R in that figure. An equivalent circuit in so far as the output is concerned has been drawn in Fig. P3-9. For a dc voltage of 50 V across $R_L = 1,000 \ \Omega$, determine V_m and C if $V_R = 1$ V ($f = 60$ Hz).

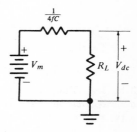

FIG. P3-9 The circuit for Prob. 3-9.

3-10. The characteristics of a Zener diode have been approximated by the straight lines indicated in Fig. P3-10. Determine an equivalent circuit for the diode in the reverse-breakdown region ($v_{ak} \leq 0$ V).

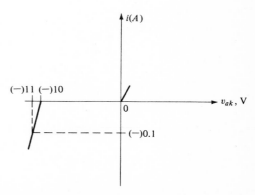

FIG. P3-10 A Zener-diode characteristic for Prob. 3-10.

3-11. An equivalent circuit for the Zener-diode regulator of Fig. 3-16 is drawn in Fig. P3-11.

(a) If the load current varies from 50 to 100 mA, determine the maximum value R_S may have for regulation.

(b) If $R_S = 100\ \Omega$, determine the output voltage change for 50 mA $\leq I_L \leq$ 100 mA.

(c) If the input voltage varies from 20 to 22 V, $R_S = 100\ \Omega$, and 50 mA $\leq I_L \leq$ 100 mA, determine the maximum load voltage change.

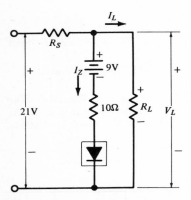

FIG. P3-11   A regulator equivalent circuit for Prob. 3-11.

3-12. Determine the dc output voltage of the circuit in Fig. 3-19 if the wiper of R_2 is adjusted midway. Determine the maximum and minimum possible values of V_{out}.

3-13. Fig. P3-13 shows a tracking regulated power supply. Discuss values for R_6 and R_7 so that $|-V_{\text{out}}| = |+V_{\text{out}}|$.

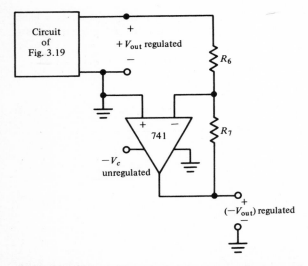

FIG. P3-13   A tracking regulated power supply for Prob. 3-13.

3-14. The input to a diode, resistor, and battery type of waveshaping circuit is given in Fig. P3-14a. Design circuits to yield the output-voltage waveforms shown in Fig. P3-14b and c. Note how the triangular wave could be modified to simulate a sine wave.

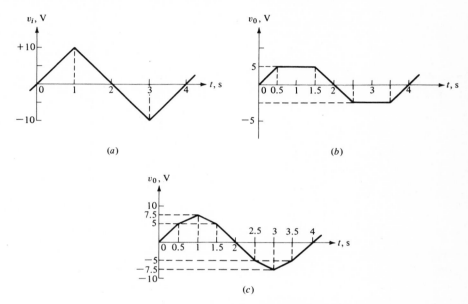

(a)

(b)

(c)

FIG. P3-14   Waveforms for use in Prob. 3-14.

3-15. The diode D in Fig. 3-26 can be represented by the equivalent circuit of Fig. P3-15. For $R = 10^6$ Ω, $C = 10^{-6}$ F (farads), $T = 2$ (seconds), $V_m = 5$ V, sketch and label the output waveform. Start at $t = 0$, and assume the initial capacitor charge to be zero.

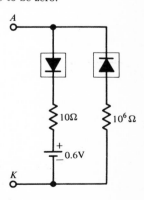

FIG. P3-15   A diode equivalent circuit for Prob. 3-15.

3-16. Repeat Prob. 3-15 if $T = 0.1$ s. Note that clamping is much better. Why?

3-17. If the resistor in Fig. 3-26 is replaced by a capacitor C_2, the capacitor charges to $v_o = -2V_m$, and the output is double the peak value of the input voltage. The circuit is called a voltage doubler. Show that the output v_{o3} of the circuit of Fig. P3-17 is $3V_m$ for the same input as in Fig. 3-26. Add another diode and capacitor to this circuit to form a voltage quadrupler. The circuit of Fig. P3-17 will double or triple V_m when the input is sinusoidal and V_m is the peak value of the sinusoid.

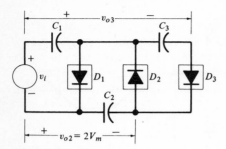

FIG. P3-17   A voltage multiplier for Prob. 3-17.

4

Transistors

INTRODUCTION

Transistors having the capability of amplifying electrical signals may be used for switching purposes similar to the diode. The present chapter attempts to describe the operation of transistors qualitatively, to indicate their terminal electrical characteristics, to include some applications of these characteristics, and to develop equivalent circuits for the devices. The material presented is basic to electronic circuit analysis and design.

The two basic types of transistors are the two-junction bipolar transistor and the unipolar field effect transistor (FET). These devices are fabricated from germanium or silicon with silicon being the predominant material, but other semiconductor materials may be employed and will undoubtedly be developed in the future.

4-1 BIPOLAR TRANSISTORS

A bipolar transistor has two *pn* junctions in close proximity as shown in Fig. 4-1 where the three regions separated by the junctions have been designated emitter, base, and collector. For signal amplification, the emitter-to-base junction will be forward-biased so that emitter majority carriers are injected into the base region where they now become minority carriers. The collector-to-base output junction is reverse-biased so as to extract or collect the minority charge carriers which diffuse or drift across the base. There are two types of construction, *pnp* and *npn*, as indicated in Fig. 4-1, and since transistors are not constructed exactly as shown, the structures should be considered as schematic.

The base region is very narrow, less than 0.001 in., so that most of the excess minority carriers in the base move across to the collector without recombining with majority carriers. A thin base region minimizes the recombination of charge carriers, and thus only a small current is required at the base lead connection. Current flow between the emitter and the collector is controlled by the base-emitter voltage and the base current. The electrical symbols for the two types of transistors are also given in Fig. 4-1, and we see that the arrow is always in the emitter lead, pointing in the direction of conventional forward bias current.

The division of current between elements is shown in Fig. 4-2 where the arrow width is proportional to current magnitude. By Kirchhoff's current law and the convention that all currents enter the device,

$$I_E + I_B + I_C = 0 \tag{4-1a}$$

or

$$I_E = -I_B - I_C \tag{4-1b}$$

In Fig. 4-2a, I_C is a negative number and I_B could be positive or negative depending on the magnitude of the reverse collector-base diode leakage current

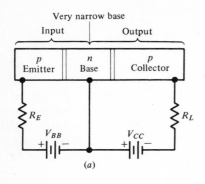
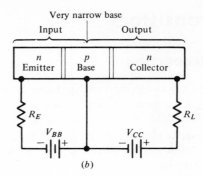

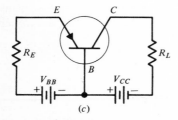
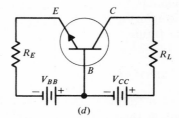

FIG. 4-1 Transistor schematic diagrams (bipolar). Note that the arrows point in the direction of conventional current flow at emitters in (c) and (d). Arrows also point toward p regions or away from n regions. (a) Representative pnp transistor; (b) representative npn transistor; (c) symbolic diagram of the pnp transistor; (d) symbolic diagram of the npn transistor.

I_{CO}. In normal operation I_B will be negative for pnp transistors. Both the I_B and I_C are positive quantities for an npn transistor as may be seen in Fig. 4-2b. Equation (4-1b) states that the sum of the base and collector currents is equal to the emitter current. A fraction of the emitter current designated by αI_E reaches the collector. The part of the base current designated by $(1 - \alpha)I_E$ contains two components:

1. Current due to majority carriers in the base that are injected into the emitter
2. Current due to minority carriers in the base that recombine with majority carriers

For most transistors, the value of α is between 0.95 and 0.995. The symbol I_{co} for leakage current of the reverse-biased collector-base diode has been changed from the symbol I_S of Eq. (3-1) because the thin base yields a different value than that found for pn diodes.

Transistor-collector characteristics, i_C versus v_C, are presented with the base terminal as reference (common base) or with the emitter terminal as reference (common emitter). Characteristics of an npn transistor are sketched in Figs. 4-3 and 4-4 where the collector-base junction is reverse-biased, and the emitter-base junction is forward-biased, as in Fig. 4-1. The collector leakage currents I_{co} and

I_{CEO} are further defined in Fig. 4-5 where the term $I_{CO} \equiv I_{CBO} \equiv I_{CO}(I_E = 0)$ represents the collector current with the emitter open circuit, and the term $I_{CEO} \equiv I_{CE}(I_B = 0)$ is similarly defined as the collector current when the base is open. We see that the subscript BO is equivalent to (base open) and EO equivalent to (emitter open). The current I_{CEO} is always larger than I_{CO}, and both currents are temperature sensitive. As temperature increases, all the curves shown move upward by the amount of increase of I_{CO} or I_{CEO}. This condition must be considered in biasing circuits, especially for germanium transistors.

By definition, an incremental current gain is given by

$$\alpha_f = -\left.\frac{\Delta i_C}{\Delta i_E}\right|_{V_{CB}=\text{constant}}$$

and it can be determined from the incremental values indicated in Fig. 4-3; $\alpha_f = -(0.99)(10^{-3})/(-10^{-3}) = 0.99$. From this figure, the collector current at point Q is given by $i_C = i_{C1} + I_{CO}$. Because the collector current is nearly

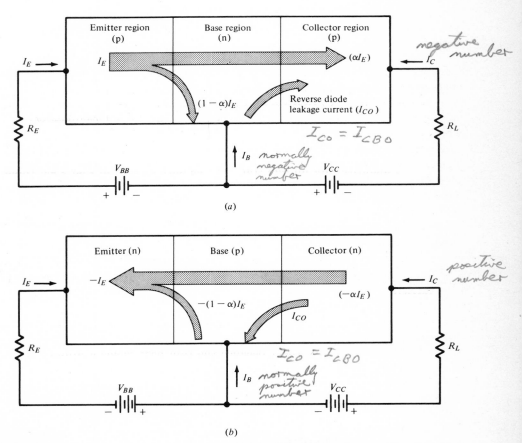

(a)

(b)

FIG. 4-2 The division of current in a transistor.

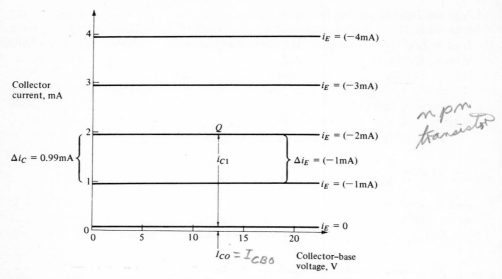

FIG. 4-3 The common-base collector characteristics (*npn*) with emitter current as a parameter.

independent of collector-emitter voltage, an equation can be written for the characteristic curve passing through Q as

$$i_C = -\alpha_0 i_{E1} + I_{co}$$

$$\alpha_o = -\left.\frac{I_C}{I_E}\right|_{V_{CB}=\text{constant}} \quad (4\text{-}2)$$

where i_{E1} would be -2 mA in Fig. 4-3. Equation (4-2) also represents any of the characteristic curves of Fig. 4-3 when the appropriate value of i_E is employed. The term α_0 is a large-signal parameter, and α_f is applied for incremental signals.

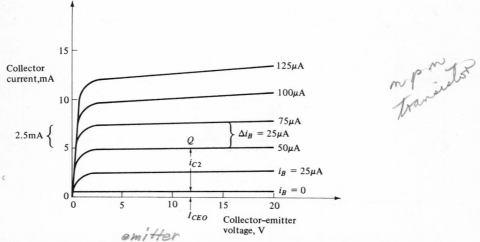

FIG. 4-4 The common-collector characteristics (*npn*) with base current as a parameter.

The common-emitter equation is written as

$$i_C = \beta_0 i_B + I_{CEO}$$

$$\beta_0 = \left.\frac{I_C}{I_B}\right| \qquad (4\text{-}3)$$

$$V_{CE} = \text{constant}$$

where $\beta_0 = I_C/I_B\,|_{V_{CE}=\text{constant}}$.

The static characteristics presented for transistors in manuals and data sheets are normally for the common-emitter case. Thus, it is desirable to develop an expression for the collector current in terms of the base current. Substituting Eq. (4-1) into Eq. (4-2) and applying instantaneous total value symbols,

$$i_C = \alpha_0(i_B + i_C) + I_{CBO}$$

or

$$i_C = \frac{\alpha_0}{1 - \alpha_0} i_B + \frac{1}{1 - \alpha_0} I_{CBO} \qquad (4\text{-}4)$$

Comparison of Eq. (4-4) with Eq. (4-3) yields,

$$\boxed{\beta_0 = \frac{\alpha_0}{1 - \alpha_0}} \qquad I_{CEO} = \frac{1}{1-\alpha_0} I_{CBO} = (\beta_0+1)I_{CBO} \qquad (4\text{-}5)$$

We can rearrange this expression to show that $(\beta_0 + 1) = 1/(1 - \alpha_0)$ or Eq. (4-4) becomes

$$i_C = \beta_0 i_B + (\beta_0 + 1)I_{CO} \qquad (4\text{-}6)$$

A definition for incremental changes may be written as

$$\beta_f = \left.\frac{\Delta i_C}{\Delta i_B}\right|_{V_{CE}=\text{constant}}$$

from which we can obtain numerical values for β_f by using the characteristic curves. As an example, the incremental values of Δi_C and Δi_B indicated in Fig. 4-4 yield $\beta_f = (2.5)(10^{-3})/(25)(10^{-6}) = 100$. There is a basic difference in definition, but, approximately, $|\beta_f| = |\beta_0|$. Since $I_{CEO} = 101 I_{CO}$, we see why I_{CEO} is always larger than I_{CO}. If Δi_b is caused by a signal, then Δi_c is β_f times larger, and a current gain has been obtained. We later show that voltage and

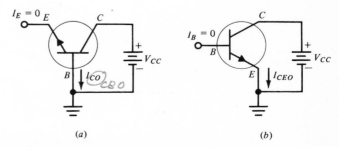

(a)   (b)

FIG. 4-5 (a) The circuit for defining the leakage current I_{CO}; (b) the circuit for defining the leakage current I_{CEO}.

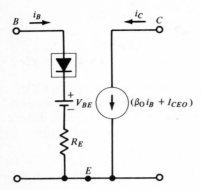

FIG. 4-6 A simple dc equivalent circuit for an *npn* transistor.

absolute maximum ratings: (25°C) (unless otherwise specified)

		2N4424	2N4425	
Voltages				
Collector to Emitter	V_{CEO}	40	40	V
Emitter to Base	V_{EBO}	5	5	V
Collector to Base	V_{CBO}	60	60	V
Current				
Collector (Steady State) *	I_c	500	500	mA
Dissipation				
Total Power (Free Air at 25°C) **	P_T	360	560	mW
Total Power (Free Air at 65°C) **	P_T	250	380	mW
Total Power (Heatsink at 25°C) ***	P_T	—	900	mW
Temperature				
Storage	T_{stg}	−55 to +150		°C
Operating	T_j	+150		°C
Lead soldering, $\frac{1}{16}'' \pm \frac{1}{32}''$ from case for 10 sec. max.	T_L	+260		°C

*Determined from power limitations due to saturation voltage at this current.
**Derate 2.88mW/°C increase in ambient temperature above 25°C.
***Derate 7.2 mW/°C for rise in heatsink temperature above 25°C.

electrical characteristics: (25°C) (unless otherwise specified)
DC CHARACTERISTICS

		Min.	Max.	
Collector Cutoff Current ($V_{CB} = 40$V)	I_{CBO}		30	nA
($V_{CB} = 40$V, $T_A = 100$°C)	I_{CBO}		10	μA
($V_{CB} = 40$V)	I_{CES}		30	nA
Emitter Cutoff Current ($V_{EB} = 5$V)	I_{EBO}		100	nA
Forward Current Transfer Ratio				
($V_{CE} = 4.5$V, $I_C = 2$ mA)	h_{FE}	180	540	
Collector Emitter Breakdown Voltage ($I_C = 10$ mA)	$V_{(BR)CEO}$	40		V
Collector Base Breakdown Voltage ($I_C = 10$ μA)	$V_{(BR)CBO}$	60		V
Emitter Base Breakdown Voltage ($I_E = 0.1$ μA)	$V_{(BR)EBO}$	5		V
Collector Saturation Voltage				
($I_B = 3$ mA, $I_C = 50$ mA)	$V_{CE(sat)}$		.30	V
Base Saturation Voltage ($I_B = 3$ mA, $I_C = 50$ mA)	$V_{BE(sat)}$		.85	V

SMALL SIGNAL CHARACTERISTICS

		Min.		
Forward Current Transfer Ratio Collector Voltage				
($V_C = 4.5$V, $I_C = 2$ mA, $f = 1$ kHz)	h_{fe}	180		

		Typical	
Forward Current Transfer Ratio	h_{fe}	180	
Input Impedance	h_{ie}	5100	ohms
Output Admittance	h_{oe}	14	μmhos
Voltage Feedback Ratio	h_{re}	.27	$\times 10^{-3}$

($V_{CE} = 10$V, $I_C = 1$ mA, $f = 1$ kHz, $T_A = 25$°C)

FIG. 4-7 Transistor data sheets. (*Courtesy General Electric Semiconductor Products Department, Syracuse, N. Y.*)

power gain is also obtained so that the transistor acts as a control device changing the power from the supply voltages to increased signal power.

A simple equivalent circuit useful for solving dc problems can be developed from Eq. (4-6) and a piecewise linear model for the base-emitter diode as in Fig. 3-9. The resulting circuit is indicated in Fig. 4-6 for the bias conditions of Fig. 4-4. For a *pnp* transistor, the polarity of the V_{BE} battery and the diode direction reverse and I_{CEO} is negative. The $\beta_0 i_B$ controlled current generator direction remains unchanged.

4-1.1 BIPOLAR-TRANSISTOR DATA

Manufacturers' data sheets for a general purpose *npn* transistor are presented in Fig. 4-7. We shall use the information given in Fig. 4-7 to determine what parameters and characteristics are important for the bipolar transistor.

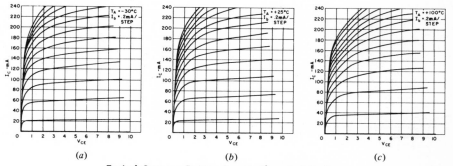

(a)   (b)   (c)

Typical Common Emitter Current Characteristic Curves

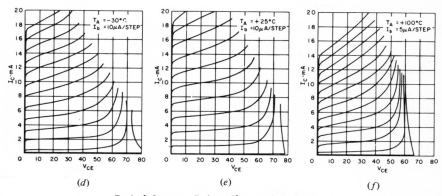

(d)   (e)   (f)

Typical Common Emitter Characteristic Curves

FIG. 4-7 Continued

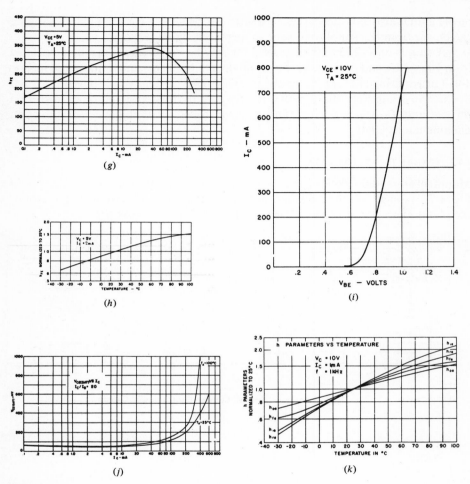

FIG. 4-7 Continued

Starting with Fig. 4-7a through f, the effect of ambient temperature and collector-to-emitter voltage on the common-emitter characteristics are shown. For this silicon transistor, the changes of I_{CEO} cannot be seen because of the current scale factors. However, it will be noted that each constant base-current curve moves upward as temperature increases.

In Fig. 4-7d, e, and f, another phenomenon occurs as seen where the curves start turning upward. This curvature is caused by avalanche breakdown at the junction between the collector and base. The pn-junction diode as discussed in Chap. 3 had a similar breakdown. The reverse bias between the collector and base results in a high field at the junction and the collision of high-energy carriers with fixed lattice electrons causes further multiplication of carriers.

The maximum reverse collector-base bias applied as in Fig. 4-7d with $I_B = 0$ before breakdown, is defined as $BV_{CBO} \equiv V_{(BR)CBO}$ and is a property of the transistor construction only. Defining M as the multiplying factor caused by avalanching of carriers, the current becomes MI_{CO}, where I_{CO} is the low-voltage value of collector current. At BV_{CBO}, the factor M becomes infinite, and the region of breakdown is attained. The M factor is found to follow the empirical expression

$$M = \cfrac{1}{1 - \left(\cfrac{V_{CB}}{BV_{CBO}}\right)^n} \qquad (4\text{-}7)$$

where

$$2 \leq n \leq 10$$

and where n is chosen to match the steepness of the curves near breakdown.

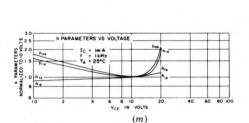

(m)

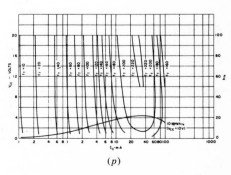

(n)

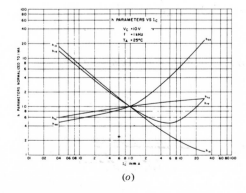

(o)

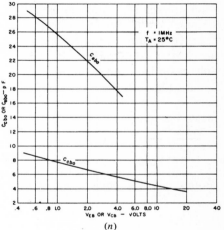

(p)

FIG. 4-7   Continued

4-1.2 COMMON-EMITTER BREAKDOWN

We defined β_f as a general term for current gain or transfer ratio; e.g.,

$$\beta_f = \frac{\Delta i_C}{\Delta i_B}\bigg|_{V_{CE}=\text{constant}}$$

The literature further breaks the definitions for common-emitter current gain into two other symbols:

$$(1) \quad h_{FE} = \frac{I_C}{I_B}\bigg|_{V_{CE}=\text{constant}}$$

$$(2) \quad h_{fe} = \frac{\partial i_c}{\partial i_b} = \frac{\Delta i_C}{\Delta i_B}\bigg|_{V_{CE}=\text{constant}} = \frac{i_c}{i_b}\bigg|_{V_{ce}=0} = \frac{di_c}{di_B}\bigg|_{V_{CE}=\text{constant}} = \frac{\partial i_C}{\partial i_B} = \beta_f$$

(See preface for notations regarding upper- and lower-case symbols and subscripts.) The dc current gain from base to collector is h_{FE} or β_0, and the ac or signal current gain from base to collector is h_{fe}. As noted previously, h_{FE} and h_{fe} have nearly the same magnitude. The dc term, h_{FE}, is often measured under pulse voltage and current conditions.

In Fig. 4-2, the major part of I_C is $|\alpha I_E|$. When multiplication occurs, the current becomes $M|\alpha I_E|$, and it appears that the common-base current-gain factor becomes

$$\alpha' = \left(\frac{I_C}{I_E}\right) = \frac{M|\alpha I_E|}{I_E} = M\alpha \tag{4-8}$$

Assuming h_{FE} to be defined as β in Eq. (4-5),

$$h_{FE}' = \frac{\alpha'}{1 - \alpha'} = \frac{M\alpha}{1 - M\alpha} \tag{4-9}$$

Since $M \to \infty$ at some point, the factor $M\alpha \to 1$ at which point $h_{FE} \to \infty$, and we conclude that breakdown occurs. When $M\alpha = 1$, from Eq. (4-7),

$$\alpha = 1 - \left(\frac{V_{CB}}{BV_{CBO}}\right)^n = \frac{1}{M} \qquad \text{the breakdown condition}$$

or

$$V_{CB} = BV_{CBO}(1 - \alpha)^{1/n} \qquad \text{for the condition } M\alpha = 1 \tag{4-10}$$

Since $h_{FE} = \alpha/(1 - \alpha) \simeq 1/(1 - \alpha)$, ($\alpha$ is near unity) and with the base an open circuit, the total potential drop is from collector to emitter so that Eq. (4-10) is

$$BV_{CEO} = BV_{CBO}\left(\frac{1}{h_{FE}}\right)^{1/n} \qquad \text{this is for the breakdown condition} \tag{4-11}$$

In Eq. (4-11), BV_{CEO} is the collector-to-emitter breakdown or sustaining voltage with the base open and with $h_{FE} \gg 1$, $BV_{CEO} < BV_{CBO}$.

When a circuit element is connected between base and emitter, the collector-to-emitter breakdown-voltage changes. The different breakdown voltages and i_C–v_{CE} curves are indicated in Fig. 4-8. When a resistor is between the base and emitter, the breakdown is defined as BV_{CER}; with the base short-circuited to the emitter, the breakdown is defined as BV_{CES}; and finally, when the base-emitter junction is reverse-biased, the breakdown is defined as BV_{CEX}. The value of BV_{CEX} can be changed by the magnitude of the base-emitter reverse bias.

If a transistor is operated in the breakdown region, it is designated as an avalanche transistor, and it has limited applications in switching networks.

From Fig. 4-7d, e, and f, BV_{CEO} is about 70 V at $-30°C$ and $+25°C$ temperature but has decreased to 60 V at $+100°C$. The absolute maximum rating as specified on the first data page of Fig. 4-7 is $V_{CEO} = BV_{CEO}$ min. (minimum) = 40 V which accounts for variation in manufacturing and maintains operation below the avalanche region.

In some applications, the emitter-base junction will be reverse-biased. It is important to note that the emitter-base breakdown voltage is low and is approximately 5 V for most transistors.

4-1.3 DEFINITIONS OF SMALL-SIGNAL ANALYSIS AND OPERATION

There are regions of nonlinearity in the characteristics of Figs. 4-3 and 4-4. If we attempt to model the entire characteristics, a nonlinear model is required. We will consider this problem in Sec. 4-1.9. A computer is usually required for electronic circuit analysis involving nonlinear models. Graphical techniques are useful, and these techniques will be considered later in the chapter. When the range of transistor currents and voltages lies within a region that is linear, the transistor can be modelled by a linear equivalent circuit. The combination of the linear equivalent circuit with other linear circuit elements may be analyzed by our conventional electrical-circuit analysis techniques. We refer to the transistor linear-equivalent circuits as small-signal equivalent circuits because the range

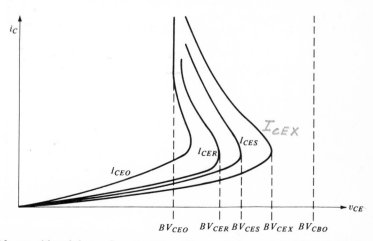

FIG. 4-8 A possible set of breakdown characteristics. The symbols are defined in the text.

over which the voltage and current change is often small. In the measurement of small-signal parameters, the voltages are usually in the millivolt range, while the currents are in the microampere range. We shall show that small-signal analysis is often applicable over much larger ranges of these voltages and currents.

We can obtain linear equivalent circuits from three different sources: (1) from the graphical characteristics (as in Sec. 4-1), (2) from the two-port or four-pole parameters of electrical-circuit analysis (as will be described in Sec. 4-1.4), and (3) or from physical electronic analysis of the transistor (as is shown in Sec. 4-1.6).

4-1.4 SMALL-SIGNAL *h* PARAMETERS

The small-signal *h* parameters were defined in Prob. 1-1, but the subscripts applied to transistors are literal rather than numerical. A second subscript, *e*, *b*, or *c*, is applied indicating the terminal that is common or reference. The first literal subscript evolves from the definitions as:

$$h_i = \left.\frac{V_1}{I_1}\right|_{V_2=0} \qquad \text{or subscript } i \text{ for } input \text{ impedance}$$

$$h_f = \left.\frac{I_2}{I_1}\right|_{V_2=0} \qquad \text{or subscript } f \text{ for } forward \text{ current gain}$$

$$h_r = \left.\frac{V_1}{V_2}\right|_{I_1=0} \qquad \text{or subscript } r \text{ for } reverse \text{ voltage gain}$$

$$h_o = \left.\frac{I_2}{V_2}\right|_{I_1=0} \qquad \text{or subscript } o \text{ for } output \text{ admittance}$$

The common-emitter *h* parameters for small signals at low frequencies are given on the first data page of Fig. 4-7. Note that these values are measured at a given bias point, $V_{CE} = 10$ V, $I_C = 1$ mA; at a frequency $f = 1$ kHz (kilohertz); and $T_A = 25°C$. An equivalent circuit with the given parameters is shown in Fig. 4-9.

The *h* parameters vary with temperature, collector-emitter voltage V_{CE}, and collector current I_C, as shown in Fig. 4-7*k*, *m*, and *o*. These curves are normalized to the typical values given above.

Example 4-1 The transistor of Fig. 4-7 is to be operated at a different bias point and temperature than that given for the *h*-parameter measurements. Operation is desired at $V_{CE} = 4$ V, $I_C = 4$ mA, and $T_A = 60°C$. Determine the *h* parameters for this new operating condition.

Referring to Fig. 4-7*k*, *m*, and *o*, each *h* parameter will be multiplied by three factors as determined from these curves. From Fig. 4-7*k*, the factors are: (approximately) $h_{oe} = 1.3$, $h_{fe} = 1.35$, $h_{re} = 1.4$, and $h_{ie} = 1.5$. From Fig. 4-7*m*, the factors are: $h_{oe} = 1.2$, h_{fe} and $h_{ie} = 0.95$, and $h_{re} = 1.15$. From Fig. 4-7*o*, the factors are: $h_{oe} = 3$, $h_{fe} = 1.2$, $h_{re} = 0.5$, and $h_{ie} = 0.35$.

The typical h parameters given in Fig. 4-7 at the new bias and temperature change to:

$$h_{oe} = (1.3)\,(1.2)\,(3)\,(14)\,(10^{-6}) = 65.5 \times 10^{-6}\,\text{mho}$$
$$h_{fe} = (1.35)\,(0.95)\,(1.2)\,(180) = 278$$
$$h_{re} = (1.4)\,(1.15)\,(0.5)\,(2.7 \times 10^{-4}) = 2.18 \times 10^{-4}$$
$$h_{ie} = (1.5)\,(0.95)\,(0.35)\,(5,100) = 2,540\ \Omega$$

Example 4-1 indicates that the equivalent circuit to be applied in any particular case may differ numerically from the typical parameters given in data sheets. It can be seen that if many variations are to be expected in a design problem, the use of a computer will be desirable.

The h parameters for bipolar transistors become complex when operated at signal frequencies in the MHz range. One reason for complex parameters can be seen from Fig. 4-7n. The emitter-base and collector-base capacitances are plotted versus the reverse bias voltage across the junctions. Since the emitter-base junction is forward-biased in amplifiers, a different value of capacitance than the plot indicates will be effective. The collector-base junction is normally reverse-biased, and the $C_{cbo} = C_{ob}$ values will be effective. The use of the computer in evaluating complex h or y parameters will be demonstrated in Chap. 9.

4-1.5 ALPHA AND h_{fe} VARIATIONS

The transit time of carriers across the base becomes an appreciable fraction of a cycle of operation at high frequencies. This transit-time effect causes significant variation of alpha and h_{fe} with frequency. Defining α_{fo} as the low-frequency value, a good approximation to α_f as a function of frequency is

$$\alpha_f = \frac{\alpha_{fo}}{(1 + jf/f_\alpha)} \tag{4-12}$$

Comparing Eq. (4-12) with Eq. (1-12), we see that a plot of α_f versus f would be of the form shown in Fig. 1-8, and f_α is the frequency at which $|\alpha_f| = 0.707\,\alpha_{fo}$. The term f_α is defined as the *alpha-cutoff frequency*. We remember that alpha variations are of importance in common-base operation, and f_α would be the frequency at which the current gain drops to 0.707 of the low-frequency value when the output load is zero ohms.

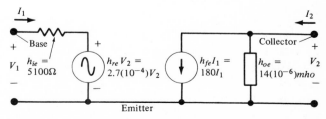

FIG. 4-9 The equivalent circuit for the typical common-emitter h parameters of the first data page of Fig. 4-7.

In the common-emitter configuration, we are interested in h_{fe} as a function of frequency. Writing h_{fe} in the form of Eq. (4-12),

$$h_{fe} = \frac{h_{feo}}{1 + j\dfrac{f}{f_\beta}} \tag{4-13}$$

where f_β is defined as the *beta-cutoff frequency*, and at $f = f_\beta$, $|\, h_{fe}\,| = 0.707\, h_{feo}$. It is of interest to relate f_α and f_β. Since β_f and h_{fe} are the same,

$$h_{fe} = \frac{\alpha_f}{1 - \alpha_f} \tag{4-14}$$

and we employ Eq. (4-12) to obtain,

$$h_{fe} = \frac{\dfrac{\alpha_{fo}}{1 + jf/f_\alpha}}{1 - \dfrac{\alpha_{fo}}{1 + jf/f_\alpha}} = \frac{\alpha_{fo}}{1 + jf/f_\alpha - \alpha_{fo}} \tag{4-15}$$

Equation (4-15) can be made in the form of Eq. (4-13) as

$$h_{fe} = \frac{\dfrac{\alpha_{fo}}{1 - \alpha_{fo}}}{1 + j\dfrac{f}{(1 - \alpha_{fo})f_\alpha}} = \frac{h_{feo}}{1 + j\dfrac{f}{(1 - \alpha_{fo})f_\alpha}} \tag{4-16}$$

and by comparison we see that

$$f_\beta = (1 - \alpha_{fo})f_\alpha = \frac{f_\alpha}{1 + h_{feo}} \tag{4-17}$$

For $\alpha > 0.9$, $f_\beta \leq 0.1\, f_\alpha$. The current and voltage gains of the common-emitter amplifier are much higher than that of the common-base circuit so that most high-frequency amplifiers are operated in the common-emitter configuration.

Measurements indicate that Eq. (4-17) is in some error, and the equation is modified to

$$f_\beta = K_\theta(1 - \alpha_{fo})f_\alpha \tag{4-18}$$

where $0.6 < K_\theta \leq 1$. The value of K_θ is often furnished on data sheets.

A current-gain bandwidth product or transition frequency defined as f_T is applied to the common-emitter stage. This parameter is the frequency at which $|\, h_{fe}\,|$ is unity. From Eq. (4-13) at the frequency $f = f_T$,

$$|\, h_{fe}\,| = \frac{h_{feo}}{[1 + (f_T/f_\beta)^2]^{1/2}} = 1 \tag{4-19}$$

It is always true that $f_T > f_\beta$ so that $(f_T/f_\beta)^2 \gg 1$. Thus Eq. (4-19) yields

$$f_T \cong h_{feo} f_\beta \tag{4-20}$$

Inserting Eq. (4-20) into Eq. (4-17), we find that f_T is only slightly smaller than f_α.

The value of f_T is a function of V_{CE} and I_C as shown in Fig. 4-7, and in general f_T is largest for large values of V_{CE} and I_C. The value of $|h_{fe}|$ at 10 MHz is also plotted in this figure.

Example 4-2 Determine the magnitude of h_{fe} and the reactance of C_{CBO} at 1 MHz for the transistor of Fig. 4-7. The transistor is biased at $V_{CE} = 10$ V, $I_C = 1$ mA so that $h_{feo} = 180$.

From Eqs. (4-13) and (4-20),

$$h_{fe} = \frac{h_{feo}}{1 + j\dfrac{h_{feo} f}{f_T}}$$

or

$$|h_{fe}| = \frac{h_{feo}}{\left[1 + \left(\dfrac{h_{feo} f}{f_T}\right)^2\right]^{1/2}}$$

Figure P4-7 shows that $f_T = 60$ MHz, and

$$|h_{fe}| = \frac{180}{\left[1 + \left(\dfrac{180 \times 10^6}{60 \times 10^6}\right)^2\right]^{1/2}} = 18$$

The magnitude of the common-emitter current gain is reduced by a factor of 10 in this example at $f = 1$ MHz. The value of h_{fe} would be approximately constant at a value of 180 to $f = 10^5$ Hz for this bias point.

The capacitive reactance for $C_{cbo} = 4$ pF is $X_C = 1/2\pi f C_{cbo} = [1/2\pi(10^6)4(10^{-12})] = 40,000$ Ω at 1 MHz and its effect on amplifier performance is negligible if the load resistance is 4,000 Ω or less.

4-1.6 THE HYBRID-π MODEL

A transistor model for small signals (where the device acts linearly) having linear-circuit parameters has been developed by Giacolletto [2]. This model can also be developed from the physical electronics of devices. The resulting hybrid-π circuit has elements which are not functions of frequency, and the circuit is applicable for frequencies less than $f_T/3$.

A form of the hybrid-π model is given in Fig. 4-10 for the useful common-emitter case. The reasons for the parameters and the expressions for determining their values may be qualitatively explained as follows:

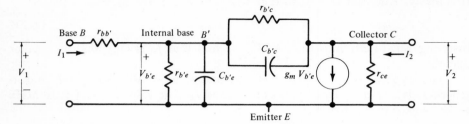

FIG. 4-10 The hybrid-π transistor model. The subscripts relate the nodes between which elements are connected.

$r_{bb'}$: The base-spreading resistance $r_{bb'}$ accounts for the resistivity of the base material and the geometry of the base region. Its value varies from tens to a few hundred ohms and may be determined from a known value of h_{ie}.

B': Because of the potential drop due to base current in $r_{bb'}$, the transistor acts as if there were <u>an internal base connection at B'</u> which cannot be reached externally.

$r_{b'e}$: The forward-biased base-emitter diode has an <u>incremental or slope resistance $r_{b'e}$</u>, and its value can be determined from the sketch of Fig. 4-11. With this junction forward-biased at point (V_{DC}, I_{DC}), the incremental change of slope is given by use of Eq. (3-1), assuming $I_S = I_O$ and the exponential much greater than unity, as

$$\frac{1}{r_e} = \frac{di_E}{dv_{EB}} = \frac{q}{kT} I_o e^{qv_{EB}/kT} \cong \frac{qI_{DC}}{kT} = \frac{I_{DC}}{.026}\bigg|_{T=300^\circ \mathrm{K}} \tag{4-21}$$

where I_{DC} = current where $v_{EB} = V_{DC}$

From Fig. 4-2 and neglecting I_{CO}, $|I_B| = (1 - \alpha)|I_E|$ and $\Delta i_B = \Delta i_E(1 - \alpha)$. By definition,

$$r_{b'e} = \frac{\Delta v_{EB}}{\Delta i_B} = \frac{1}{1-\alpha}\frac{\Delta v_{EB}}{\Delta i_E} = \frac{r_e}{1-\alpha} \cong h_{fe}r_e \tag{4-22}$$

Hence, knowing the bias point and h_{fe} (or β), $r_{b'e}$ can be calculated.

$C_{b'e}$: With the emitter-base junction biased as in Fig. 4-11, there is a charge Q in the base region which is directly proportional to I_{DC}. Thus $Q = Q_o(e^{qv_{EB}/kT} - 1)$. Approximately,

$$C_{b'e} = \frac{\Delta Q}{\Delta v_{EB}} \cong \frac{Qq}{kT} = I_{DC}t_D \cdot \frac{q}{kT}$$

where t_D = average diffusion time of carriers in the base. We see that $r_{b'e}C_{b'e}$ is independent of I_{DC} (except where α or β is a function of I_{DC}), and it will be shown that $C_{b'e} = g_m/2\pi f_T$.

g_m: Since the circuit is linear, the short-circuit collector current will be linearly related to the voltage $V_{b'e}$, and the proportionality constant is defined as a transconductance g_m. We may write

$$g_m = \frac{di_{C(\text{short-circuit})}}{dv_{B'E}} = \frac{d(\alpha i_E)}{dv_{B'E}} \cong \frac{\alpha}{r_e} \tag{4-23}$$

Also, g_m may be related to h_{fe} as follows:

$$h_{fe} = \left.\frac{I_2}{I_1}\right|_{v_2=0} = \frac{g_m V_{b'e}}{I_1} = g_m r_{b'e} \tag{4-24}$$

Since $\alpha \cong 1$, from Eq. (4-23)

$$g_m \cong \frac{1}{r_e} = \left(\frac{1}{.026}\right) I_E \cong 38.5 I_C \tag{4-25}$$

where collector current $I_C \cong I_E$. To a good approximation, g_m is independent of the type of transistor.

$C_{b'c}$: The capacitance between collector and base $C_{b'c}$ is the same as $C_{cbo} = C_{ob}$ as discussed previously.

$r_{b'c}$: This is a large resistance in the order of megohms which is due to changes in effective base width caused by signal fluctuations in the collector-base voltage. We normally omit $r_{b'c}$ for approximate calculations, but it can easily be included in computer analysis.

r_{ce}: The collector-to-emitter resistance is in the tens of thousands of ohms range and is also caused by the effective base-width modulation due to collector-signal voltage variation. Its value is most easily determined from h parameters.

4-1.7   RELATING $C_{b'e}$ TO f_T

Since the f_T values are normally published in transistor data sheets, it is desirable to relate $C_{b'e}$ to f_T so that the hybrid-π model can be determined. We refer to Fig. 4-12, and write an expression for input admittance under steady-state sinusoidal excitation as,

$$\frac{I_i}{V_i} = \frac{1}{r_{b'e}} + j\omega(C_{b'e} + C_{b'c}) \tag{4-26}$$

(Note that $r_{b'c}$ is large and has been omitted.) From the definition of h_{fe} and

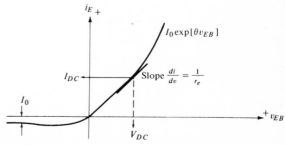

FIG. 4-11   The base-emitter diode characteristics for the definition of r_e (npn).

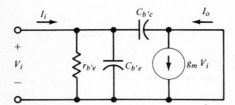

FIG. 4-12 A simplified hybrid-π circuit with a short circuit load.

Eq. (4-26),

$$h_{fe} = \frac{I_o}{I_i} \cong \frac{g_m V_i}{I_i} = \frac{g_m r_{b'e}}{1 + j\omega r_{b'e}(C_{b'e} + C_{b'c})} \tag{4-27}$$

Equation (4-27) is in the same form as Eq. (4-13), and we see that

$$f_\beta = \frac{1}{2\pi r_{b'e}(C_{b'e} + C_{b'c})} \tag{4-28}$$

Solving for $C_{b'e} + C_{b'c}$,

$$C_{b'e} + C_{b'c} = \frac{1}{2\pi f_\beta r_{b'e}} \cong \frac{1}{2\pi f_\beta h_{fe} r_e}$$

and applying Eq. (4-23) with $\alpha = 1$,

$$C_{b'e} = \frac{g_m}{2\pi f_\beta h_{fe}} - C_{b'c} = \frac{g_m}{2\pi f_T} - C_{b'c} \tag{4-29}$$

where Eq. (4-20) has been used. Thus, knowing f_T, h_{fe}, and the bias current which yields g_m, we can calculate $C_{b'e}$ from Eq. (4-29).

Example 4-3 Determine all the hybrid-π parameters possible for the transistor of Fig. 4-7 at the bias point $V_{CE} = 10$ V, $I_C = 1$ mA.

From Eq. (4-25), $g_m = 38.5 I_C = 38.5(10^{-3}) = 38.5$ mmho (millimhos), and since $f_T = 60$ MHz, with $C_{b'c} = C_{cbo} = 4$ pF from Fig. 4-7, Eq. (4-29) yields

$$C_{b'e} = \frac{g_m}{2\pi f_T} - C_{b'c} = \frac{0.0385}{2\pi(6)(10^7)} - 4(10^{-12}) = 98 \text{ pF}$$

From Eq. (4-22),

$$r_{b'e} = h_{fe} r_e \doteq \frac{h_{fe}}{g_m} = \frac{180}{0.0385} = 4,670 \ \Omega$$

Thus far, the above parameters are all that we can calculate. Table 4-1 indicates approximate relations between h and hybrid-π parameters (see

TABLE 4-1 Approximate Relations between h and Hybrid-π Parameters with a Summary of Parameter Expressions

$$r_{b'e} = h_{ie} - r_{bb'}$$

$$r_{b'c} = \frac{h_{ie} - r_{bb'}}{h_{re}}$$

$$g_m = \frac{h_{fe}}{h_{ie} - r_{bb'}}$$

$$\frac{1}{r_{ce}} = h_{oe} - \frac{h_{fe} h_{re}}{h_{ie} - r_{bb'}}$$

$$r_{b'e} = h_{fe}(kT)/qI_C$$

$$g_m = 38.5 \, I_C$$

$$C_{b'e} = g_m/2\pi f_T - C_{b'c}$$

$$C_{b'c} = C_{cbo}$$

Prob. 4-15). Applying this table, we obtain the following values: with $h_{ie} = r_{bb'} + r_{b'e}$,

(1) $$r_{bb'} = h_{ie} - \frac{h_{fe}}{g_m} = 5{,}100 - \frac{180}{0.0385} = 425 \, \Omega$$

(2) $$r_{b'c} = \frac{h_{ie} - r_{bb'}}{h_{re}} = \frac{5{,}100 - 425}{2.7 \times 10^{-4}} = 17.3 \, \text{M}\Omega$$

(3) $$1/r_{ce} = h_{oe} - \frac{h_{fe}h_{re}}{h_{ie} - r_{bb'}} = 14 \times 10^{-6} - \frac{180(2.7)(10^{-4})}{4{,}670}$$

$$= 3.6 \times 10^{-6} \, \text{mmho}$$

or

$$r_{ce} = 278{,}000 \, \Omega$$

Figure 4-13 summarizes the results.

Example 4-3 shows that $r_{b'c}$ and r_{ce} can be omitted from the circuit as a first approximation because of their size in relation to other parameters including normal load values, but for computer analysis or design, the circuit can contain all the circuit components. However, calculations with a simplified circuit will

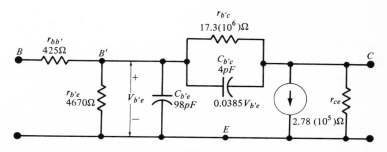

FIG. 4-13 The results of Example 4-3.

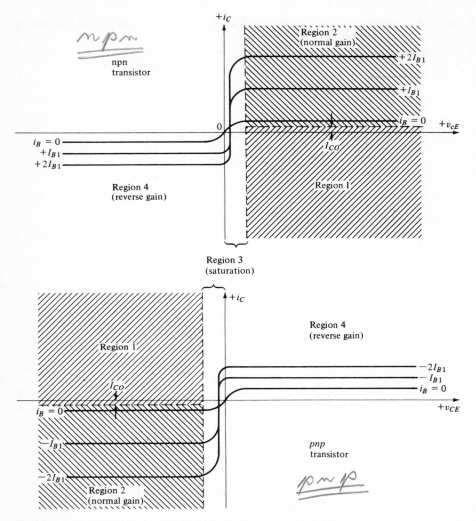

FIG. 4-14 The defined regions of operation for *npn* and *pnp* transistors.

often yield satisfactory answers more quickly than the time necessary to place
the complex circuit in the computer.

4-1.8 REGIONS OF OPERATION

Transistors have applications over wide ranges of currents and voltages. It is
convenient to define operating regions so that analysis and design may be
simplified.

Four regions are shown in Fig. 4-14 and may be defined by the following
bias conditions:

Region 1:   emitter- and collector-base junctions reverse-biased

Region 2:   emitter junction forward-biased and collector junction reverse-biased
Region 3:   emitter- and collector-base junctions forward-biased
Region 4:   collector-base junction forward-biased and emitter-base junction reverse-biased

A further definition of these regions is indicated in Fig. 4-14; the normal gain, the saturation, and reverse gain regions. In region 2, normal amplifier gain is accomplished, and the h and hybrid-π parameters were determined for region 2 in the preceding section.

4-1.9 EBERS-MOLL TRANSISTOR MODEL

For large-signal or switching problems and computer applications, a model of the transistor which applies to all regions of Fig. 4-14 has been developed by J. J. Ebers and J. L. Moll [7].

The defining of voltages and currents for the Ebers-Moll model are given in Fig. 4-15. The junction voltages v_C and v_E have positive values if the p region is positive with respect to the n region. The terminal voltages differ from v_C and v_E by ohmic voltage drops across the volume resistances of the transistor, especially the base spreading resistance.

With the v–i characteristic of any junction given by Eq. (3-1), $i_C = +I_{CO}(e^{qv_C/kT} - 1)$ when i_E is zero; and in region 2, the expression for current is given by Eq. (4-3). Defining $\theta = q/\eta kT$, we may rewrite Eq. (4-3) as

$$i_C = -\alpha_F i_E + I_{CO}(e^{\theta v_C} - 1) \qquad (4\text{-}30)$$

(+) sign due to forward polarity of v_C

If v_C is negative, the case of a reverse-biased junction, Eq. (4-30) reduces to Eq. (4-2). The F subscript on alpha represents the α value in the forward gain region 2. The α_F is the same as α and α_0 used earlier.

We next consider that the transistor could be operated in the reverse mode in which the collector is the input terminal and the emitter the output terminal, i.e., operation is in the reverse gain area of Fig. 4-14. We may write an expression similar to Eq. (4-30) as

$$i_E = -\alpha_R i_C + I_{EO}(e^{\theta v_E} - 1) \qquad (4\text{-}31)$$

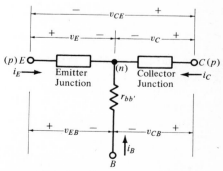

FIG. 4-15   Voltages and currents applied in the Ebers-Moll model (*pnp* transistor).

where I_{EO} is the reverse saturation current of the emitter junction. Here, α_R is the reverse common-base current gain defined as $\Delta i_E / \Delta i_C$ in region 4. Some references indicate α_F by the symbol α_N, and the reverse alpha α_R by α_I.

It is assumed that superposition of transistor currents applies; i.e., the current injected at the emitter junction is independent of the current injected at the collector junction and vice versa. Actually, the emitter and collector currents depend upon the excess holes (or electrons) injected into the base as

$$i_E = a_{11} p_E + a_{12} p_C$$
$$i_C = a_{21} p_E + a_{22} p_C \qquad \Big\} \quad pnp \ transistor$$

for pnp transistors, where the a's are constants and p_E and p_C are the excess hole densities injected into the base from the emitter and collector, respectively. These are linear equations so that superposition may be used; the nonlinearities arise from the law of the junction; that is, $p_E = p_n(e^{\theta v_{EB}} - 1)$. Thus, solving Eqs. (4-30) and (4-31) for i_C and i_E,

$$i_E = \frac{-\alpha_R I_{CO}}{1 - \alpha_R \alpha_F} (e^{\theta v_C} - 1) + \frac{I_{EO}}{1 - \alpha_F \alpha_R} (e^{\theta v_E} - 1) \qquad (4\text{-}32)$$

$$i_C = \frac{-\alpha_F I_{EO}}{1 - \alpha_R \alpha_F} (e^{\theta v_E} - 1) + \frac{I_{CO}}{1 - \alpha_F \alpha_R} (e^{\theta v_C} - 1) \qquad (4\text{-}33)$$

If $|\theta v_C| \ll 1$ and $|\theta v_E| \ll 1$, the power series expansion for the exponentials in Eqs. (4-32) and (4-33) can be represented by the first two terms or $(e^{\theta v_C} - 1) = 1 + \theta v_C - 1 = \theta v_C$, etc. Equations (4-32) and (4-33) become

$$i_E = - \left[\frac{\alpha_R I_{CO} \theta}{(1 - \alpha_F \alpha_R)} \right] v_C + \left[\frac{I_{EO} \theta}{(1 - \alpha_F \alpha_R)} \right] v_E \qquad (4\text{-}32a)$$

$$i_C = + \left[\frac{I_{CO} \theta}{(1 - \alpha_F \alpha_R)} \right] v_C - \left[\frac{\alpha_F I_{EO} \theta}{(1 - \alpha_F \alpha_R)} \right] v_E \qquad (4\text{-}33a)$$

Reciprocity has to apply for small values of current and voltage to avoid the possibility that net power can be supplied from the transistor. For reciprocity to apply,

$$\alpha_R I_{CO} = \alpha_F I_{EO} \qquad (4\text{-}34)$$

and we find Eq. (4-34) has to be valid to avoid a violation of the laws of thermodynamics. Only three of the parameters α_R, I_{CO}, α_F and I_{EO} are independent at low currents. For better curve fitting at normal operating currents, Eq. (4-34) may not be valid.

Another form for Eqs. (4-32) and (4-33) is sometimes convenient. If we define

$$I_{ES} = \frac{I_{EO}}{(1 - \alpha_F \alpha_R)} \qquad (4\text{-}35)$$

and

$$I_{CS} = \frac{I_{CO}}{(1 - \alpha_F \alpha_R)} \tag{4-36}$$

Equations (4-32) and (4-33) may be rewritten as:

$$i_C = -\alpha_F I_{ES}(e^{\theta v_E} - 1) + I_{CS}(e^{\theta v_C} - 1) \tag{4-37}$$

$$i_E = I_{ES}(e^{\theta v_E} - 1) - \alpha_R I_{CS}(e^{\theta v_C} - 1) \tag{4-38}$$

The magnitude of the emitter current when the emitter junction is reverse-biased by more than about $3/\theta$ V, with a short circuit from collector to base, is given by I_{ES}, and a similar definition applies for I_{CS}.

4-1.10 COMMON-EMITTER COLLECTOR CHARACTERISTICS FROM THE EBERS-MOLL MODEL

We can become familiar with the use of the Ebers-Moll transistor model by developing the common-emitter i–v characteristics. For region 1 of Fig. 4-14 with v_C and v_E much less than zero (negative), Eq. (4-37) becomes

$$i_C = -I_{CS} + \alpha_F I_{ES}$$

and since $\alpha_F I_{ES} = \alpha_R I_{CS}$ by reasoning similar to that for Eq. (4-34),

$$i_C = -(1 - \alpha_R)I_{CS} \tag{4-39}$$

The emitter current from Eq. (4-38) becomes

$$i_E = -I_{ES} + \alpha_R I_{CS} = -(1 - \alpha_F)I_{ES} \tag{4-40}$$

and the base current, from Kirchhoff's law is

$$i_B = -(i_E + i_C) = (1 - \alpha_R)I_{CS} + (1 - \alpha_F)I_{ES} \tag{4-41}$$

For region 2, with $v_C \ll 0$ and $v_E \gg 0$ in Eq. (4-30), $i_C = -I_{CO} - \alpha_F i_E$ $= -I_{CO} + \alpha_F(i_B + i_C)$, or

$$i_C = \frac{\alpha_F}{1 - \alpha_F} i_B - \frac{I_{CO}}{1 - \alpha_F} = \beta_F i_B - (\beta_F + 1)I_{CO} \tag{4-42}$$

where $\beta_F = \alpha_F/(1 - \alpha_F)$

For region 4, with $v_E \ll 0, v_C \gg 0$ in Eq. (4-31), $i_E = -(i_B + i_C) = -\alpha_R i_C$ $- I_{EO}$, so that the collector current is

$$i_C = -\frac{i_B}{1 - \alpha_R} + \frac{I_{EO}}{1 - \alpha_R} \tag{4-43}$$

Expanding the region near the voltage axis (not to scale) in order to show the salient features, Fig. 4-16 summarizes the results of the above analysis. The common point, through which all curves pass, is the cutoff value of collector current given by Eq. (4-39). At this common point, Eqs. (4-37) and (4-39)

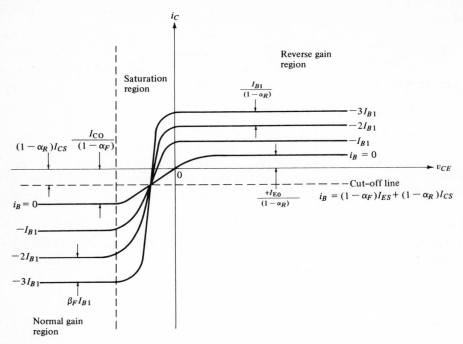

FIG. 4-16 Common-emitter i_C versus v_{CE} characteristics from the Ebers-Moll model of a *pnp* transistor.

yield,

$$-(1 - \alpha_R)I_{CS} = -\alpha_F I_{ES}(e^{\theta v_E} - 1) + I_{CS}(e^{\theta v_C} - 1)$$

With $\alpha_F I_{ES} = \alpha_R I_{CS}$ and $v_{CE} = v_C - v_E$, this last expression leads to

$$v_{CE} = -\frac{1}{\theta}\ln\frac{1}{\alpha_R} \tag{4-44}$$

Current values for i_C in the saturation area can be computed from Eqs. (4-32) and (4-33).

The principal difference between the curves of Fig. 4-16 and the $i_C(v_{CE}, i_B)$ curves for a commercial transistor is that the collector-current lines for constant i_B have nonzero slope in region 2 where the collector junction is reverse-biased. When the magnitude of this bias increases, the effective base width narrows due to widening of the collector junction barrier or depletion layer, α_F increases causing larger β_F. The breakdown regions are not included in the models of Fig. 4-16, but modifications can be made to account for most variations encountered with actual transistors.

Table 4-2 presents some actual Ebers-Moll parameters for four transistor models [8, 9]. In this table, θ_E is $q/\eta kT$ for the emitter-base junction while θ_C is $q/\eta kT$ for the collector-base junction, the values being determined at room

TABLE 4-2 Table of Transistor Model Parameters

	2N404	2N916	2N955A	2N1016B	
α_F	0.9949	0.9892	0.994	0.9896	
α_R	0.88	0.2593	0.2857	0.8333	
I_{ES}	3.0×10^{-6}	4.95×10^{-12}	1.81×10^{-4}	2.73×10^{-7}	mA
I_{CS}	3.51×10^{-6}	1.93×10^{-10}	8.05×10^{-3}	8.76×10^{-7}	mA
θ_e	39.	39.	37.	38.	V^{-1}
θ_c	39.	35.	36.	35.	V^{-1}
R_b	0.12	0.042		0.04	kΩ
R_c	0.0044	0.033		0.6×10^{-3}	kΩ

temperature. Other parameters not defined in the table are: $R_b = r_{bb'}$ and R_c is a series resistance added to the collector circuit.

The transistor ratings and types are:

2N404	*pnp*	150 mW	germanium
2N916	*npn*	360 mW	silicon
2N955A	*npn*	150 mW	germanium
2N1016B	*npn*	150 W	silicon

4-1.11 DETERMINING REGIONS OF OPERATION

We shall indicate a procedure for determining the region of transistor operation by an example. In the circuits of Fig. 4-17 neglect the leakage currents, assume all forward-biased junctions to have zero voltage across them, and $\beta_F = 50$, $\beta_R = \alpha_R/(1 - \alpha_R) = 2$. We wish to find the dc values of V_{CE}, I_C, and I_B, and define the region of operation.

From Fig. 4-17a, the polarity of the 2-V source indicates that the base-emitter junction is reverse-biased. Since $I_B = I_C = 0$, $V_{CB} = 10 + 2 = 12$ V reverse bias so that $V_{CE} = 10$ V, and the transistor is in the cutoff region.

For Fig. 4-17b, the base-emitter junction is now forward-biased with $I_B = 2/10^4 = 0.2$ mA. If operation is in the normal gain region, $I_C = \beta_F I_B = 10$ mA. However, this current would cause a 20-V drop in the collector resistance, which is not possible; therefore, the transistor is saturated with $V_{CE} = 0$ V, and $I_C = 10/2{,}000 = 5$ mA.

For Fig. 4-17c, the base-emitter junction is forward-biased and $I_B = 0.2$ mA with $I_C = \beta_F I_B = 10$ mA. The output voltage is $V_{CE} = 30 - I_C(2 \text{ k}) = 30 - 20 = 10$ V. Here the operation is in the normal gain region.

In Fig. 4-17d, the base-emitter junction appears to be reverse-biased while the collector-base junction is forward-biased. Therefore, we assume operation is in the reverse gain region where $I_C = -(I_B + I_E) = -(\beta_R + 1)I_B$ since $I_E = \beta_R I_B$. Writing a voltage equation around the collector-base loop, $10 - 2 = I_B(10 \text{ k}) - I_C(2 \text{ k}) = (10 \text{ k} + (\beta_R + 1)(2 \text{ k}))I_B$ or $I_B = 0.5$ mA, and $I_C = -(3)I_B = -15$ mA. For these values, $V_{CE} = -10 + 3 = -7$ V. The transistor has been reversed so that the base-collector junction is forward-biased and the

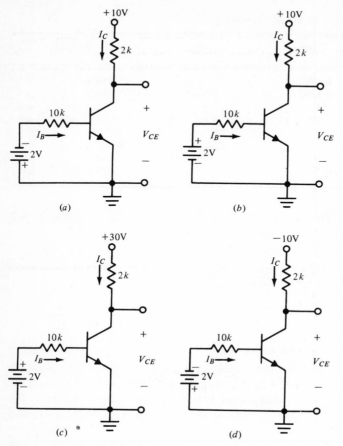

FIG. 4-17 Example circuits for determining regions of operation.

base-emitter reverse-biased. The calculations are similar to those for the normal gain region except for the use of β_R in place of β_F, and the roles of the collector and emitter currents have been interchanged.

4-1.12 JUNCTION-VOLTAGE EXPRESSIONS

We solve for junction voltages in terms of the currents from Eqs. (4-30) and (4-31). The results are:

$$v_E = \frac{1}{\theta} \ln \left| 1 + \frac{i_E + \alpha_R i_C}{I_{EO}} \right| \tag{4-45}$$

and

$$v_C = \frac{1}{\theta} \ln \left| 1 + \frac{i_C + \alpha_F i_E}{I_{CO}} \right| \tag{4-46}$$

These equations may be applied as follows. For the 2N916 transistor in Table 4-2, $\alpha_F = 0.9892$, $\alpha_R = 0.2593$, $I_{ES} = 4.95(10^{-12})$ mA and $I_{CS} = 1.93(10^{-10})$ mA. From Eq. (4-35) and (4-36), $I_{EO} = 3.68(10^{-12})$ mA and $I_{CO} = 1.43(10^{-10})$ mA. We may estimate the cutin voltage from Eq. (4-45) by assuming that the cutin voltage occurs when $I_C = 100\ I_{CO}/(1 - \alpha_F)$ or 100 I_{CEO} (see Fig. 4-16) or $I_C = (100)(1.43)(10^{-10})/(0.0108) = (1.32)(10^{-6})$ mA. This is the approximate value of I_E since $I_B = 0$, and from Eq. (4-45), with current directions for an *npn* transistor, $\eta = 2$ for silicon,

$$v_{BE}(\text{cutin}) = \frac{1}{\theta} \ln \left| 1 + \frac{(-1.32)(10^{-6}) + (0.2593)(1.32)(10^{-6})}{(3.68)(10^{-12})} \right|$$

$$= 2(0.026)\ln(265687) = 0.649 \text{ V}$$

in good agreement with the value obtained for a silicon transistor such as given in Fig. 4-7*i*.

If we consider the same transistor in the saturation region operating with $I_C = 10$ mA, $I_B = 1.0$ mA, so that $I_E = -11$ mA, the junction saturation voltages are obtained from Eqs. (4-45) and (4-46) as:

$$v_{E(sat)} = 2(0.026)\ln\left| 1 + \frac{(-11) + (0.2593)(10)}{(3.68)(10^{-12})} \right| = 1.48 \text{ V}$$

and

$$v_{C(sat)} = 2(0.026)\ln\left| 1 + \frac{(10) + (0.9892)(-11)}{(1.43)(10^{-10})} \right| = 1.17 \text{ V}$$

Thus, for an *npn* saturated transistor, $v_{CE(sat)} = -v_C - (-v_E) = -1.17 + 1.48 = 0.31$ V. From Fig. 4-7*j*, $v_{CE(sat)}$ is about 0.3 V indicating reasonable agreement. The value of $v_{E(sat)}$ agrees favorably with $v_{BE(sat)}$ as specified in the electrical characteristics of this same figure.

4-1.13 BIAS CIRCUIT

Transistor equivalent circuits are useful for small signals where device operation is nearly linear, but for large-signal operation, graphical or computer techniques are necessary.

An *npn* transistor is drawn in a preferred bias circuit in Fig. 4-18. It can be shown that this bias arrangement yields excellent stability of Q point when β and temperature variations are considered. The coupling capacitors C_1 and C_2 are included to isolate the load R_L and source v_s and R_s from the dc voltages. The effect of R_E and C_E on signal gain will be considered later.

4-1.14 DC BIAS

The dc bias point for the transistor is defined by specified values for I_C, V_{CE}, and I_B. In Fig. 4-18, these bias values are controlled by V_{CC}, R_1, R_2, R_C, and R_E. In Fig. 4-19, V_{CC}, R_1, and R_2 have been combined into a Thévenin equivalent

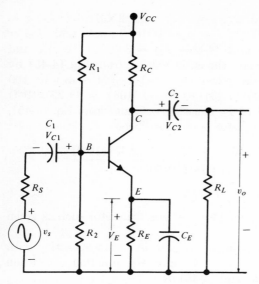

FIG. 4-18 A preferred bias circuit with a load and generator coupled by capacitors. The capacitor C_E is selected for low impedance at moderate-to-high signal frequencies.

source. The bias point will be determined from this figure. We suggest that the transistor be replaced by the circuit of Fig. 4-6 in order to follow the calculations ($I_{CEO} = 0$).

For a particular application, the bias point is known, that is, I_C and V_{CE} have been chosen to yield a specific output voltage and current variation under signal. The supply voltage V_{CC} will be specified by available sources in the system. A "rule of thumb" for reasonable bias stability is to allow $I_E R_E \simeq I_C R_E$ to be 0.1 V_{CC}, or

$$R_E = \frac{0.1\ V_{CC}}{I_C} \tag{4-47}$$

Since R_{TH} shunts the input to the transistor, it should be much greater than $r_{b'e}$ so that most of the input signal current goes into the base. A satisfactory design starting point is,

$$R_{TH} = 10 r_{b'e} = 10\ \frac{\beta}{g_m} = \frac{10\beta}{40\ I_C} = \frac{0.25\beta}{I_C} \tag{4-48}$$

Considering the base-emitter loop of Fig. 4-19,

$$V_{TH} = I_B R_{TH} + V_{BE} - I_E R_E \cong I_B R_{TH} + V_{BE} + I_C R_E$$

or

$$V_{TH} = I_C \left(\frac{R_{TH}}{\beta} + R_E\right) + V_{BE} \tag{4-49}$$

Knowing R_{TH} from Eq. (4-48) and V_{TH} from (4-49), R_1 and R_2 may be determined. The collector resistance R_C is found from the collector-emitter loop as,

$$V_{CC} = I_C R_C + V_{CE} - I_E R_E \cong I_C (R_C + R_E) + V_{CE}$$

or

$$R_C = \frac{V_{CC} - V_{CE}}{I_C} - R_E \tag{4-50}$$

Example 4-4 Design the bias circuit of the amplifier in Fig. 4-18 for a transistor having characteristics as in Fig. 4-7e. The bias point is to be set at $I_C = 10$ mA, $V_{CE} = 20$ V, and V_{CC} is 40 V.

From Fig. 4-7e, $\beta = h_{FE} = I_C/I_B = (10 \times 10^{-3})/(50 \times 10^{-6}) = 200$ at the bias point. From Eq. (4-47),

$$R_E = \frac{0.1(40)}{10^{-2}} = 400 \ \Omega$$

and from Eq. (4.48),

$$R_{TH} = \frac{(0.25)(200)}{(0.01)} = 5,000 \ \Omega$$

We note from Fig. 4-7i that $V_{BE} \cong 0.6$ V so that from Eq. (4-49),

$$V_{TH} = 10^{-2} \left(\frac{5 \times 10^3}{200} + 400 \right) + 0.6 = 4.85 \ \text{V}$$

From Fig. 4-19, $R_{TH} = R_1 R_2/(R_1 + R_2)$ and $V_{TH} = (V_{CC} R_{TH})/R_1$, so that $R_1 = (V_{CC} R_{TH})/V_{TH} = [40(5)(10^3)]/4.85 = 41,200 \ \Omega$, and a nominal value of R_1 would be selected (39,000 Ω). From $V_{TH} = V_{CC} R_2/(R_1 + R_2)$,

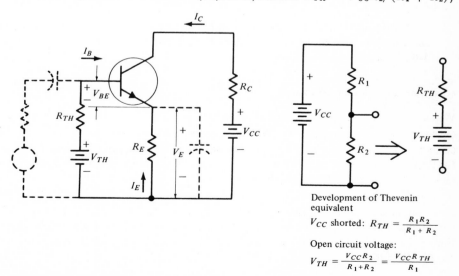

Development of Thévenin equivalent

V_{CC} shorted: $R_{TH} = \dfrac{R_1 R_2}{R_1 + R_2}$

Open circuit voltage:

$V_{TH} = \dfrac{V_{CC} R_2}{R_1 + R_2} = \dfrac{V_{CC} R_{TH}}{R_1}$

FIG. 4-19 The bias circuit of Fig. 4-18 converted to Thévenin equivalent form.

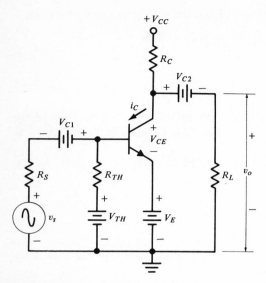

FIG. 4-20 The circuit for drawing dc and ac (or dynamic) load lines. Capacitors C_1, C_2, and C_E have been charged to voltages V_{C1}, V_{C2}, and V_E, respectively, that are assumed constant for the frequencies applied.

$R_2 = [R_1/(V_{CC}/V_{TH}) - 1] = [41,200/(40/4.85) - 1] = 5,700 \ \Omega$ or a nominal value of 5,600 Ω. Finally from Eq. (4-50),

$$R_C = \frac{40 - 20}{10^{-2}} - 400 = 1,600 \ \Omega$$

or nominally, 1,500 Ω. The resulting bias point is accurate enough considering the variability of transistor characteristics. (See Probs. 4-2 through 4-6 for further bias considerations.)

4-1.15 AC OR DYNAMIC LOAD LINE

When a time-varying signal is applied as in Fig. 4-18, the transistor voltages and currents also vary with time. We shall show that the operating path on the collector characteristics is a straight line passing through the bias point. Assume that the frequency of v_s is such that all capacitors maintain the dc voltage (or charge) developed by the bias circuit. From Fig. 4-18 with $v_s = 0$, $V_{C1} = V_{BE} + V_E$, $V_{C2} = V_{CE} + V_E$, and the voltage across C_E is $V_E = 0.1 \ V_{CC}$. The resulting circuit is shown in Fig. 4-20. So that the path of operation on the collector characteristics can be plotted, an expression of i_C versus v_{CE} will be determined assuming $v_s = V_s \sin \omega t$. Applying Thévenin's theorem to Fig. 4-20, the circuit simplifies to that of Fig. 4-21. From this figure,

$$V_{CTH} - V_E = i_C(R_{CTH}) + v_{CE}$$

or

$$i_C = \frac{V_{CTH} - V_E - v_{CE}}{R_{CTH}} \tag{4-51}$$

Equation (4-51) is the equation of a straight line having a slope of $(-1/R_{CTH})$, and the operating path is linear.

We can show that the line passes through the bias point by solving Eq. (4-51) for v_{CE}, substituting $V_{C2} = V_{CC} - I_{CQ}R_C$ in V_{CTH} and noting from Fig. 4-19 that the collector bias current is

$$I_{CQ} = \frac{V_{CC} - V_E - V_{CEQ}}{R_C} \tag{4-52}$$

The algebraic manipulations yield $v_{CE} = V_{CEQ}$, the bias voltage.

The resulting ac or dynamic load line and Q point are shown in Fig. 4-22 with the transistor characteristics suppressed. To draw this load line, it is only necessary to sketch a line having a slope of $(-1/R_{CTH})$ through the bias point. As the base current varies with signal above and below the Q point, operation

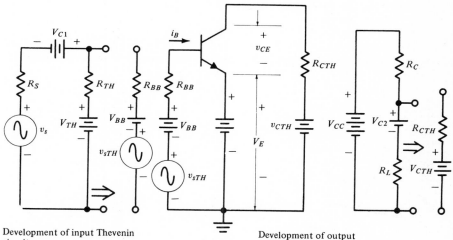

Development of input Thevenin circuit

Open circuit voltage:

$$v_{BB} = V_{TH} + [(v_s + V_{C1} - V_{TH})R_{TH}]/(R_S + R_{TH})$$
$$\text{BY SUPERPOSITION THEOREM,}$$
$$V_{BB} = V_{TH} + [V_{C1}R_{TH} - V_{TH}R_{TH}]/(R_S + R_{TH})$$
$$= [V_{TH}R_S + V_{C1}R_{TH}]/(R_S + R_{TH})$$

$$v_{sTH} = v_s R_{TH}/(R_S + R_{TH})$$

Voltages short circuit:

$$R_{BB} = R_S R_{TH}/(R_S + R_{TH})$$

Development of output Thevenin circuit

Open circuit voltage:

$$V_{CTH} = V_{C2} + [(V_{CC} - V_{C2})R_L/(R_C + R_L]$$
$$= (V_{CC}R_L + V_{C2}R_C)/(R_C + R_L)$$

With batteries shorted:

$$R_{CTH} = R_C R_L/(R_C + R_L)$$

FIG. 4-21 The result of applying the Thévenin and superposition theorems to Fig. 4-20.

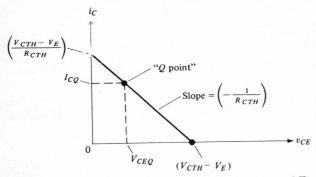

FIG. 4-22 The dynamic or ac load line for the circuit of Fig. 4-18.

follows the ac load line. Since V_E and V_{C2} do not change, the changes in v_O are exactly the same as the changes in v_{CE}, and the signal load is R_{CTH} or $R_C \parallel R_L$.

Example 4-5   Sketch the ac load line for Example 4-4 and determine the transfer characteristic of i_C and v_{CE} versus i_B. Assume $R_S = R_L = 5,000\ \Omega$, and estimate the current, voltage, and power gains.

Solution   Since $R_C = 1,500\ \Omega$ and $R_L = 5,000\ \Omega$, $R_{CTH} = (1.5 \times 5)(10^6)/$ $(6.5)(10^3)$ or $R_{CTH} = 1,150\ \Omega$. The ac load line is sketched in Fig. 4-23, and points for the transfer characteristics are read at intersection points of the

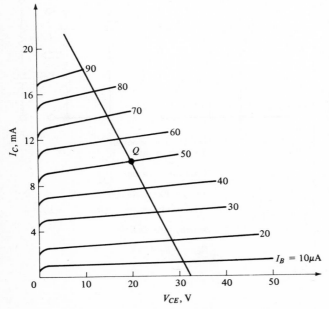

FIG. 4-23   The ac load line for Example 4-5.

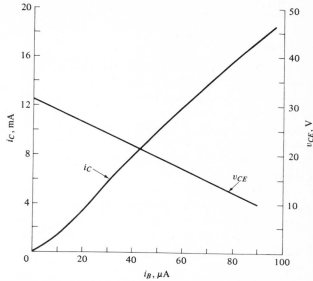

FIG. 4-24 The i_C and v_{CE} transfer characteristics for Example 4-5.

load line and the constant i_B curves. The transfer characteristics shown in Fig. 4-24 are reasonably linear for i_B between 20 and 80 μA, and it appears that a linear equivalent circuit can be applied over this range of i_B. Note that the input signal causes i_B to vary above and below the Q point of $I_{BQ} = 50$ μA. Referring to the input of the circuit, as shown in Fig. 4-21, $i_B = I_{BQ} + i_b$, $I_{BQ} = (V_{BB} - V_{BE} - V_E)/R_{BB} = 50$ μA, $i_b = v_{sTH}/R_{BB} = v_s/R_S = v_s/5{,}000$, and $i_B = 50(10^{-6}) + (2)(10^{-4})v_s$ (amperes).

From Fig. 4-24, operation is approximately linear for 30 μA on each side of the bias point of $I_B = 50$ μA. The slope of the collector-current transfer curve in this Δi_B range yields a current gain of $\Delta i_C/\Delta i_B = (16.2 - 3.3)(10^{-3})/60(10^{-6}) = 215$. The peak-to-peak input voltage causing the $\Delta i_B = 60$-μA change is found from the i_B equation as $\Delta i_B = (2)(10^{-4})\,\Delta v_s$, or $\Delta v_s = 60(10^{-6})/2(10^{-4}) = 0.3$ V, and the change in collector-to-emitter voltage corresponding to this same Δi_B change is $\Delta v_{CE} = (13 - 27) = -14$ V from Fig. 4-24. The voltage gain is $\Delta v_{CE}/\Delta v_s = -14/0.3 = -48$, and the power gain is $\Delta v_{CE}\Delta i_C/\Delta v_s\Delta i_B \cong 10{,}000$. We see that as i_B and i_C increase, v_{CE} decreases in Fig. 4-24, and for sinusoidal signals, the input voltage and base current will be in phase so that the output or collector-emitter voltage will be 180° out of phase with the input voltage.

4-2 FIELD-EFFECT TRANSISTORS (FET)

The field-effect transistor (unipolar) has two structures, both of which differ from that of the bipolar or two-junction transistor. The FET has high input resistance while the bipolar transistor has relatively low input resistance. The main applications of the FET have been in high-input-impedance circuits, high-frequency circuits and low-power integrated logic circuits.

4-2.1 MOSFET

The insulated gate (IGFET) or metal-oxide semiconductor (MOSFET) transistor structure is shown in Fig. 4-25. The three-device terminals are designated as source, gate, and drain as indicated, and these terminals, for reference purposes, can be roughly compared to the emitter, base, and collector, respectively, of the bipolar transistor. The layer between the gate and substrate is an insulator so that the metal-oxide-semiconductor layers form a capacitor. The charge carriers in the semiconductor side of this capacitor control the operation of the device.

To understand the operation of the MOSFET, assume a fixed value of drain-source voltage V_{DS}, and refer to Fig. 4-25a. When the gate-source voltage V_{GS} is zero, no current flows between the source and drain p regions since one of these junctions formed with the n substrate is always reverse-biased. Application of a negative voltage on the gate with respect to the source induces a positive charge in the n substrate near the SiO_2 layer and between the source and drain. This positive charge is made up of holes from the minority carriers in the n region and holes injected from the source. A p type channel forms between the p type

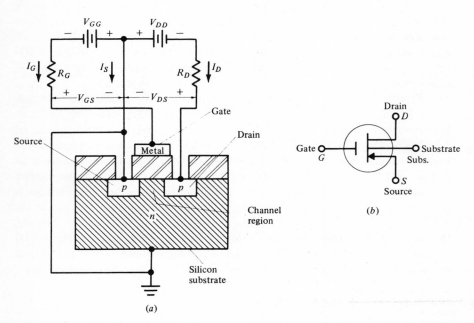

FIG. 4-25 (a) A cross-sectional sketch of a p-channel MOSFET with normal biasing polarities; (b) a symbol for the MOSFET. The arrow direction is shown for a p-channel device, reversing direction for an n-channel type.

drain and source, and current can flow. As the gate-source voltage becomes more negative, more holes appear between the drain and source, and more drain current flows. Because of electric field conditions in the channel, the thickness of the injected channel "pinches off" near the drain when the gate-source voltage reaches a particular value. If the gate-source-voltage magnitude is increased beyond this pinchoff value, the drain current remains nearly constant. The resultant drain characteristics with v_{GS} as a parameter are shown in Fig. 4-26a. The negative values of current and voltage occur because of the assumed direction of current and polarities of voltages of Fig. 4-25a. These current directions and voltage polarities are standard as is the case for the bipolar transistor.

There are n channel MOSFETs manufactured as well as the p channel model discussed above. The characteristics are similar to Fig. 4-26a except that both voltage polarities and the current direction are of opposite sign.

*mechanical data

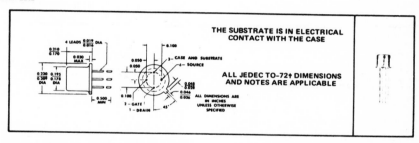

†TO-72 outline is same as TO-18 outline with the addition of a fourth lead.

handling precautions

Curve-tracer testing and static-charge buildup are common causes of damage to insulated-gate devices. Permanent damage may result if either gate-voltage rating is exceeded even for extremely short time periods. Each transistor is protected during shipment by a gate-shorting device, which should be removed only during testing and after permanent mounting of the transistor. Personnel and equipment, including soldering irons, should be grounded.

*absolute maximum ratings at 25°C free-air temperature (unless otherwise noted)

Drain-Gate Voltage	–25 V
Drain-Source Voltage	–25 V
Forward Gate-Source Voltage	–25 V
Reverse Gate-Source Voltage	25 V
Continuous Drain Current	–125 mA
Continuous Device Dissipation at (or below) 25°C Free-Air Temperature (See Note 1)	360 mW
Continuous Device Dissipation at (or below) 25°C Case Temperature (See Note 2)	1.8 W
Storage Temperature Range	–65°C to 200°C
Lead Temperature 1/16 Inch from Case for 10 Seconds	300°C

NOTES: 1. Derate linearly to 175°C free-air temperature at the rate of 2.4 mW/deg.

2. Derate linearly to 175°C case temperature at the rate of 12 mW/deg.

*Indicates JEDEC registered data

†Enhancement-mode operation entails the use of a forward gate-source voltage to increase drain current from I_{DSS}, the drain current at $V_{GS} = 0$, as opposed to depletion-mode operation wherein a reverse gate-source voltage is used to decrease drain current. An enhancement-type transistor is in the "off" state at $V_{GS} = 0$ and hence will not operate normally in the depletion mode.

FIG. 4-26 Typical characteristics of the 3N160 MOSFET.(*By permission Texas Instruments Inc., Dallas, Texas.*)

*electrical characteristics at 25°C free-air temperature (unless otherwise noted)

	PARAMETER	TEST CONDITIONS†	MIN	TYP	MAX	UNIT
I_{GSSF}	Forward Gate-Terminal Current	V_{GS} = –25 V, V_{DS} = 0		<-1	–10	pA
		V_{GS} = –25 V, V_{DS} = 0, T_A = 100°C		–10	–50	pA
I_{GSSR}	Reverse Gate-Terminal Current	V_{GS} = 25 V, V_{DS} = 0		<1	10	pA
I_{DSS}	Zero-Gate-Voltage Drain Current	V_{DS} = –15 V, V_{GS} = 0		<1	–10	nA
		V_{DS} = –25 V, V_{GS} = 0			–10	µA
$V_{GS(th)}$	Gate-Source Threshold Voltage	V_{DS} = –15 V, I_D = –10 µA	–1.5		–5	V
V_{GS}	Gate-Source Voltage	V_{DS} = –15 V, I_D = –8 mA	–4.5		–8	V
$I_{D(on)}$	On-State Drain Current	V_{DS} = –15 V, V_{GS} = –15 V, See Note 3	–40		–120	mA
$|y_{fs}|$	Small-Signal Common-Source Forward Transfer Admittance		3.5		6.5	mmho
$|y_{os}|$	Small-Signal Common-Source Output Admittance	f = 1 kHz			0.25	mmho
C_{iss}	Common-Source Short-Circuit Input Capacitance	V_{DS} = –15 V, I_D = –8 mA			10	pF
C_{rss}	Common-Source Short-Circuit Reverse Transfer Capacitance	f = 1 MHz			4	pF

TYPICAL CHARACTERISTICS

COMMON-SOURCE DRAIN CHARACTERISTICS †

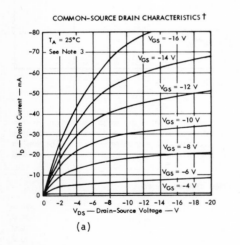

(a)

SMALL-SIGNAL COMMON-SOURCE FORWARD TRANSFER ADMITTANCE vs DRAIN CURRENT †

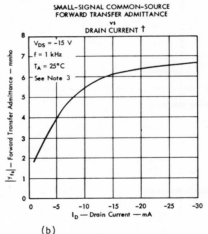

(b)

NOTE: 3. These parameters must be measured using pulse techniques. t_p ≈ 100 ms, duty cycle ≤ 10%

*Indicates JEDEC registered data

†All measurements are made with the third lead (case and substrate) connected to the fourth lead (source).

FIG. 4-26 Continued

Figure 4-26a shows that a negative V_{GS} value is necessary to cause significant drain current flow. This type of operation is defined as the *enhancement mode* since an applied gate-source voltage is necessary to enhance or cause a channel to form. The voltage V_{GS} at which drain current begins is defined as the *threshold* voltage V_T. In Fig. 4-26a, $V_{GS(th)} \equiv V_T \cong -3$ V.

Theoretically, it can be shown that [3]

$$i_D = \pm K \left[(v_{GS} - V_T)v_{DS} - \frac{v_{DS}^2}{2} \right] \quad \text{for } v_{DS} < (v_{GS} - V_T) \tag{4-53}$$

$$i_D = 0 \quad \text{for } v_{GS} < V_T$$

and

$$i_D = \pm \frac{K V_T^2}{2} \left[\frac{v_{GS}}{V_T} - 1 \right]^2 \quad \text{for } v_{DS} \geq (v_{GS} - V_T) \tag{4-54}$$

The $+$ sign in these equations applies for n-channel devices, whereas the negative sign applies for p-channel devices. Equation (4-53) is valid in the ohmic region before the channel pinches off, and Eq. (4-54) is valid in the constant current region after the channel pinches off. For constant v_{GS}, Eq. (4-54) states that i_D is constant in the defined range. We see from Fig. 4-26a that i_D increases continually so that Eq. (4-54) is an approximation. Modified models will be given to account for this increase in drain current.

The value of K in Eqs. (4-53) and (4-54) can be found experimentally or by curve matching. As an example, from Fig. 4-26a we evaluate K by choosing the boundary between the constant current and ohmic regions given by $v_{DS} = v_{GS} - V_T$. At $V_{GS} = -8$ V, and $V_T = -3$ V, $V_{DS} = -5$ V, and $I_D = -14$ mA, Eq. (4-54) yields $K = 1.12 \times 10^{-3}$ A/V^2. With this value of K in Eqs. (4-53) and (4-54), the resulting curve is sketched on the measured characteristic in Fig. 4-27. The agreement with measured values is good in the ohmic region but is in error in the constant-current region. The amount of error depends upon the construction of the IGFET. A linear correction can be made by adding a term

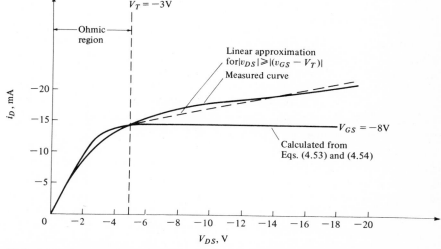

FIG. 4-27   A comparison of measured and calculated curves for the device of Fig. 4-26.

to Eq. (4-54) so that it becomes

$$i_D = \pm \frac{K(V_T)^2}{2}\left(\frac{v_{GS}}{V_T} - 1\right)^2 + \frac{v_{DS} - (v_{GS} - V_T)}{r_{ds}} \quad \text{for } v_{DS} \geq (v_{GS} - V_T).$$

$$(4\text{-}54a)$$

The value of r_{ds} may be determined by a linear approximation to the actual curve in the region $v_{DS} \geq (v_{GS} - V_T)$. With the dashed-line approximation in Fig. 4-27, $r_{ds} = 1.83$ kΩ.

For computer-aided design, Eqs. (4-53) and (4-54) can be expressed in terms of v_{GS}, v_{GD}, and V_T so as to be similar to the Ebers-Moll equations applied to the bipolar transistor. Noting that $v_{GD} = v_{GS} - v_{DS}$ or $v_{DS} = v_{GS} - v_{GD}$, we obtain the following relations:

$$i_D = K\left[\left(\frac{v_{GD}^2}{2} - V_T v_{GD}\right) - \left(\frac{v_{GS}^2}{2} - V_T v_{GS}\right)\right]$$

$$\text{for} \quad v_{GS} < V_T \quad \text{and} \quad v_{GD} < V_T \quad (4\text{-}55)$$

$$i_D = K\left[\frac{V_T^2}{2} - \left(\frac{v_{GS}^2}{2} - V_T v_{GS}\right)\right] \quad \text{for} \quad v_{GS} < V_T \quad \text{and} \quad v_{GD} > V_T$$

$$(4\text{-}56)$$

$$i_D = K\left[\left(\frac{v_{GD}^2}{2} - V_T v_{GD}\right) - \frac{V_T^2}{2}\right] \quad \text{for} \quad v_{GS} > V_T \quad \text{and} \quad v_{GD} < V_T$$

$$(4\text{-}57)$$

$$i_D = K[V_T^2] \quad \text{for} \quad v_{GD} \geq V_T \quad \text{and} \quad v_{GS} \geq V_T \qquad (4\text{-}58)$$

4-2.2 MOSFET SMALL-SIGNAL PARAMETERS

The data sheets of Fig. 4-26 should be studied to learn other parameters of interest. These parameters are defined under given test conditions, and we shall comment on a few of them.

A small-signal equivalent circuit for the MOSFET that applies over the normal frequency range is shown in Fig. 4-28. We see that this model is similar to the hybrid-π circuit for the bipolar transistor with the exception that the input resistance at low frequencies is infinite in this case.

At a particular bias point in region 2, the amplifying region, four measurements are made to determine two y parameters and two capacitor values. These small-signal measurements yield (See Prob. 1-1 for general y-parameter equations):

1. y_{fs}, the common-source short-circuit transadmittance,

$$\left.\frac{I_d}{V_{gs}}\right|_{\text{with } V_{ds}=0}$$

2. y_{os}, the common-source short-circuit output admittance,

$$\frac{I_d}{V_{ds}}\bigg|_{\text{with }V_{gs}=0}$$

3. C_{iss}, the common-source input capacitance with $V_{ds} = 0$.
4. C_{rss}, the common-source reverse-transfer capacitance with $V_{gs} = 0$

We note that these values appear on page two of data sheet Fig. 4-26.

From the measurements defined above, parameters of Fig. 4-28 may be determined. From the circuit, we see that

$$y_{fs} = \frac{I_d}{V_{gs}}\bigg|_{V_{ds}=0} = g_m - j\omega C_{gd} \tag{4-59}$$

However, a good approximation for g_m can be obtained from the vertical spacing between drain characteristic curves at the operating point. Thus,

$$g_m = \frac{\Delta i_D}{\Delta v_{GS}}\bigg|_{V_{DS}=\text{constant}} \tag{4-60}$$

Since amplifier operation occurs in the constant-current region, we may also find g_m from Eq. (4-54) as,

$$g_m = \frac{di_D}{dv_{GS}}\bigg|_{V_{DS}=\text{constant}} = \frac{\partial i_C}{\partial v_{GS}} = K(V_{GS} - V_T) \tag{4-61}$$

$$= \sqrt{2KI_D} \tag{4-62}$$

where
   V_{GS} = bias gate-source voltage
   I_D = bias drain current.
From the definition of C_{rss}, $V_{gs} = 0$, and from Fig. 4-28,

$$C_{rss} = C_{gd} \tag{4-63}$$

Similarly for C_{iss} with $V_{ds} = 0$,

$$C_{iss} = C_{gd} + C_{gs}$$

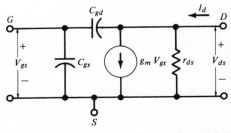

FIG. 4-28 A small-signal MOSFET equivalent circuit.

or

$$C_{gs} = C_{iss} - C_{gd} \qquad (4\text{-}64)$$

Finally,

$$\frac{1}{r_{ds}} = y_{os} = \frac{I_d}{V_{ds}}\bigg|_{V_{gs}=0} \qquad (4\text{-}65)$$

Example 4-6 Determine the circuit parameters of Fig. 4-28 for the maximum device parameters given on data-page 2 of Fig. 4-26.

From Fig. 4-26, $y_{fs} = 6.5$ mmho, $y_{os} = 0.25$ mmho, $C_{iss} = 10$ pF, $C_{rss} = 4$ pF at $V_{DS} = -15$ V, $I_D = -8$ mA, and $V_{GS} = -6$ V. From Eq. (4-63), $C_{gd} = C_{rss} = 4$ pF. From Eq. (4-59), $g_m = y_{fs} + j\omega C_{gd}$, and y_{fs} is measured at $\omega = 2\pi(10^3)$ rad, so that ωC_{gd} is negligible, and $g_m \cong y_{fs} = 6.5$ mmho. If we apply Eq. (4-62) with $K = 1.12 \times 10^{-3}$ from our previous calculations, $g_m = 4.23$ mmhos which is about the minimum y_{fs} specified on the data sheet. From Eq. (4-64), $C_{gs} = C_{iss} - C_{gd} = (10 - 4)(10^{-12}) = 6$ pF. Finally, Eq. (4-65) shows that $r_{ds} = 1/y_{os} = 4,000$ Ω. The linear slope of the $V_{GS} = -6$ V curve of Fig. 4-26a in the constant-current region also gives about the same value of r_{ds}.

All the parameters of Fig. 4-28 have a range of values for a particular MOSFET, and the parameters are also a function of temperature and bias point. Approximate calculations are usually made with typical values, and if extremes of voltage and current ranges are needed, all the necessary information can be included in a computer solution.

4-3 JFET

The second type of FET is schematically shown in Fig. 4-29a along with the circuit symbol in Fig. 4-29b. A double-junction FET is drawn in Fig. 4-29a where two pn junctions are formed on opposite sides of the channel semiconductor material. A single-junction structure is also possible.

Referring to Fig. 4-29a, the two junctions formed by the p regions on the n-type conducting channel control the electrical characteristics. Normally, the junctions are reverse-biased resulting in a depletion layer, barrier layer, or layer of essentially no movable charge which extends into the conducting channel. The cross-sectional area A, through which current flows in the conduction channel, is controlled by the gate-source voltage. Since the channel resistance is proportional to $\rho L/A$ where ρ is the channel resistivity and L the effective length, the channel resistance increases with decreasing A and channel current decreases, assuming fixed drain voltage V_{DS}. By varying the junction reverse bias, the channel current may be varied, and little current flows across the gate junctions due to the reverse bias.

The drain characteristics for an n-channel JFET are shown in Fig. 4-30. A p-channel device has similar characteristic curves, but the current direction and voltage polarities are reversed. We see that the gate-source voltage v_{GS} must increase in magnitude to decrease the drain current. This mode of operation is defined as being a *depletion mode* because the channel is depleted of movable charge carriers as $|v_{GS}|$ increases. Certain MOSFETS can be operated in the depletion mode, or both depletion and enhancement modes, but all JFETS operate in the depletion mode.

The defining equations [3] for the drain current of an *n*-channel JFET are:

$$i_D = G_o v_{DS} \left[1 + \frac{2}{3} \frac{V_p}{v_{DS}} \left(\frac{v_{GS} - v_{DS}}{V_p} \right)^{3/2} - \frac{2}{3} \frac{V_p}{v_{DS}} \left(\frac{v_{GS}}{V_p} \right)^{3/2} \right]$$

$$\text{for} \quad v_{GS} - v_{DS} = v_{GD} \geq V_p \quad \text{and} \quad v_{GS} \geq V_p \quad (4\text{-}66)$$

and

$$i_D = G_o \left\{ -\frac{V_p}{3} + v_{GS} \left[1 - \frac{2}{3} \left(\frac{v_{GS}}{V_p} \right)^{1/2} \right] \right\}$$

$$\text{for} \quad v_{GS} > V_p \quad \text{and} \quad v_{DS} > v_{GS} - V_p \quad (4\text{-}67)$$

Pinchoff is attained when the two depletion layers of Fig. 4-29a meet. Below the knee of the curves of Fig. 4-30, the channel is open; and above the knee, the channel is pinched off by the depletion layers. The boundary points between the two regions are shown by the dashed line in Fig. 4-30. The voltage V_p is defined as the *pinchoff voltage*, and it is the value of $v_{GS} - v_{DS}$ at pinchoff. The value of V_p is obtained by extrapolation from drain currents greater than zero.

Equation (4-66) applies to the ohmic region between $v_{DS} = 0$ and $v_{DS} = (v_{GS} - V_p)$, whereas Eq. (4-67) applies to the constant-current region for $v_{DS} > (v_{GS} - V_p)$. For amplifying purposes (region 2), Eq. (4-67)—which is

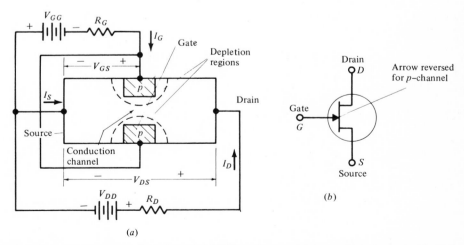

FIG. 4-29 A cross-sectional sketch of a JFET structure with normal biasing polarity indicated; (b) a symbol for an *n*-channel JFET.

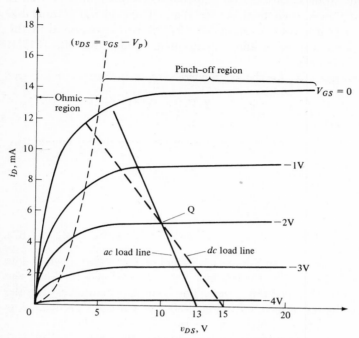

FIG. 4-30 The drain characteristics of an n-channel JFET and the load lines for Example 4-7.

approximate—is replaced by the parabolic expression

$$i_D = I_{DSS}\left(-\frac{v_{GS}}{V_p} + 1\right)^2$$

(4-68)

for $v_{GS} > V_p$, $v_{DS} > (v_{GS} - V_p)$

The term I_{DSS} is the drain current (called saturation drain current) with $V_{GS} = 0$ and $v_{DS} > (-V_p)$ but below breakdown, BV_{DSS}.

The transfer characteristics i_D versus v_{GS} for the FET of Fig. 4-30 at $V_{DS} = 15$ V has been plotted in Fig. 4-31, and from this plot, we can estimate V_p. The slope of the curve can be found from Eq. (4-68) as

$$g_m = \frac{-\Delta i_D}{\Delta v_{GS}} = \frac{2I_{DSS}}{V_p}\left(-\frac{v_{GS}}{V_p} + 1\right)$$

(4-69)

At $v_{GS} = 0$,

$$\frac{\Delta i_D}{\Delta v_{GS}} = \frac{2I_{DSS}}{V_p} = \frac{I_{DSS}}{(V_p/2)}$$

(4-70)

implying that a line tangent to the i_D versus v_{GS} curve at $v_{GS} = 0$ will intersect the v_{GS} axis at $V_p/2$. This line is shown in Fig. 4-31 and indicates that $V_p/2 = -2.5$ or $V_p = -5$ V.

To see how Eq. (4-68) fits the curve of Fig. 4-31, circled points are shown that have been calculated from $i_D = 13.8(10^{-3})[(-v_{GS}/-5) + 1]^2$. The calculated points appear to be as accurate as the experimental plot.

4-3.1 JFET PARAMETERS

Since a designer must work from data sheets describing devices, we show typical information given for a JFET in Fig. 4-32.

4-3.2 BREAKDOWN VOLTAGE

Since $v_{DG} = v_{DS} - v_{GS}$, a larger voltage appears across the drain-gate diode than the gate-source diode and the former diode will break down first. The breakdown voltage from drain to source with the gate shorted to the source is defined as $BV_{DSS} \equiv V_{BR(GSS)}$, and the minimum value BV_{DSS} is given in the data of Fig. 4-32 as -30 V.

4-3.3 GATE CUTOFF CURRENT

Connecting the drain to the source and reverse-biasing the gate-source diode, a measurement of the gate current I_{GSS} can be obtained. This measurement indicates the quality of the diode. Resulting resistance values are to be found in the thousands of megohms range at 0°C. According to Fig. 4-32, $|I_{GSS}| = 0.5$ nA (nanoampere) at $T = 25$°C and from the test conditions, $r_{GS} = [20/(0.5) \times 10^{-9}] = 4 \times 10^{11}$ Ω.

4-3.4 SATURATION OR PINCHOFF DRAIN CURRENT

The IEEE standard symbol for the drain current at zero gate-to-source voltage is I_{DSS}. Since the current increases slightly in the pinchoff region, I_{DSS} or $(I_{D \text{ on}})$ is stated at a particular drain-source voltage. Note that there is a ratio of 5:1

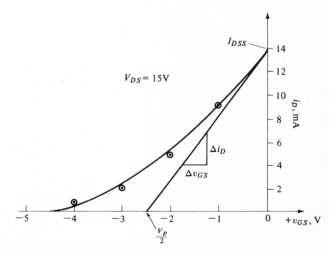

FIG. 4-31 The transfer characteristics of i_D versus v_{GS} plotted from Fig. 4-30 at $V_{DS} = 15$ V.

**2N3823 SYMMETRICAL N-CHANNEL FIELD-EFFECT TRANSISTOR
FOR VHF AMPLIFIER AND MIXER APPLICATIONS**

- **Low Noise Figure: $\leq$ 2.5 db at 100 Mc**
- **Low C_{rss}: $\leq$ 2 pf**
- **High y_{fs}/C_{iss} Ratio (High-Frequency Figure-of-Merit)**
- **Cross Modulation Minimized by Square-Law Transfer Characteristic**

*mechanical data

†TO-72 outline is same as TO-18 except for addition of a fourth lead.

* absolute maximum ratings at 25°C free-air temperature (unless otherwise noted)

Drain-Gate Voltage . . . . 30 v
Drain-Source Voltage . . . . 30 v
Reverse Gate-Source Voltage . . . . −30 v
Gate Current . . . . 10 ma
Continuous Device Dissipation at (or below) 25°C Free-Air Temperature (See Note 1) . . . 300 mw
Storage Temperature Range . . . . −65°C to + 200°C
Lead Temperature ¹⁄₁₆ Inch from Case for 10 Seconds . . . . 300°C

*electrical characteristics at 25°C free-air temperature (unless otherwise noted)

	PARAMETER	TEST CONDITIONS‡	MIN	MAX	UNIT
$V_{(BR)GSS}$	Gate-Source Breakdown Voltage	$I_G = -1\ \mu a,\quad V_{DS} = 0$	−30		v
I_{GSS}	Gate Cutoff Current	$V_{GS} = -20\ v,\quad V_{DS} = 0$		−0.5	na
		$V_{GS} = -20\ v,\quad V_{DS} = 0,\quad T_A = 150°C$		−0.5	μa
I_{DSS}	Zero-Gate-Voltage Drain Current	$V_{DS} = 15\ v,\quad V_{GS} = 0,\quad$ See Note 2	4	20	ma
V_{GS}	Gate-Source Voltage	$V_{DS} = 15\ v,\quad I_D = 400\ \mu a$	−1	−7.5	v
$V_{GS(off)}$	Gate-Source Cutoff Voltage (V_P)	$V_{DS} = 15\ v,\quad I_D = 0.5$ na		−8	v
$\|y_{fs}\|$	Small-Signal Common-Source Forward Transfer Admittance	$V_{DS} = 15\ v,\quad V_{GS} = 0,\quad f = 1\ kc,$ See Note 2	3500	6500	μmho
$\|y_{os}\|$	Small-Signal Common-Source Output Admittance	$V_{DS} = 15\ v,\quad V_{GS} = 0,\quad f = 1\ kc,$ See Note 2		35	μmho
C_{iss}	Common-Source Short-Circuit Input Capacitance	$V_{DS} = 15\ v,$		6	pf
C_{rss}	Common-Source Short-Circuit Reverse Transfer Capacitance	$V_{GS} = 0,$ $f = 1\ Mc$		2	pf
$\|y_{fs}\|$	Small-Signal Common-Source Forward Transfer Admittance	$V_{DS} = 15\ v,$	3200		μmho
$Re(y_{is})$	Small-Signal Common-Source Input Conductance	$V_{GS} = 0,$		800	μmho
$Re(y_{os})$	Small-Signal Common-Source Output Conductance	$f = 200\ Mc$		200	μmho

NOTES: 1. Derate linearly to 175°C free-air temperature at the rate of 2 mw/C°.
  2. These parameters must be measured using pulse techniques. PW = 100 msec, Duty Cycle $\leq$ 10%.

*Indicates JEDEC registered data.
‡The fourth lead (case) is connected to the source for all measurements.

FIG. 4-32   *(By permission Texas Instruments Inc., Dallas, Texas.)*

between the minimum and maximum values of I_{DSS} in Fig. 4-32 due to manufacturing tolerances.

4-3.5 SMALL-SIGNAL PARAMETERS

The small-signal equivalent circuit for the JFET can be the same as in Fig. 4-28.
Equations (4-59), (4-63), (4-64), and (4-65) along with data given in Fig. 4-32

are applied to obtain circuit values. Equation (4-69) also yields an estimate for g_m. Data are given in Fig. 4-32 for a complete common-source y-parameter equivalent circuit where the y parameters are a function of frequency.

Example 4-7   Figure 4-33 shows one possible FET bias circuit. For the bias point $V_{GS} = (-2)$ V, $V_{DS} = 10$ V, $I_{DS} = 5.4$ mA, or point Q in Fig. 4-30, find R_D, R_S, draw the dc and ac load lines, and find the voltage gain in the frequency range where X_{C_1} and X_{CS} are approximately zero. Determine the low frequency at which the voltage gain has dropped by 3 dB.

* operating characteristics at 25°C free-air temperature

PARAMETER		TEST CONDITIONS‡	MAX	UNIT
NF	Common-Source Spot Noise Figure	$V_{DS} = 15$ v, $V_{GS} = 0$, f = 100 Mc, $R_G = 1$ kΩ	2.5	db

TYPICAL CHARACTERISTICS‡

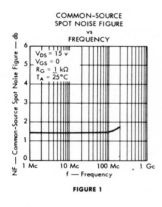

COMMON-SOURCE
SPOT NOISE FIGURE
vs
FREQUENCY

FIGURE 1

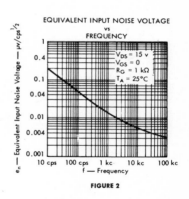

EQUIVALENT INPUT NOISE VOLTAGE
vs
FREQUENCY

FIGURE 2

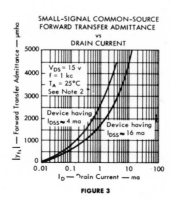

SMALL-SIGNAL COMMON-SOURCE
FORWARD TRANSFER ADMITTANCE
vs
DRAIN CURRENT

FIGURE 3

NOTE 2: These parameters must be measured using pulse techniques. PW = 100 msec, Duty Cycle ≤ 10%.

*Indicates JEDEC registered data.
‡The fourth lead (case) is connected to the source for all measurements.

FIG. 4-32   Continued

TYPICAL CHARACTERISTICS ‡

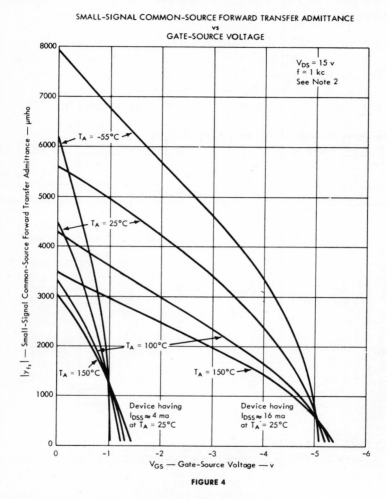

SMALL-SIGNAL COMMON-SOURCE FORWARD TRANSFER ADMITTANCE
vs
GATE-SOURCE VOLTAGE

FIGURE 4

NOTE 2: These parameters must be measured using pulse techniques. PW = 100 msec, Duty Cycle ≤ 10%.

‡The fourth lead (case) is connected to the source for all measurements.

FIG. 4-32 Continued

We find the dc bias elements R_S and R_D first. Since the gate current is very small, ($I_{GSS} < 1$ nA), the dc voltage drop in R_G is negligible, the voltage appearing across R_S is V_S, and $V_{GS} = -V_S$ is a negative value as required. Therefore, $V_{GS} = -V_S = -I_D R_S$ since $I_D = -I_S$, and $R_S = (-V_{GS}/I_D) = -(-2)/(0.0054) = 370$ Ω, or a nominal value of 390 Ω. Since $V_{DS} = 10$ V, the voltage across R_D is $V_{DD} - V_{DS} - V_S = 15 - 10 - 2 = 3$ V, and $R_D = V_{RD}/I_D = 3/(0.0054) = 556$ Ω or a nominal value of 560 Ω.

TYPICAL CHARACTERISTICS‡

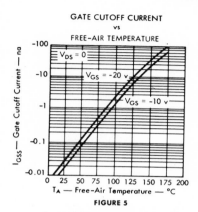

GATE CUTOFF CURRENT
vs
FREE-AIR TEMPERATURE

FIGURE 5

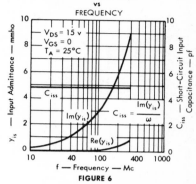

SMALL-SIGNAL COMMON-SOURCE
INPUT ADMITTANCE
vs
FREQUENCY

FIGURE 6

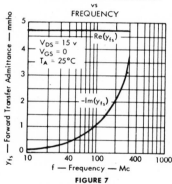

SMALL-SIGNAL COMMON SOURCE
FORWARD TRANSFER ADMITTANCE
vs
FREQUENCY

FIGURE 7

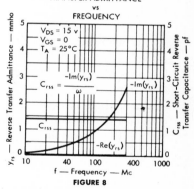

SMALL-SIGNAL COMMON-SOURCE REVERSE
TRANSFER ADMITTANCE
vs
FREQUENCY

FIGURE 8

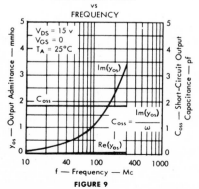

SMALL-SIGNAL COMMON-SOURCE
OUTPUT ADMITTANCE
vs
FREQUENCY

FIGURE 9

‡The fourth lead (case) is connected to the source for all measurements.

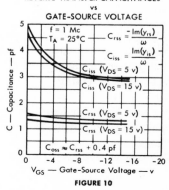

COMMON-SOURCE SHORT-CIRCUIT INPUT AND
REVERSE-TRANSFER CAPACITANCES
vs
GATE-SOURCE VOLTAGE

FIGURE 10

FIG. 4-32 Continued

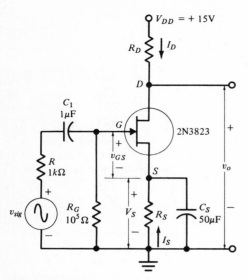

FIG. 4-33   The circuit for Example 4-7.

If we draw a dc load line on Fig. 4-30, it passes through the points $i_D = 0$, $v_{DS} = 15$ V, and $i_D = 5.4$ mA, $v_{DS} = 10$ V. By Kirchhoff's voltage law around the drain-source circuit, the dc load line is the path followed by the drain current if the capacitor C_S is removed from Fig. 4-33 and the gate-source voltage is varied, as shown by the dashed line in Fig. 4-30. At frequencies in the constant-gain region where C_S acts as a short circuit to signals, the ac load line passes through the Q point, $i_D = 5.4$ mA, $v_{DS} = 10$ V, but has a slope $(-1/R_D)$ in the same manner as that shown for the bipolar transistor in Fig. 4-22 and drawn by a solid line in Fig. 4-30. Sketching the ac load line is important because it shows graphically how large an output signal can be obtained without clipping the peaks of the waveform or entering into regions that will distort the signal. The signal could cause v_{GS} to vary from $-V_p$ to zero volts due to signal amplitude and the output voltage is obtained from the projection of the ac load line on the v_{DS} axis. The ac load line passes through $i_D = 0$, $v_{DS} = V_{DD} - V_S = 15 - 2 = 13$ V, and the peak value of the output signal with distortion is the difference between the voltage at this intersection point and the v_{DS} bias voltage of 10 V, or $13 - 10 = 3$ V. If the bias point had been selected at $i_D = 5$ mA, $v_{DS} = 5$ V, the ac load line would have the same slope (R_D unchanged), but now the peak output signal is distorted or clipped because operation enters the ohmic region. The importance of locating the bias point to accommodate the largest-required signal output voltage can be seen from this discussion.

If v_{GS} is caused to vary from (-3) V to (-1) V, from the intersection points of the ac load line and the characteristic curves, we read the values of v_{DS} to be 11.6 and 8.1 V, respectively. Therefore, the voltage gain is $\Delta v_{DS}/\Delta v_{GS} = (11.6 - 8.1)/[-3 - (-1)] = -1.75$, the negative sign indicating 180° phase shift of the output sinusoidal signal with respect to the input.

We wish to check the graphically calculated value of voltage gain by use of the equivalent circuit shown in Fig. 4-34, assuming the capacitive reactances to be zero as in the preceding calculation. We have applied the equivalent circuit of Fig. 4-28 assuming frequencies to be low enough that capacitors C_{gs} and C_{gd} act as open circuits. With $X_{CS} = 0$, the source is connected to ground, and we also see that R_D is connected from drain to ground. The reason for R_D being connected as shown may be argued as follows. If we apply an ac signal source to a battery of zero internal resistance, by the superposition principle, the battery is replaced with a short circuit and the ac signal current found in this short circuit; the signal source is replaced by its internal resistance, and the dc current is found due to the battery; the total current is the sum of these two values, and we see that, if the signal only is considered, the battery appears as a short circuit. If the battery has an internal resistance, it must remain in the circuit. We assume ideal voltage supplies.

We need numerical values for g_m and r_{ds} of Fig. 4-34, and we can estimate these from the curves of Fig. 4-30 or from y_{fs} and $y_{os} = 1/r_{ds}$ in Fig. 4-32. From Fig. 4-30, at the bias point,

$$g_m = \frac{\Delta i_D}{\Delta v_{GS}}\bigg|_{v_{DS}=\text{constant}}$$

$$= \frac{(9 - 2.5)(10^{-3})}{(-1) - (-3)}\bigg|_{v_{DS}=10\text{ V}}$$

$$= 3.25(10^{-3})\text{ mho}$$

where we have found i_D at the intersection points of a vertical line through

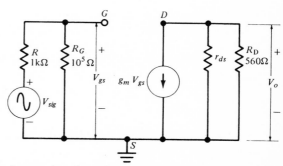

FIG. 4-34 The equivalent circuit at frequencies where the capacitors of Fig. 4-33 are short circuits. The voltages are rms.

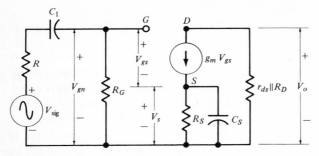

FIG. 4-35 The equivalent circuit of Fig. 4-33 for determining the low-frequency response function of Example 4-7.

$v_{DS} = 10$ V and $v_{GS} = (-3)$ and (-1) V, to obtain Δi_D. From Fig. 4-32, the typical value of y_{fs} is 3,200 μmho (micromhos) agreeing with g_m above. The value of r_{ds} is obtained from the slope of the characteristic curve for $v_{GS} = -2$ V and passing through Q. We estimate this slope to be 5,000 Ω, agreeing with $1/y_{os}$ of Fig. 4-32.

We are now ready to calculate the gain for Fig. 4-34. With R_G much larger than R, $V_{gs} = V_{sig}$, and

$$V_o = -g_m V_{gs} \frac{1}{(1/r_{ds}) + (1/R_D)}$$

or

$$\frac{V_o}{V_{sig}} = -(3.25)(10^{-3}) \frac{1}{(1/5,000) + (1/560)}$$

$$= -1.64$$

agreeing well with our graphically determined value of -1.75.

To investigate the low-frequency point where the voltage gain is 3 dB down, the equivalent circuit including capacitors is drawn in Fig. 4-35. The battery has been replaced by a short circuit as in Fig. 4-34. This type of problem is much easier to solve by the computer methods of Chap. 9, but we shall apply circuit analysis and Bode plots to obtain an answer. (See Reference [10] for another method.) From Fig. 4-35, $V_{gs} = V_{gn} - V_s$, and we see that it is important to label the figure as shown to verify where V_{gs} occurs. Defining R_S in parallel with C_S as an impedance Z_S, Kirchhoff's law around the drain-source loop yields

$$V_s = g_m V_{gs} Z_S = g_m (V_{gn} - V_s) Z_S$$

or

$$V_s = \frac{g_m V_{gn}}{g_m + (1/Z_S)}$$

Since $Z_S = R_S/(1 + j\omega C_S R_S)$,

$$V_s = \frac{g_m R_S V_{gn}}{(1 + g_m R_S) + j\omega C_S R_S}$$

and

$$V_{gs} = V_{gn} - V_s = V_{gn} - \frac{g_m R_S V_{gn}}{1 + g_m R_S + j\omega C_S R_S}$$

$$= V_{gn} \frac{1 + j\omega C_S R_S}{1 + g_m R_S + j\omega C_S R_S}$$

Finally, from Fig. 4-35,

$$V_{gn} = \frac{V_{\text{sig}} R_G}{(R + R_G + 1/j\omega C_1)}$$

so that

$$V_{gs} = \frac{V_{\text{sig}} R_G}{R + R_G + 1/j\omega C_1} \frac{1 + j\omega C_S R_S}{1 + g_m R_S + j\omega C_S R_S}$$

and

$$\frac{V_{gs}}{V_{\text{sig}}} = \frac{R_G}{(R + R_G)(1 + g_m R_S)}$$

$$\times \frac{1 - j(1/\omega C_S R_S)}{\{1 - j[1/\omega C_1(R + R_G)]\}\{1 - j[1/\omega C_S R_S(1 + g_m R_S)]\}}$$

Now, $V_o = -g_m V_{gs}(r_{ds} \| R_D)$, and using the V_{gs}/V_{sig} expression with the

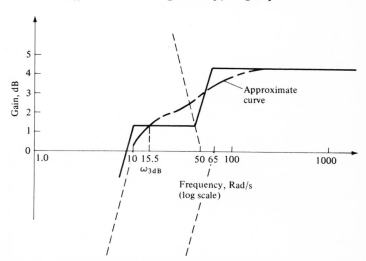

FIG. 4-36 The asymptotic frequency-response plot for Example 4-7.

numerical values as specified and calculated previously, we obtain

$$\frac{V_o}{V_{\mathrm{sig}}} = -1.64 \frac{1}{1 - j(10/\omega)} \frac{1 - j(50/\omega)}{1 - j(65/\omega)}$$

In asymptotic form,
$$20 \log(V_o/V_{\mathrm{sig}}) = [4.3 - 20 \log(10/\omega) + 20 \log(50/\omega) - 20 \log(65/\omega)] \, \mathrm{dB}$$
and from Fig. 4-36, we estimate from the approximate curve sketch that $f_{3\mathrm{dB}}$ is 2.5 Hz. We now see why it is desirable to solve even a simple circuit as shown in Fig. 4-33 by the computer.

Several important factors have been demonstrated in Example 4-7, and it should be studied carefully. The same techniques apply for the bipolar and MOS transistors.

REFERENCES

1. Fitchen, "Transistor Circuit Analysis and Design," 2d ed., D. Van Nostrand Company, Inc., Princeton, N. J., 1966.
2. Giacolletto, L. L., Study of PNP Alloy Junction Transistor from DC through Medium Frequencies, *RCA Review*, vol. 15, no. 4, December, 1954.
3. Lindmayer, J., and C. Y. Wrigley, "Fundamentals of Semiconductor Devices," D. Van Nostrand Company, Inc., Princeton, N. J., 1965.
4. Millman, J. and H. Taub, "Pulse Digital and Switching Waveforms," McGraw-Hill Book Company, New York, 1965.
5. General Electric Co., Semiconductor Data Handbook, 2d ed., 1973.
6. Smith, Ralph J., "Circuits, Devices and Systems," 3d ed., John Wiley & Sons, Inc., New York, 1971.
7. Ebers, J. J. and J. L. Moll, Large-Signal Behavior of Junction Transistors, *Proc. IRE*, December, 1954.
8. Cordwell, W. A., Transistor and Diode Model Handbook, *Technical Report No. AFWL-TR-69-44*, Air Force Weapons Laboratory, Kirtland AFB, N. M., October, 1969.
9. Bowers, J. C. and S. R. Sedore, "Sceptre: A Computer Program for Circuit and System Analysis," Prentice-Hall, Inc., Englewood Cliffs, N. J., 1971.
10. Gray, P. E. and C. L. Searle, "Electronic Principles, Physics, Models, and Circuits," John Wiley & Sons, Inc., New York, 1969.

PROBLEMS

4-1. Some of the curves of Fig. 4-7 have been linearized in Fig. P4-1 for different ranges of collector and base currents. The line slopes have not been drawn accurately, but their values have been indicated where necessary. Since the transistor is made of silicon, I_{CEO} is negligible. Determine the parameters for the circuit of Fig. 4-6 for (a) data from the curves of Fig. P4-1a and b; (b) data from the curves of Fig. P4-1c and d. Add a resistance in parallel with the controlled current generator of Fig. 4-6 to account for the line slopes of Figs. P4-1a and c.

4-2. The standard bias circuit for discrete transistors is drawn in Fig. P4-2. To make calculations easier, transform the external base-to-ground circuit to a Thévenin equivalent as shown in Fig. P4-2b. Determine expressions for V_{TH} and R_{TH} in terms of R_1, R_2, and V_{CC}.

4-3. For the circuit of Fig. P4-2, $V_{CC} = 24$ V, $R_E = 600 \, \Omega$, $R_C = 2.2 \, \mathrm{k}\Omega$, $R_1 = 60 \, \mathrm{k}\Omega$, and $R_2 = 20 \, \mathrm{k}\Omega$. The transistor is made of silicon so that I_{CEO} is negligible, and $V_{BE} = 0.7$ V when forward-biased. Determine I_C, I_B, I_E, and V_{CE} if (a) $\beta = 50$; (b) $\beta = 100$; (c) $\beta = 200$.

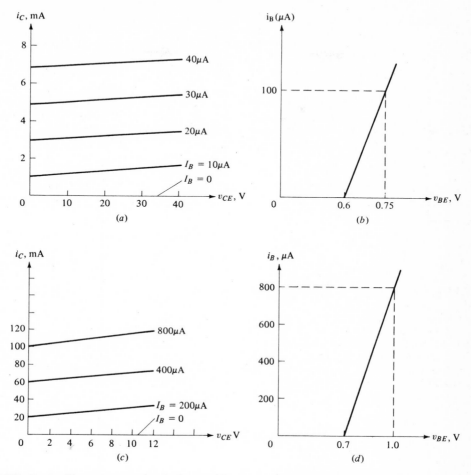

FIG. P4-1 Characteristics for Prob. 4-1 (a) slope: each line equals 1.25×10^{-5} A/V; (c) slope: each line equals 5×10^{-4} A/V.

4-4. Repeat Prob. 4-3 if $R_1 = 6$ kΩ and $R_2 = 2$ kΩ. Note that saturation occurs for $\beta = 100$.

4-5. In Prob. 4-4 we found that saturation occurs if we are not careful. However, it should be noted that for small values of R_{TH} as defined in Prob. 4-1, the bias voltage V_{CE} changes only slightly. For Fig. P4-2, assume $R_1 = 6$ kΩ, $R_2 = 2$ kΩ, and $R_C = 2.2$ kΩ. Determine a new value for R_E so that $I_C = 5$ mA when $\beta = 50$, keeping $V_{CC} = 24$ V. Then recalculate I_C, I_E, and V_{CE} when $\beta = 100$ and 200. If R_{TH} is as small as practical, we conclude that bias stability is good for β and temperature variations (see Prob. 4-6).

4-6. We noted in Prob. 4-5 that R_{TH} of the equivalent bias circuit of Prob. 4-2 should be as small as practical for good bias stability. We now consider factors which control the size of R_{TH}. Assume the circuit to be driven by a Norton source as shown in Fig. P4-6. Assume $i_s = I_S \sin(\omega t)$, C_C and C_E large so that they may be considered a short circuit at the signal frequency applied, R_S to be infinite, $\beta = h_{fe} = 100$, and the transistor to be at room temperature, $T = 300°$K. Determine:

    (a) the current gain I_c/I_s, the ratio of rms values;

    (b) the base current signal I_b in terms of I_s;

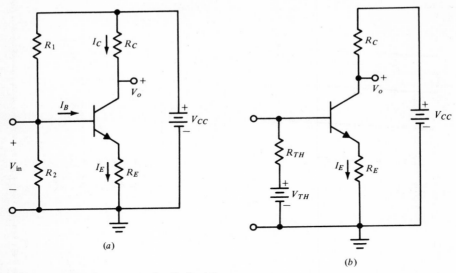

FIG. P4-2 The bias circuit for Prob. 4-2.

    (c) the voltage gain V_o/V_b;
    (d) the quiescent power loss in the circuit ($i_s = 0$);
    (e) the values of (a) through (c) if $h_{fe} = 200$;
    (f) Summarize the effects of a low value of R_{TH}.
(Note: Use a hybrid-π equivalent circuit with $r_{bb'} = 0$ and $r_{b'c} = r_{ce} \rightarrow \infty$; omit $C_{b'c}$ and $C_{b'e}$ at the frequencies of interest.)

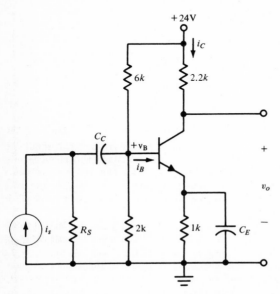

FIG. P4-6 The circuit for Prob. 4-6.

4-7. The common-emitter h-parameter circuit for small signals has been rearranged for use in a common-collector circuit in Fig. P4-7a. The general h-parameter circuit for the collector as reference is shown in Fig. P4-7b, where $V_1 = I_1 h_{ic} + h_{rc}V_2$, and

$$I_2 = I_1 h_{fc} + h_{oc}V_2$$

For the two circuits to be equivalent, find expressions for h_{ic}, h_{rc}, h_{fc}, and h_{oc} in terms of h_{fe}, h_{ie}, h_{re} and h_{oe}.

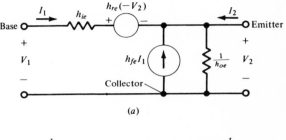

(a)

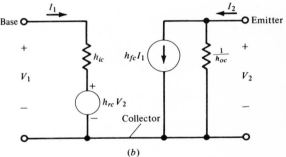

(b)

FIG. P4-7   The h-parameter circuits for Prob. 4-7.

4-8. Each transistor in Fig. P4-8 can be represented by the common-collector h-parameter circuit shown in Fig. P4-7b. Because the bias currents of the transistors are not the same, the h parameters of Q_1 and Q_2 differ. (a) Draw the complete equivalent circuit for small signals. (b) Derive expressions for V_o/V_i, $R_i = V_i/I_i$, and $R_o = V_o/I_{o'}$. (c) Evaluate the expressions of part (b) if: $R_E = 1 \text{ k}\Omega$, $R_B = 50 \text{ k}\Omega$, $h_{fc1} = -50$, $h_{ic1} = 2 \text{ k}\Omega$, $h_{rc1} = 1$, $h_{oc1} = 2(10^{-5})$ mho, $h_{fc2} = -60$, $h_{ic2} = 200 \ \Omega$, $h_{rc2} = 1$, $h_{oc2} = 2(10^{-5})$ mho, and $R_A = 100 \text{ k}\Omega$.

4-9. Consider the low-frequency characteristics of Fig. 4-18. Assume $h_{fe} = \beta = 100$, $r_{bb'} = 0$, $r_{ce} = r_{b'c} \rightarrow \infty$, $R_C = 2.2\text{k}$, $R_1 = 6 \text{ k}$, $R_2 = 2\text{k}$, $R_E = 1 \text{ k}$, and $R_S = 1 \text{ k}$. (a) Assuming C_E very large, determine the value for C_1 so that the low-frequency half-power point for the voltage gain V_o/V_s occurs at $f = 100$ Hz. (b) Assuming C_1 very large, determine C_E for the same break frequency as in part (a). (c) Compare the effect of C_1 and C_E on the low-frequency breakpoint.

4-10. The circuit of Fig. P4-10a is biased so that the h parameters of Fig. 4-9 apply. The resulting equivalent circuit for signals is given in Fig. P4-10b. Determine: (a) V_2/V_1; (b) I_2/I_i; (c) $R_i = V_1/I_i$; (d) $R_o = V_2/I_{2'}$. Assume a signal source applied to the output for determining R_o. (e) Draw a Thévenin equivalent circuit at the output terminals where the Thévenin generator is controlled by V_1.

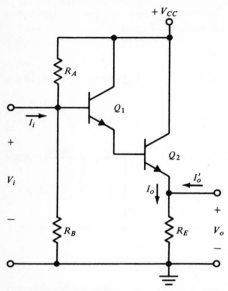

FIG. P4-8 The Darlington transistor circuit for Prob. 4-8.

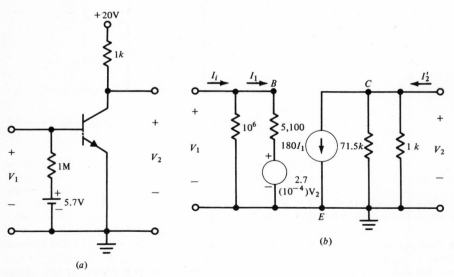

FIG. P4-10 The circuits for Prob. 4-10.

4-11. In Prob. 4-10, we found that $h_{re}V_2 = (2.7)(10^{-4})V_2$ and $h_{oe} = 1/71.5$ k, had negligible effect on the calculated results. Thus, a first-order approximation to the circuit of Fig. P4-10b is as in Fig. P4-11. Assume that $h_{fe} = (180)/[1 + j(f/10^4)]$. Sketch and label the asymptotic frequency response of V_2/V_1.

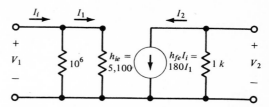

FIG. P4-11   The h-parameter circuit for Prob. 4-11.

4-12. Apply the hybrid-π equivalent circuit to the amplifier of Fig. P4-12. The $+20$-V supply acts as a short circuit to signal frequencies. Omit $r_{b'c}$ and r_{ce} from your diagram since they have negligible effect on calculations. Determine V_2/V_1, and I_2/I_1 at low frequencies where effects of $C_{b'e}$ and $C_{b'c}$ are negligible.

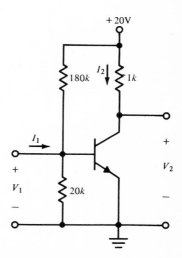

FIG. P4-12   The amplifier for Prob. 4-12.

4-13. For Fig. P4-13, determine V_O, I_C, I_B, and define the region of operation for the transistor. Assume that all forward-biased junctions have zero voltage drop, $\beta_F = 50$, and $\beta_R = \alpha_R/(1 - \alpha_R) = 2$.

4-14. A 3N160 IGFET whose characteristics are given by Fig. 4-26 is applied in the common-source circuit of Fig. P4-14. (a) Draw a dc load line, and determine V_{GS}, V_{DS}, and I_D. (b) An input signal causes v_{GS} to vary from -10 to -6 V. Determine the change in drain-source voltage and the gain, $\Delta v_{DS}/\Delta v_{GS} = \Delta v_{DS}/4$. (c) From part (b), the gate-source voltage may be written $v_{GS} = -8 - 2\sin(\omega t)$. Sketch and label the instantaneous value of v_{DS} versus ωt. Note the amplitude distortion of the output signal and the phase inversion of the output signal compared to the input signal.

4-15. In Fig. P4-15a, the h-parameter circuit is drawn with $h_{re} = 0$. The hybrid-π is shown in Fig. P4-15b where the capacitors and $r_{b'c}$ have been omitted. These circuits are useful approximations at low frequencies. Verify the expressions of Table 4-1.

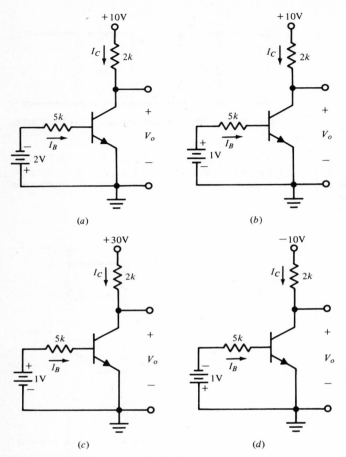

FIG. P4-13 The circuits for Prob. 4-13.

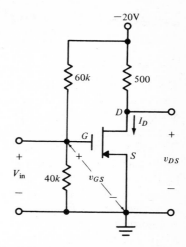

FIG. P4-14 The MOSFET circuit for Prob. 4-14.

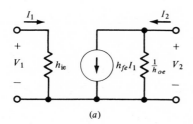

(a)

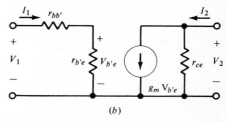

(b)

FIG. P4-15 The *h*-parameter circuit for Prob. 4-15.

5

Linear Integrated Circuits

INTRODUCTION

The equivalent circuits for all transistor types which we have presented are applicable to the analysis of integrated circuits. Our purpose is not to present the design and fabrication of ICs. For our purposes, an IC is made of several transistors, resistors, and a few capacitors (no inductors) in a small package. We will analyze some ICs to become familiar with the operation and terminal characteristics. For electronic circuit design, terminal characteristics, i.e., input and output i–v properties, are sufficient.

Analysis of typical linear ICs leads us to the understanding of the operation of devices in circuits. In addition, we see why devices are utilized in different configurations. We shall start with complex-looking circuits but break them into simpler sections and analyze the operation. In this manner, we shall learn how the component parts make up the whole system.

Most linear ICs have a differential amplifier as an input stage. We begin by defining some terms which were not considered in Chap. 2.

5-1 DIFFERENTIAL AMPLIFIER

There are several specifications which indicate the quality of a differential amplifier (diff-amp), and we introduce these factors before analyzing the actual circuits.

The output voltage of a diff-amp is proportional to the difference of two input voltages. The basic diff-amp will have input and output terminals as shown in Fig. 5-1, and the specifications we find in data sheets often refer to only three of the four terminals, e.g., terminals 1, 2, and 3. The input and output can be dc or time-varying voltages, and if both inputs are at the same voltage, $v_1 = v_2$, the ideal diff-amp output is zero. In real amplifiers, when $v_1 = v_2$, the output is nonzero and related to v_1. For small signals, the output voltage is linearly related to the input and a common-mode gain for $v_1 = v_2$ is defined as $A_c = v_3/v_1$ or $v_3 = A_c v_1$. If v_1 and v_2 are not equal, the common-mode gain still exists, and a common-mode signal is defined as $v_c = \frac{1}{2}(v_1 + v_2)$ so that if $v_1 = v_2$, $v_3 = A_c v_c = A_c v_1$ as before. The difference-mode gain is defined

$$v_3 = A_d(v_1 - v_2) = A_d v_d \tag{5-1}$$

where v_d = difference signal. The purpose of the amplifier is to amplify the difference of the input signals, and we desire A_d to be large with $A_c = 0$. Obtaining $A_c = 0$ is impossible, and a quality factor due to this imperfection is the common-mode rejection ratio defined as

$$\text{CMRR} = |A_d/A_c| \tag{5-2}$$

The ideal quality factor is infinite. The CMRR is expressed in decibels as $\text{CMRR(dB)} = 20 \log |A_d/A_c|$. For the μA741 of Fig. 2-6, the CMRR is typi-

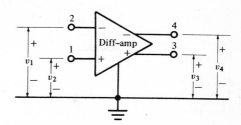

FIG. 5-1   The differential amplifier terminal designations.

cally 90 dB, and the signal voltage gain A_d is $2(10^5)$ so that the common-mode gain A_c is

$$A_c = \frac{A_d}{\log^{-1}[\text{CMRR}(\text{dB})/20]} = \frac{2(10^5)}{\log^{-1}(90/20)} = 6.3$$

By the superposition theorem, $v_3 = A_d v_d + A_c v_c = A_d(v_1 - v_2) + 0.5A_c(v_1 + v_2)$. If $v_1 = 0.1$ and $v_2 = 0$, then for the μA741, $v_3 = 2(10^5)(0.1) + 0.5(6.3)(0.1)$ or $v_3 = 2(10^4) + 0.315$, and we see that the effect of the common-mode gain is negligible. A CMRR in the range of 70-to-90 dB is usually satisfactory.

In many applications, the differential input resistance of a differential amplifier is of interest, and we shall calculate this resistance in several examples to follow. There is also a common-mode input resistance which may be important in measurements such as performed in the biomedical field. The common-mode input resistance is measured with both inputs connected together and a signal applied, yielding large values of resistance. If both inputs are connected to a capacitive probe and the voltage increased, the settling time—or time for the input to reach a steady voltage—can be long.

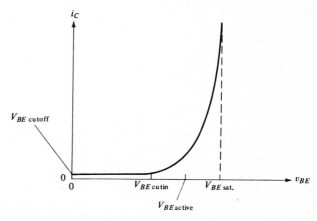

FIG. 5-2   The definitions of the base-emitter voltages for Table 5-1. An *npn* transistor has been assumed.

TYPICAL ELECTRICAL CHARACTERISTICS

PARAMETER	CONDITIONS ($T_A = 25°C$, $V^+ = 12.0$ V, and $V_{CM} = 3.5$ V Unless otherwise specified)	VALUE	UNITS
Input Offset Voltage	$R_S \leqslant 50\Omega$	1	mV
Input Offset Current		0.5	μA
Input Bias Current		3.5	μA
Input Resistance		20	$k\Omega$
Differential Voltage Gain	$R_L \geqslant 100$ kΩ	145	
Differential Distortion	$R_L \geqslant 100$ kΩ	80	mVpp
Bandwidth		1.5	MHz
Single-Ended Output Resistance		70	Ω
Output Voltage Swing	$R_L \geqslant 100$ kΩ	8.0	Vpp
Supply Current	$R_L \geqslant 100$ kΩ	9.5	mA
Power Consumption	$R_L \geqslant 100$ kΩ	114	mW
Input Voltage Range		3.5-5.2	V
Common Mode Rejection Ratio	$R_S \leqslant 50\Omega$, $f \leqslant 1$ kHz, $+3.5$ V $\leqslant V_{CM} \leqslant +5.2$ V	85	dB

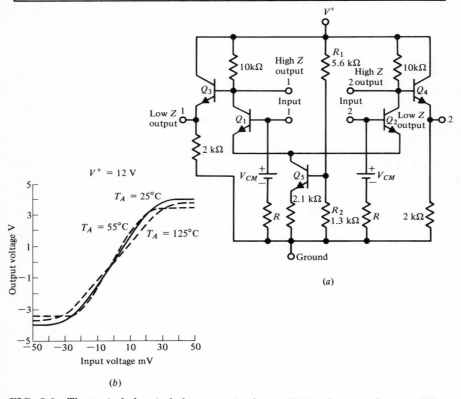

FIG. 5-3 The typical electrical characteristics for a μA730, a basic single-stage differential amplifier intended for use as a gain block in general purpose dc and ac applications. (a) The schematic diagram; (b) the transfer characteristics. (*By permission Fairchild Semiconductor Corporation, Mountain View, Calif.*)

5-1.1 DIFFERENTIAL-AMPLIFIER ANALYSIS

One convenient factor in ICs is found to be the matching of transistors. The ICs are made on silicon, and all transistors are made at the same time. The transistors are in close physical proximity and experience nearly the same tem-

TABLE 5-1 Average Values for Silicon

$V_{BE\text{cutoff}}$	$V_{BE\text{cutin}}$	$V_{BE\text{active}}$	$V_{BE\text{sat}}$	$V_{CE\text{sat}}$
0 V	0.5 V	0.6–0.7 V	0.7–0.85 V	0.1–0.3 V

perature environment. Thus, the i–v characteristics of similarly made transistors are nearly equal, and the characteristics change with temperature in the same manner.

We shall approximate the base-emitter junction of transistors as a battery for dc calculations having voltage values given in Table 5-1 and defined in Fig. 5-2. Table 5-1 lists the average values for silicon. We apply the active region values in this chapter, but the saturation values will be useful later.

Figure 5-3 sketches the circuit diagram of a basic IC, a single-stage differential amplifier. We note that this IC has two input terminals, two low-impedance output terminals, and two high-impedance output terminals. We wish to verify as many of the typical characteristics given as possible.

The first problem is to calculate all the dc voltages and currents when $V^+ = 12$ V and the input bias voltage $V_{CM} = 3.5$ V. To solve for these values, replace all transistors by the simplest dc equivalent circuit possible from Fig. 4-6, assuming I_{CEO} negligible.

There is a constant current generator or its equivalent connected to the emitters of the two input transistors. In Fig. 5-3 transistor Q_5 is the constant current generator furnishing bias current to transistors Q_1 and Q_2. No signal is applied to the base or emitter of Q_5 indicating that it is only performing a dc function.

Little error occurs from assuming that base currents are negligible with respect to collector and emitter currents when dc voltages are determined. The

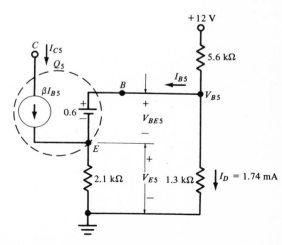

FIG. 5-4 The bias current source Q_5 for determining I_{C5} in the circuit of Fig. 5-3.

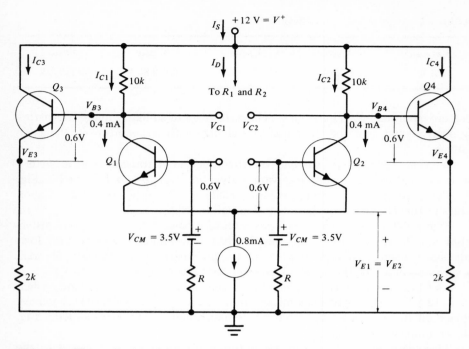

FIG. 5-5 The resulting values after calculating the bias currents of Fig. 5-4. Transistor Q_5 has been replaced by a constant current source.

first approximation assumes that base currents are zero. This is a good approximation for $\beta > 20$.

Calculations begin with the constant current source shown in Fig. 5-4. Since base current I_{B5} is negligible, $V_{B5} = 12(1.3)/6.9 = 2.26$ V, by voltage division across the 5.6-kΩ and 1.3-kΩ resistors. Assuming all base-emitter voltages to be 0.6 V, the voltage across the 2.1-kΩ emitter resistor is $V_{E5} = V_{B5} - V_{BE5} = 2.26 - 0.6 = 1.66$ V. Thus, $I_{E5} = I_{C5} \simeq V_{E5}/2.1$ k $= (1.66/2.1)10^{-3} = 0.79$ mA. The current I_{C5} divides equally between transistors Q_1 and Q_2 because they are assumed to be identical. Thus $I_{E1} = I_{E2} = I_{C1} = I_{C2} = 0.79/2$ mA $\cong 0.4$ mA.

We refer to Fig. 5-5 for the remaining dc calculations.

$$V_{E1} = V_{E2} = V_{CM} - V_{BE} = 3.5 - 0.6 = 2.9 \text{ V}$$

$$V_{B3} = V_{B4} = V_{C1} = V_{C2} = V^+ - I_{C1}(10 \text{ k}\Omega) = 12 - 0.4(10) = 8 \text{ V}$$

$$V_{E3} = V_{E4} = V_{B3} - V_{BE} = 8 - 0.6 = 7.4 \text{ V}$$

$$I_{C3} = I_{C4} = \frac{V_{E3}}{2 \text{ k}} = \frac{7.4}{2} \text{ mA} = 3.7 \text{ mA}$$

$$I_D = \frac{12}{5,600 + 1,300} = 1.74 \text{ mA}$$

$$I_S \equiv \text{Supply current} = I_{C3} + I_{C4} + I_{C1} + I_{C2} + I_D = 9.9 \text{ mA}$$

We see that the supply current agrees with the specified value of 9.5 mA with reasonable accuracy. The calculated power consumption is $V^+I_S = (12)(9.9)(10^{-3}) = 119$ mW compared to the specified value of 114 mW.

The input base bias current to transistors Q_1 and Q_2 can be approximated from $I_C = \beta I_B$. Since $I_{C1} = I_{C2} = 0.4$ mA, $I_{B1} = I_{B2} = (0.4 \text{ mA})/100 = 4$ μA assuming $\beta = 100$. The typical value from Fig. 5-3 is 3.5 μA, and we see why base currents may be neglected in the previous calculations.

5-2 SMALL-SIGNAL CALCULATIONS

For the approximate calculations of signal characteristics, we assume simple equivalent circuits with a h_{fe} of 100 as a typical value. The output transistors Q_3 and Q_4 will be analyzed first.

5-2.1 EMITTER-FOLLOWER OR COMMON-COLLECTOR AMPLIFIER

Figure 5-6 shows the circuit used for either transistor Q_3 or Q_4. This configuration has been named the emitter follower, and we see that the collector potential cannot vary with the signal because of the fixed V^+ voltage. The small-signal, low-frequency h-parameter circuit is drawn in Fig. 5-6b with $h_{re} = h_{oe} = 0$. V_1 and V_2 are the rms signal input and output voltages, respectively. To determine the characteristics of an emitter follower, we shall obtain expressions for voltage gain, input resistance, current gain, and output resistance.

5-2.2 EMITTER-FOLLOWER VOLTAGE GAIN

From Fig. 5-6b, V_2 and V_1 are:

$$V_2 = (h_{fe} + 1)I_1 R_E$$

and

$$V_1 = [h_{ie} + (h_{fe} + 1)R_E]I_1 \qquad (5-3)$$

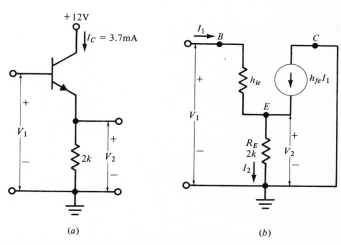

FIG. 5-6   (a) The circuit for Q_3 or Q_4 of Fig. 5.3; (b) the low-frequency equivalent circuit for (a).

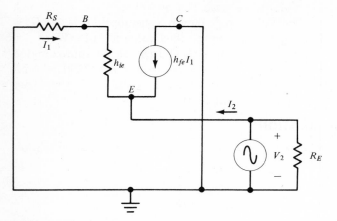

FIG. 5-7   The circuit for finding the output resistance of an emitter follower.

The voltage gain may be expressed as

$$A_v = \frac{V_2}{V_1} = \frac{(h_{fe} + 1)R_E}{h_{ie} + (h_{fe} + 1)R_E} \qquad (5\text{-}4)$$

Equation (5-4) shows that the voltage gain is less than unity, the output and input voltages are in phase, and for h_{fe} large, the voltage gain is near unity.

5-2.3   EMITTER-FOLLOWER INPUT RESISTANCE

From Eq. (5-3), the input resistance is

$$R_i = \frac{V_1}{I_1} = h_{ie} + (h_{fe} + 1)R_E \qquad (5\text{-}5)$$

Equation 5-5 shows that the input resistance can be large; approximately, $R_i \cong h_{fe}R_E$, an approximation useful as a "rule of thumb."

5-2.4   EMITTER-FOLLOWER CURRENT GAIN

By inspection of Fig. 5-6b, $I_2 = (h_{fe} + 1)I_1$ or the current gain is

$$A_i = \frac{I_2}{I_1} = h_{fe} + 1 \qquad (5\text{-}6)$$

The current gain is large, and even though A_v is less than unity, a power gain $A_v A_i$ is obtained.

5-2.5   EMITTER-FOLLOWER OUTPUT RESISTANCE

As shown in Fig. 5-7, we connect a signal source across the emitter resistance, and the input is replaced by a resistor equal to the driving-source resistance R_S. This method of measurement is one which might be applied in the laboratory. The currents I_1 and I_2 are found from Fig. 5-7 as,

$$I_1 = \frac{-V_2}{R_S + h_{ie}}$$

and

$$I_2 = -(h_{fe} + 1)I_1 = -\frac{-V_2}{R_s + h_{ie}}(h_{fe} + 1)$$

The output resistance omitting R_E becomes

$$R_o' = \frac{V_2}{I_2} = \frac{R_s + h_{ie}}{1 + h_{fe}} \tag{5-7}$$

Equation (5-7) indicates a low output resistance. The actual resistance R_o is R_o' in parallel with R_E.

Summarizing the low-frequency emitter-follower characteristics: (1) voltage gain is near but less than unity, (2) current gain is $(h_{fe} + 1)$, (3) input resistance is high, (4) the output resistance is low, and (5) the input and output voltages are in phase. The emitter follower is applied as a matching or buffering element between a high- and a low-impedance circuit. We see why the low-output-impedance points are so named in Fig. 5-3.

Calculating numerical values for the circuit of Fig. 5-6, from Table 4-1 assuming $r_{bb'} \ll r_{b'e}$ so that $h_{ie} = r_{b'e} = h_{fe}/g_m = h_{fe}/38.5I_C$ and assuming $h_{fe} = \beta = 100$, $h_{ie} = 700$ Ω for $I_C = 3.7$ mA. The calculations follow: From Eq. (5-5), $R_i = 700 + (101)(2,000) = 2.03(10^5)$ Ω. From Eq. (5-4), $A_v = (101)(2,000)/[700 + (101)(2,000)] = 0.997$. From Eq. (5-6), $A_i = 101$. From Eq. (5-7) with $R_S = 10^4$ Ω, $R_o' = 10,700/101 = 106$ Ω. The stated value of R_o in Fig. 5-2 is 70 Ω, in satisfactory agreement with our calculations.

5-2.6 DIFFERENTIAL-AMPLIFIER CALCULATIONS

The input resistance R_i of the emitter followers Q_3 and Q_4 of Fig. 5-3 will be a load connected to the collectors of Q_1 and Q_2. Referring to Fig. 5-8b, $R_L' = 10^4 \parallel R_i$ Ω, and because the signal voltage cannot change across V^+, the 10-kΩ resistors are effectively connected to ground. The 0.8-mA bias current source has been replaced by an infinite resistance assuming an ideal current generator. An improvement on the equivalent circuit would be to place a resistance of $1/h_{oe}$ across the current generator. The value of $1/h_{oe}$ for all transistors has been assumed to be infinite.

Referring to Fig. 5-8b, if I_1 increases, I_2 decreases by the same amount, the increase in $h_{fe}I_1$ is offset by the decrease in $h_{fe}I_2$, and there is no net change in the current to node E because of the controlled generators. Between terminals 1 and 2, the differential input resistance is found to be,

$$R_i = \frac{V_1 - V_2}{I_1 - I_2} = 2h_{ie} \tag{5-8}$$

The input signal can be applied to one or both terminals 1 and 2, and the output

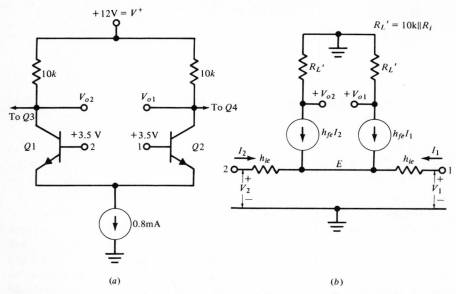

FIG. 5-8 (a) The $Q_1 - Q_2$ differential amplifier of Fig. 5-3; (b) the small-signal equivalent circuit of (a).

can be obtained from one collector or between collectors. The various combinations lead to the following expressions.

5-2.7 DIFFERENTIAL GAIN, SINGLE-ENDED INPUT, SINGLE-ENDED OUTPUT (SISO)

Assuming $V_2 = 0$ in Fig. 5-8, $I_1 = -I_2 = V_1/2h_{ie}$ and $V_{o2} = -h_{fe}I_2R'_L = h_{fe}V_1R'_L/2hie$ so that,

$$A_v = \frac{V_{o2}}{V_1} = \frac{h_{fe}R'_L}{2h_{ie}} \tag{5-9}$$

and similarly,

$$A_v = \frac{V_{o1}}{V_1} = \frac{-h_{fe}R'_L}{2h_{ie}} \tag{5-10}$$

We see that one output voltage is in phase with the input while the second output voltage is 180° out of phase with the input.

5-2.8 SINGLE-ENDED INPUT, DOUBLE-ENDED OUTPUT (SIDO)

Assuming $V_2 = 0$,

$$V_o = V_{o2} - V_{o1} \tag{5-11}$$

and from Eqs. (5-8), (5-9), and (5-10),

$$A_v = \frac{V_o}{V_1} = \frac{h_{fe}R'_L}{h_{ie}} \tag{5-12}$$

The phase relation of Eq. (5-12) is arbitrary since the reference point is undefined for the output voltage.

5-2.9 DOUBLE-ENDED INPUT, DOUBLE-ENDED OUTPUT (DIDO)

From Fig. 5-8b, $I_2 = -I_1$, $V_1 - V_2 = 2h_{ie}I_1$, and $V_{o2} - V_{o1} = h_{fe}(-I_2 + I_1)R'_L = 2h_{fe}I_1R'_L$, or

$$A_v = \frac{V_{o2} - V_{o1}}{V_1 - V_2} = \frac{h_{fe}R'_L}{h_{ie}} \tag{5-13}$$

5-2.10 DOUBLE-ENDED INPUT, SINGLE-ENDED OUTPUT (DISO)

For the double-ended-input single-ended-output condition, from Fig. 5-8b, $(V_1 - V_2) = 2h_{ie}I_1$ and $I_2 = -I_1$ so that $V_{o2} = -h_{fe}I_2R'_L = h_{fe}I_1R'_L$, yielding

$$A_v = \frac{V_{o2}}{V_1 - V_2} = \frac{h_{fe}R'_L}{2h_{ie}} \tag{5-14}$$

or

$$A_v = \frac{V_{o1}}{V_1 - V_2} = \frac{-h_{fe}R'_L}{2h_{ie}} \tag{5-15}$$

and the results are the same as for Eqs. (5-8) and (5-9). Using abbreviations for these differential gain connections, the SISO and DISO have the same voltage gain while SIDO and DIDO have the same gain.

5-2.11 DIFFERENTIAL-AMPLIFIER INPUT AND OUTPUT RESISTANCE

The single- or double-ended input resistance is $2h_{ie}$ as may be seen by traversing the loop from ground to terminal 1, to terminal 2, and back to ground in Fig. 5-8.

The single-ended output resistance of either transistor Q_1 or Q_2 is 10^4 Ω, the collector resistor, as measured from collector to ground, Fig. 5-8, with $V_1 = V_2 = 0$. The double-ended output resistance between collectors of Q_1 and Q_2 is $2(10^4)$ Ω. These output-resistance values would be modified slightly if h_{oe} and h_{re} were included.

The values of gains, input and output resistance can now be calculated for the μA730. The collector currents for Q_1 and Q_2 are 0.4 mA so that $h_{ie} = h_{fe}/38.5\,I_C$, and the load resistance at each collector is the 10^4-Ω resistor in parallel with the input to the emitter followers, $R'_L = 10^4 \,\|\, 2.03(10^5) \simeq 9.5$ kΩ. The differential voltage gain as defined in Eq. (5-1) is obtained from Eq. (5-14) as

$$A_v = \frac{h_{fe}R'_L}{2h_{ie}} = \frac{(0.0154)(9.5\text{ k})}{2} = 73$$

We see that the gain specified in Fig. 5-3 must be for double-ended output with either single- or double-ended input, and our calculated value becomes $(2)(73) = 146$, in reasonable agreement with 145 as specified. The emitter followers have a gain of 0.997, and the gain to either the low- or high-output impedance terminals is about 146.

The input resistance is $R_i = 2h_{ie} = 2h_{fe}/g_m = 2(100)/0.0154 = 13{,}000 \ \Omega$, agreeing with the 20,000-Ω specified value within the accuracy of the calculations.

5-2.12   TRANSFER CHARACTERISTICS

The transfer characteristics of Fig. 5-3b may be determined from the Ebers-Moll equation (4-38)

$$i_E = I_{ES}(e^{\theta v_{BE}} - 1) \simeq I_{ES}e^{\theta v_{BE}} \tag{5-16}$$

where $\theta = q/\eta kT$. Summing the two emitter currents of Q_1 and Q_2 in Fig. 5-3a,

$$I_O = i_{E1} + i_{E2} \tag{5-17}$$

or from Eq. (5-16)

$$I_O = I_{ES}e^{\theta v_{BE2}}(1 + e^{\theta(v_{BE1} - v_{BE2})}) = i_{E2}(1 + e^{\theta v_I}) \tag{5-18}$$

where $v_I = v_{BE1} - v_{BE2}$. Solving Eq. (5-18) for i_{E2},

$$i_{E2} = \frac{I_O}{1 + e^{\theta v_I}} \tag{5-19}$$

and since $i_{C1} = \alpha i_{E1}$, $i_{C2} = \alpha i_{E2}$,

$$i_{C1} = \frac{\alpha I_O}{1 + e^{\theta v_I}} \quad \text{and} \quad i_{C2} = \frac{\alpha I_O}{1 + e^{-\theta v_I}} \tag{5-20}$$

For the transfer characteristic,

$$v_O = i_{C1}R_{L1} - i_{C2}R_{L2} \tag{5-21}$$

where R_{L1} and R_{L2} are the collector resistors of Q_1 and Q_2, respectively. For $R_{L1} = R_{L2}$,

$$v_O = \alpha I_O R_L \frac{e^{\theta v_I} - e^{-\theta v_I}}{2 + e^{\theta v_I} + e^{-\theta v_I}} \tag{5-22}$$

$$= \alpha I_O R_L \frac{\sinh(\theta v_I)}{1 + \cosh(\theta v_I)} \tag{5-23}$$

Equation (5-23) or (5-22) is plotted in Fig. 5-9 for $\alpha = 1$, $I_0 = 0.8$ mA, $R_L = 10^4 \ \Omega$, $\theta = 40$. (Note: Eq. (5-23) is for double-ended input and output.)

When the collector current of transistor Q_1 is cut off by the input signal, all 0.8-mA bias current passes through transistor Q_2. For this condition, $V_{O1} = 12$ V and $V_{O2} = 4$ V. The situation is similar if transistor Q_2 is cut off. Thus V_{O1} or V_{O2} varies from 4 to 12 V or a maximum peak-to-peak voltage swing of 8 V as indicated by typical values in Fig. 5-3. The double-ended maximum peak-to-peak voltage swing is 8 V. We can see that the emitter currents of Q_1 and Q_2 are switched back and forth with input signal. The circuit is sometimes called an emitter-coupled amplifier.

For linear operation, the input signal can vary by about ± 20 mV and the output by ± 3.0 V, as seen in Fig. 5-9. Calculation of the percent of distortion will not be attempted since measurements of distortion are more meaningful. To set the input bias voltage at the center of the curve of Fig. 5-9, it is necessary to have an external bias adjustment. We note that the input signal voltage is the difference of v_{BE1} and v_{BE2}.

A comparison of calculated and typical values for the differential amplifier of Fig. 5-3 is summarized in Table 5-2. We have not calculated input offset values because these depend upon differences in transistor characteristics and construction.

The slope of Fig. 5-9 also yields the double-ended differential gain of the amplifier. An expression for this gain may be derived from Eq. (5-22) or (5-23) by taking the derivative with respect to v_I (see Prob. 5-3).

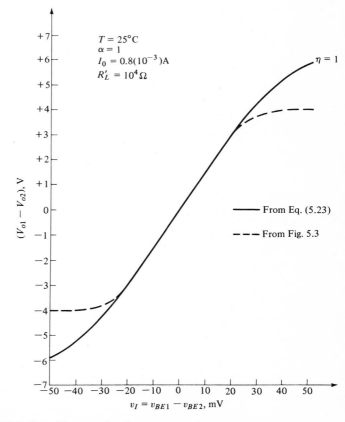

FIG. 5-9 The transfer characteristics from Fig. 5-3 and Eq. (5-23).

TABLE 5-2

Parameter	Conditions $T_A = 25°C$, $V^+ = 12$ V, $V_{CM} = 3.5$ V $\beta = h_{fe} = 100$	Typical value	Calc. value	Units
Input bias current		3.5	4	μA
Input resistance		20	13	kΩ
Differential voltage gain $R_L \geq 100$ kΩ		145	146	
Single-ended output resistance		70	106	Ω
Output voltage swing (single-ended)		8	8	V_{pp}
Supply current		9.5	9.9	mA
Power consumption		114	119	mW

5-2.13 FREQUENCY-RESPONSE ANALYSIS

The differential voltage gain as a function of frequency for the circuit of Fig. 5-3a will be determined by applying Eq. (4-13) for h_{fe}. From Eqs. (5-4), and (5-13), the overall differential gain function is

$$A_v = \frac{h_{fe}R_L'}{h_{ie}} \frac{(h_{fe} + 1)R_E}{h_{ie} + (h_{fe} + 1)R_E}$$

$$\cong \frac{h_{fe}R_L'}{h_{ie}}$$

$$= \frac{h_{feo}R_L'}{h_{ie}} \frac{1}{(1 + jf/f_\beta)} \tag{5-24}$$

From the table in Fig. 5-3 the break frequency is given as 1.5 MHz. Thus, from Eq. (5-24), the Beta cutoff frequency f_β for the transistors must be about 1.5 MHz.

5-3 LINEAR-IC OPERATIONAL-AMPLIFIER ANALYSIS

The operational amplifier is a differential amplifier without differential output. It has a high overall voltage gain by including several stages of amplification.

Three new applications of transistors will be demonstrated by analyzing the operational amplifiers shown in Figs. 5-10 and 5-11. The new subcircuits are: (1) the Darlington-transistor configuration made up of transistors Q_1 and Q_2, or Q_4 and Q_5; (2) the phase splitter at transistor Q_{11}; and (3) a push-pull output amplifier formed by transistors Q_{12} and Q_{13}. From the previous analysis, transistors Q_2 and Q_4 of Fig. 5-10 form a differential amplifier with constant-current source, Q_3. Transistors Q_6 and Q_7 form a second differential amplifier with a large resistor R_6 and $-V_{EE}$ approximating a constant current source. Transistor Q_8 is a common collector or emitter-follower stage driving the phase splitter Q_{11}, and transistor Q_{10} is for bias purposes. The bases and emitters of Q_3 and Q_{10} are connected to a fixed voltage point. No signal is to be amplified by these transistors, which indicates that they are bias transistors.

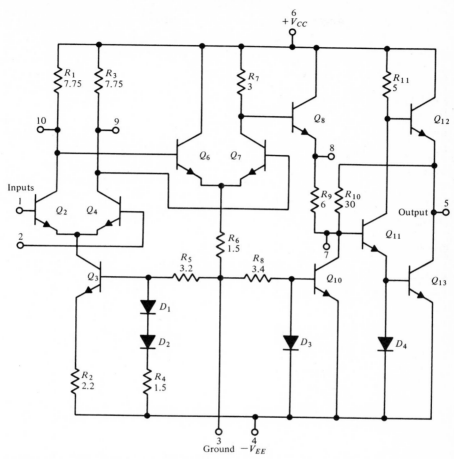

FIG. 5-10   The MC1530 operational amplifier. (All resistances are in $k\Omega$.) (*By permission Motorola Semiconductor Products, Inc., Phoenix, Ariz.*)

Two voltage supplies are needed as denoted by $+V_{CC}$ and $-V_{EE}$ in Fig. 5-10. The second terminal of these sources is connected to ground, and the complete system is designed so that if the inputs are at ground potential, the output will be at ground potential. Thus, two or more of these operational amplifiers with appropriate feedback elements can be directly connected, if desired, without changing bias conditions. It should be pointed out, however, that if the gain of each IC stage is large, any input offset voltage may saturate or latch the output making the circuit inoperative. Null balancing circuits to balance out the offset effects may be necessary. Feedback networks as discussed in Chap. 2 minimize offset effects.

The new subcircuits will be analyzed before the complete system is considered.

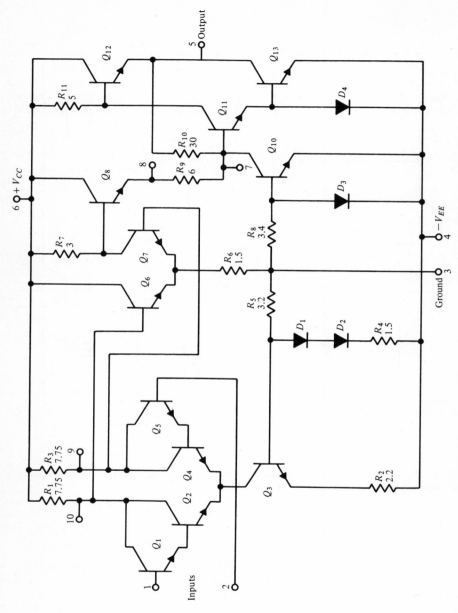

FIG. 5-11 The MC1531 op-amp with Darlington input. (All resistances are in kΩ.) (*By permission Motorola Semiconductor Products, Inc., Phoenix, Ariz*).

5-3.1 DARLINGTON TRANSISTOR PAIRS

The circuit of Fig. 5-11 differs from Fig. 5-10 in the input stage only. The circuit of Fig. 5-11 has a larger input resistance and less voltage gain than that of Fig. 5-10. To demonstrate these facts, consider the transistor pair Q_1 and Q_2 of Fig. 5-11 as redrawn in Fig. 5-12. Knowing the bias currents and h_{fe}, the parameters of the circuit of Fig. 5-12b may be determined as for the μA730 and applied at low frequencies. Since the collector or emitter current of Q_1 is the base current of Q_2, we know that the collector currents of the two transistors are unequal, and the equivalent circuits for them must be different.

To show the advantage of the combination of transistors in Fig. 5-12a, the circuit of Fig. 5-12c will be derived from Fig. 5-12b.

In Fig. 5-12c we see that

$$V_1 = I_1 h_{iec} \tag{5-25}$$

where the subscript c refers to the combination. Writing a similar expression for Fig. 5-12b,

$$V_1 = I_1 h_{ie1} + (h_{fe1} + 1)I_1 h_{ie2} \tag{5-26}$$

Therefore, comparing Eqs. (5-25) and (5-26),

$$h_{iec} = h_{ie1} + (h_{fe1} + 1)h_{ie2} \tag{5-27}$$

We see that Eq. (5-27) is of the same form as Eq. (5-5) for the emitter follower, and we expect high input resistance. To obtain h_{fec}, from Fig. 5-12c, we have

$$I_c = h_{fec} I_1 \tag{5-28}$$

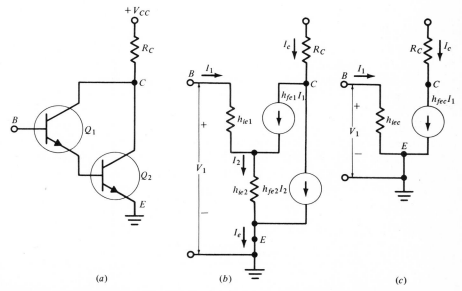

(a)   (b)   (c)

FIG. 5-12   (a) The Darlington pair of transistors at the input of Fig. 5-11; (b) the equivalent circuit for small-signal analysis of (a); (c) a composite equivalent circuit for (b).

and writing a similar equation for Fig. 5-12b,

$$I_c = h_{fe1}I_1 + h_{fe2}I_2$$

or

$$I_c = h_{fe1}I_1 + h_{fe2}(h_{fe1} + 1)I_1 \qquad (5\text{-}29)$$

Comparing Eqs. (5-28) and (5-29),

$$h_{fec} = h_{fe1} + h_{fe2}(h_{fe1} + 1) \cong h_{fe1}h_{fe2} \qquad (5\text{-}30)$$

The Darlington pair will produce high current gain and input resistance.

Example 5-1 For Fig. 5-12, $I_{C2} = 0.5$ mA and h_{fe2} is assumed to be 100. Determine the values for the equivalent transistor of Fig. 5-12c.

For Q_2, $h_{ie2} = h_{fe2}/38.5I_{C2} = 100/(38.5)(0.5)(10^{-3})$ or $h_{ie2} = 5,200$ Ω. Assuming h_{FE2} is also 100, $I_{B2} = I_{E1} = I_{C2}/h_{FE2} = 5(10^{-5})$ A. From Fig. 4-7, when collector current goes down from the rated value, h_{fe} also decreases. From this figure, we estimate that $h_{fe1} = 50$ so that $h_{ie1} = h_{fe1}/38.5I_{C1} = 50/(38.5)(5)(10^{-5}) = 26,000$ Ω. From Eq. (5-27), $h_{iec} = (26,000) + (50 + 1)5,200 = 2.9(10^5)$ Ω, and Eq. (5-30) yields,
$$h_{fec} = 50 + 100(50 + 1) = 5,150$$
If the same collector current of 0.5 mA were to go through Q_2 in Fig. 5-10, $h_{ie2} = 5,200$ and $h_{fe2} = 100$. If the voltage is 1 mV across h_{ie2} of Fig. 5-10 and h_{iec} of Fig. 5-11, we find $I_c = 19$ μA for Fig. 5-10, and $I_c = 18$ μA in Fig. 5-11. Therefore, the input resistance of the MC 1531 is higher than for the MC 1530, but the gain of the MC 1530 should be larger. The circuit to be selected will depend on the input resistance necessary for the application.

5-3.2 PHASE SPLITTER

Transistor Q_{11} in Fig. 5-10 delivers two output signals to transistors Q_{12} and Q_{13}, the two signals being obtained from the collector and the emitter of Q_{11}. To understand the operation of the phase splitter, assume transistors Q_{12} and Q_{13} have large input resistance, and replace diode D_4 by a piecewise-linear equivalent circuit, as shown in Fig. 5-13a. The small-signal equivalent circuit of Fig. 5-13b will be applied to demonstrate the circuit operation. Since the base current is small, approximately the same current flows through r_d and R_{11}. However, $V_{i13} = I_c r_d$ and $V_{i12} = -I_c R_{11}$. The *two output signals* are 180° *out of phase* with V_{i11} while I_c is in phase with the input voltage. Since the input voltage has been split into two out-of-phase voltages, the name for the circuit is self-evident.

A more exact analysis follows from Fig. 5-13b. At the input,

$$V_{i11} = I_1[h_{ie} + (h_{fe} + 1)r_d] \qquad (5\text{-}31)$$

and with $V_{i12} = -h_{fe}I_1R_{11}$, we find

$$V_{i12} = \frac{-V_{i11}R_{11}h_{fe}}{h_{ie} + (h_{fe} + 1)r_d} \qquad (5\text{-}32)$$

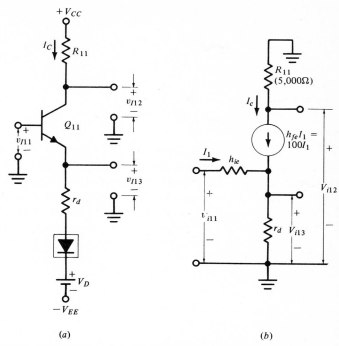

(a)   (b)

FIG. 5-13 (a) The phase-splitter circuit with base bias elements removed; (b) the small-signal equivalent circuit of (a).

Similarly,

$$V_{i13} = (h_{fe} + 1)I_1r_d = (h_{fe} + 1)\frac{V_{i11}r_d}{h_{ie} + (h_{fe} + 1)r_d} \tag{5-33}$$

The numerical values for V_{i12} and V_{i13} are developed in Example 5-2.

Example 5-2 Transistor Q_{11} in Fig. 5-13a is biased so that $I_C = 1$ mA. Determine numerically the voltages V_{i12} and V_{i13} in terms of V_{i11}.

For the diode in the emitter circuit, applying Eq. (3-2) with $\eta = 2$, $r_d = 2kT/qI_c = 0.052/10^{-3} = 52$ Ω. The transistor h_{ie} value for Fig. 5-13b with $h_{fe} = 100$ is, $h_{ie} = h_{fe}/g_m = 100/(38.5)(10^{-3})$ or $h_{ie} = 2,560$ Ω. With $R_{11} = 5,000$ Ω, Eq. (5-33) becomes

$$V_{i13} = \frac{(101)(V_{i11})(52)}{2,560 + (101)(52)} = 0.672V_{i11}$$

and Eq. (5-32) yields

$$V_{i12} = \frac{-(5,000)(100)V_{i11}}{2,560 + (101)(52)} = -64V_{i11}$$

If the loads were fixed resistors, as in this example, the unbalance in voltage would be undesirable. However, as shown later, the transistor loads on the phase splitter tend to counterbalance the uneven splitting of voltage found here.

In discrete circuit design, the values of r_d and R_C can be controlled to yield a more equal magnitude of voltages at the emitter and collector of the phase splitter if desired. The output is affected by the load which must be included to obtain a final solution.

5-3.3 PUSH-PULL OUTPUT STAGE

The output stage of Fig. 5-10 with resistor R_{10} removed, is redrawn in Fig. 5-14a. As will be shown, there is current flow in Q_{12} and Q_{13} for all signals, and the transistors operate Class A or in region 2 (linear region). From the equivalent circuit of Fig. 5-14b, we see that Q_{12} is operated as an emitter follower, and Q_{13} has its emitter connected to ground. The two input signals V_{i12} and V_{i13} are out of phase from the phase splitter. At the load,

$$I_o = (h_{fe12} + 1)I_{12} - h_{fe13}I_{13}$$

$$= \frac{(h_{fe12} + 1)(V_{i12} - V_o)}{h_{ie12}} - \frac{h_{fe13}V_{i13}}{h_{ie13}} \tag{5-34}$$

and since $I_o = V_o/R_L$, Eq. (5-34) can be rearranged so that

$$V_o = \frac{R_L h_{ie12}[(h_{fe12} + 1)V_{i12}/h_{ie12} - h_{fe13}V_{i13}/h_{ie13}]}{[h_{ie12} + (h_{fe12} + 1)R_L]} \tag{5-35}$$

For $h_{fe} \gg 1$ and $h_{fe}R_L \gg h_{ie}$, Eq. (5-35) simplifies to

$$V_o = V_{i12} - V_{i13} \tag{5-36}$$

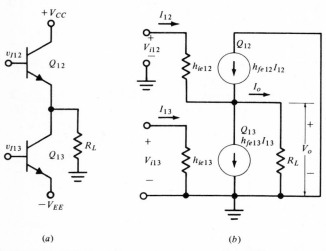

(a)   (b)

FIG. 5-14 (a) The output circuit of Fig. 5-10 with a load added; (b) the equivalent circuit of (a) at low frequencies.

To understand the operation, if V_{i12} increases, V_{i13} decreases because of the driving-source phase splitter, and V_o increases. If V_{i12} decreases, V_{i13} increases, and V_o decreases and could become negative. For sinusoidal input signals, the output will be a sine waveform centered about ground, the bias setting the ground reference point, as shown later. Since one input signal increases (*pushes*) and the other input signal decreases (*pulls*), the circuit is described as *push-pull*.

The input resistances to transistors Q_{12} and Q_{13} are not the same. From Fig. 5-14 the input resistance of Q_{13} is h_{ie13} by inspection. The input resistance of Q_{12} is approximately $(h_{fe12} + 1)R_L + h_{ie12}$ if $V_{i13} = 0$. Since the resistance looking into the collector of the phase splitter is much higher than the resistance looking into its emitter, the unbalanced inputs to the push-pull amplifier can be tolerated.

Example 5-3 The phase splitter of Example 5-2 drives the push-pull stage of Fig. 5-14. If collector bias currents of Q_{12} and Q_{13} are 3 mA, express the output voltage in terms of V_{i11} if $R_L = 1,000\ \Omega$.

For Q_{12} and Q_{13} with $h_{fe} = 100$, $g_{m12} = g_{m13} = 38.5I_C = 0.12$ mho and $h_{ie12} = h_{ie13} = h_{fe}/g_m = 830\ \Omega$. The complete circuit with numerical values is drawn in Fig. 5-15. The general procedure for the solution will be to: (1) find I_{13} from the input circuit in terms of V_{i11}, (2) write a node equation at the collector of the phase splitter to get I_{12} in terms of V_o and V_{i11}, and (3) express the output voltage in terms of the I_{12} and I_{13} found in (1) and (2). (1) At the input, $I_1 = V_{i11}/[2,560 + (101)(49)] = 1.33(10^{-4})V_{i11}$, and $I_{13} = (101)(49)I_1/830 = 7.93(10^{-4})V_{i11}$. (2) At the collector node of Q_{11}, $100I_1 + I_{12} + (V_c/5,000) = 0$. With $V_c = 830I_{12} + V_o$, we solve this node equation and find $I_{12} = -1.72(10^{-4})V_o - 0.0114V_{i11}$. (3) Writing the

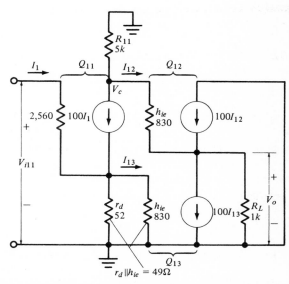

FIG. 5-15 The total equivalent circuit for the phase-splitter and class A push-pull amplifier of Fig. 5-10. This circuit is a combination of Figs. 5-13(b) and 5-14(b).

expression for output voltage in terms of I_{12} and I_{13} and employing the expressions of (1) and (2) we find

$$V_o = [(101)I_{12} - 100I_{13}]R_L$$

$$= (101)[-1.72(10^{-4})V_o - 0.0114\,V_{i11}](10^3)$$

$$- (100)[7.93(10^{-4})\,V_{i11}](10^3)$$

$$= -67V_{i11}$$

The gain of the phase-splitter push-pull output stage is (-67) where the negative sign indicates $180°$ phase shift.

The push-pull stage is also a bias or dc level shifter so that the dc output voltage will be zero when the input is zero. We shall show this fact in the next section.

5-3.4 DC ANALYSIS

The circuit of Fig. 5-10 is reproduced in Fig. 5-16 and partitioned by dashed lines to indicate the stages in the system. All diodes are made from transistors by connecting the collector to base. The importance of this connection will be shown for obtaining the bias current of Q_{10}.

The dc analysis for finding all bias currents and voltages begins with the constant-current source formed by Q_3. The supply voltages are $-V_{EE} = -6$ V and $V_{CC} = +6$ V. At the base of Q_3 (neglecting all base currents),

$$V_{B3} = \frac{(-V_{EE} + V_{D1} + V_{D2})R_5}{R_4 + R_5}$$

$$= \frac{(-6 + 1.4)3.2\,\mathrm{k}}{(1.5\,\mathrm{k} + 3.2\,\mathrm{k})} = -3.13\ \mathrm{V}$$

The current in R_2 is found to be

$$I_{R_2} = \frac{V_{B3} - V_{BE3} + V_{EE}}{R_2}$$

and

$$I_{C3} \simeq I_{R_2} = \frac{-3.13 - 0.7 + 6}{2.2\,\mathrm{k}} \simeq 1\ \mathrm{mA}$$

Since Q_2 and Q_4 have matched characteristics, the collector current of Q_3 divides between Q_2 and Q_4. Thus, assuming terminals 1 and 2 to be at ground,

$$I_{C2} = I_{C4} = \frac{I_{C3}}{2} = 0.5\ \mathrm{mA}$$

The collector voltages of Q_2 and Q_4 and the base voltages of Q_6 and Q_7 are

$$V_{C2} = V_{C4} = V_{B6} = V_{B7} = V_{CC} - I_{C4}R_3 = 6 - (0.5)(10^{-3})(7.75)(10^3)$$

$$= 6 - 3.8 = 2.2\ \mathrm{V}$$

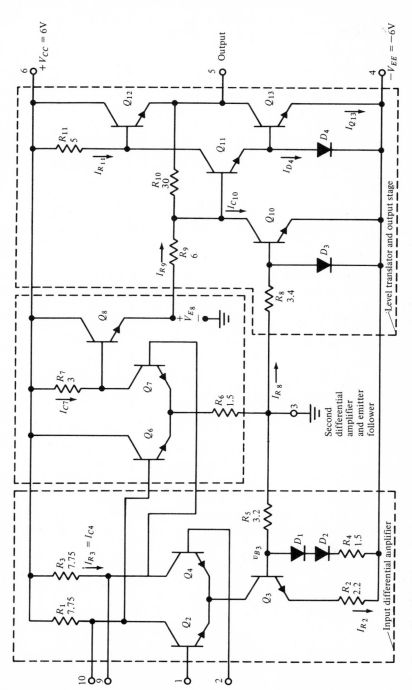

Notes: (1) All diode forward voltages and base–emitter voltages are approximately 0.7 V.
(2) All base currents are neglected.
(3) Notation: v_{B3} is read as the positive to negative voltage from base to ground for Q_3; V_{E8} is from the emitter of Q_8 to ground, etc.
(4) All resistors are in kΩ.

FIG. 5-16   The MC1530 IC amplifier sectioned for analysis. (*By permission Motorola Semiconductor Products, Inc., Phoenix, Ariz.*)

The input common-mode voltage range is determined by V_{B3} and V_{B6} or V_{B7}. So that the collector-base junctions of Q_2 and Q_4 remain reverse-biased, the common voltage on terminals 1 and 2 cannot safely exceed about $+2$ V. So that the Q_3 collector base also remains reverse-biased, and since $V_{B3} = -3.14$ V, the voltage should be greater than -2 V.

5-3.5 SECOND STAGE

The current in R_6 is found to be

$$I_{R_6} = \frac{V_{B6} - V_{BE6}}{R_6} = \frac{V_{B7} - V_{BE7}}{R_6}$$

$$= \frac{2.2 - 0.7}{1.5 \text{ k}} = 1 \text{ mA}$$

Thus, for Q_6 and Q_7 matched,

$$I_{C6} = I_{C7} = \frac{I_{R_6}}{2} = 0.5 \text{ mA}$$

The resistor R_6 is employed instead of a constant-current source because the common-mode voltage swing is small being fixed by the collector voltages of Q_2 and Q_4.

The collector voltage of Q_7 or the base voltage of Q_8 is

$$V_{C7} = V_{B8} = V_{CC} - I_{C7}R_7$$

$$= 6 - (0.5)(10^{-3})(3)(10^3)$$

$$= 4.5 \text{ V}$$

and

$$V_{E8} = V_{B8} - V_{BE8} = 4.5 - 0.7 = 3.8 \text{ V}$$

The output stage must translate from this $V_E = 3.8$ V to zero output voltage so as to maintain the zero input–zero output condition of the system.

5-3.6 THIRD STAGE

To understand the method of biasing Q_{10}, refer to Fig. 5-17 which shows that diodes are made by connecting the collector to base of a transistor. Assuming negligible base current with respect to collector or emitter current,

$$I_{CD3} = I_{ED3} = \frac{V_{EE} - V_{BED3}}{R_8}$$

$$= \frac{6 - 0.7}{3.4 \text{ k}} = 1.56 \text{ mA}$$

Since the transistor connected as D_3 matches characteristics with Q_{10} and the base-emitter voltages are the same, $I_{C10} = I_{ED3} = 1.56$ mA.

The current through R_9 (see Fig. 5-16) is:

$$I_{R_9} = \frac{V_{E8} - V_{BE11} - V_{BE13} + V_{EE}}{R_9}$$

$$= \frac{3.8 - 0.7 - 0.7 + 6}{6\,k} = 1.4\,\text{mA}$$

For the output voltage at terminal 5,

$$V_O = (I_{C10} - I_{R_9})R_{10} + V_{BE11} + V_{BE13} - V_{EE}$$

$$= (1.56 - 1.4)(10^{-3})(30)(10^3) + 1.4 - 6$$

$$= 4.8 + 1.4 - 6 = 0.2\,\text{V}$$

Thus the output is essentially zero within the accuracy of the calculations.

To complete the bias calculations, (assuming $V_O \simeq 0$) at the phase-splitter push-pull stage,

$$I_{R_{11}} = \frac{V_{CC} - V_{BE12}}{R_{11}} = \frac{6 - 0.7}{5\,k} = 1.06\,\text{mA}$$

and

$$I_{R_{11}} = I_{C11} = I_{D4} = 1.06\,\text{mA}$$

It appears that diode D_4 should act on transistor Q_{13} in the same manner as shown in Fig. 5-17 for D_3 and Q_{10}. However, Q_{13} is made with the base-emitter-junction area three times larger than the area of D_4 [8] so that

$$I_{C13} = I_{C12} = 3I_{D4} = 3.18\,\text{mA}$$

It is not always possible to tell from the schematic drawing if the transistors are made identically.

All bias currents and voltages have now been found, and the signal gain can be calculated for the system. Note again that the input voltage was assumed to be zero in the above calculations.

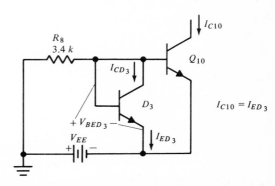

FIG. 5-17 The bias circuit for transtor Q_{10} of Fig. 5-16.

5-3.7 SIGNAL GAIN OF THE SYSTEM

From Eq. (5-12) or (5-13) the differential voltage gain of the first stage is

$$A_v = \frac{h_{fe}R_L'}{h_{ie}}$$

and since

$$I_{C2} = I_{C4} = I_{C6} = I_{C7}$$

all g_m values of Q_2, Q_4, Q_6, and Q_7 are
$g_m = 38.5I_C = (38.5)(0.5)(10^{-3}) = 0.02$ mho
Assuming $h_{fe} = 100$ and $r_{bb'}$ negligible throughout, the load on the collectors of Q_2 and Q_4 are $R_1 \parallel h_{ie6}$ and $R_3 \parallel h_{ie7}$, respectively. Therefore,

$$h_{ie6} = h_{ie7} \cong \frac{h_{fe}}{g_m} = 5,000 \ \Omega$$

and

$$R_L' = R_1 \parallel h_{ie6} = \frac{(7.75 \text{ k})(5,000)}{12,750} = 3,039 \ \Omega$$

For the first stage,

$$A_v = \frac{(100)(3,039)}{5,000} = 60.8$$

For the second stage, the input impedance of the emitter follower Q_8 will be assumed large so that it does not load Q_7. From Eq. (5-15), for double-ended input single-ended output,

$$A_{v2} = -h_{fe}R_L/2h_{ie} = \frac{(-100)(3,000)}{2(5,000)}$$

$$= -30$$

We have calculated the gain of the phase-splitter push-pull stage neglecting R_9 and R_{10} in Examples 5-2 and 5-3. However, including the feedback and input resistors, the final stage may be drawn as in Fig. 5-18. From Eq. (2-3) for an

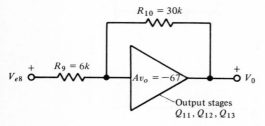

FIG. 5-18 The gain of the final stage of Fig. 5-16, calculated by use of the usual op-amp gain equation. Transistors Q_{11}, Q_{12}, and Q_{13} have a stage gain of -67 as calculated in the text.

operational amplifier,

$$A_{v3} = \frac{-R_{10}}{R_9} = -5$$

Thus the total gain of the system is

$$A_{vT} = (A_{v1})(A_{v2})(A_{v3}) = (60.8)(-30)(-5) \cong 9{,}120$$

This agrees with the manufacturer's specifications which state that the gain is between 5,000 and 12,000.

Input Resistance The input resistance between terminals 1 and 2 can be determined from the equivalent circuit of Fig. 5-8. As shown in that example,

$$R_{\text{in}} = h_{ie2} + h_{ie4}$$

Since $I_{C_2} = I_{C_4} = 0.5$ mA and with $h_{fe} = 100$,

$$h_{ie2} = h_{ie4} = \frac{h_{fe}}{g_m} = \frac{100}{0.02}$$

$$= 5{,}000 \ \Omega$$

Thus, $R_{\text{in}} = 10{,}000 \ \Omega$, which agrees favorably with the published value.

5-3.8 OUTPUT VOLTAGE SWING

The final evaluation for the circuit of Fig. 5-10 involves the output voltage swing. Referring to Fig. 5-15 and Eq. (5-34), the load current is proportional to the difference of the collector currents in Q_{12} and Q_{13}. As the collector current in Q_{12} increases, the collector current in Q_{13} decreases because of the phase-splitter drive. Since the bias current in these transistors is about 3 mA, the current in one transistor can go to zero, while the current in the other can increase to about 6 mA so that the load current can vary ±6 mA; or for a load of 1,000 Ω, the output voltage can vary ±6 V peak. However, to keep out of the saturation region of the transistors, the collector-to-emitter voltage should not become less than 0.5 V. Thus the output-voltage limits are about ±5.5 V. The data sheets indicate a variation of ±5.2 V. This limit is the same for any load greater than 1,000 Ω due to the V_{CC} and V_{EE} supplies of 6 V.

Other calculations concerning the ICs of Fig. 5-10 are most conveniently performed by use of the computer. However, the understanding of the components of the system and good approximations to the actual operation have been accomplished by the above analysis.

5.4 INTERNALLY COMPENSATED ICs

The internally frequency-compensated IC, type 741, was discussed earlier, and the characteristics of Fig. 2-6 were presented. Some differences between the design of the IC of Fig. 5-10 and of the 741 are of interest. The 741 is called a second-generation operational amplifier. Its schematic and block diagrams are reproduced in Fig. 5-19, and we see that both *npn* and *pnp* transistors are used.

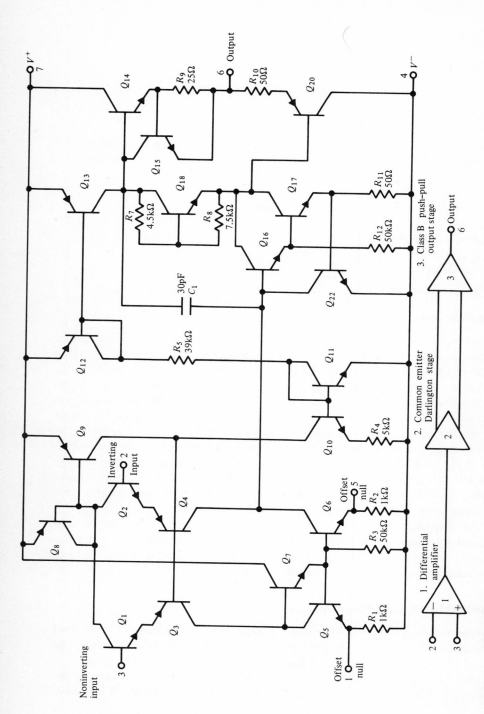

FIG. 5-19 The 741 schematic and block diagram. The total gain of the op-amp is determined by the first two stages. The push-pull output stage allows the circuit to drive a reasonable load.

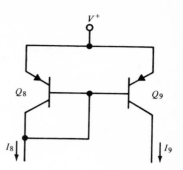

FIG. 5-20 The collector load on the input differential amplifier of Fig. 5-19. For matched transistors, $V_{BE8} = V_{BE9}$ and $I_8 = I_9$.

For such a complicated-looking circuit, the block diagram is quite simple. The input stage is a differential amplifier with a single-ended output; it is followed by a high-gain common-emitter amplifier, and the output is a push-pull amplifier.

The input-stage differential amplifier has a collector load that is a current source made up of transistors Q_8 and Q_9. Since a current source has a high impedance, the collector loads on Q_1 and Q_2 are high-impedance loads, the load being redrawn in Fig. 5-20. (Transistors Q_7 and Q_{15} are omitted in the following because they are in the circuit to prevent latching only.) Since base currents are less than collector or emitter currents, and $V_{BE8} = V_{BE9}$ for matched transistors, $I_8 \simeq I_9$. A more exact calculation will be made later. The current I_9 is fixed by a bias circuit, and I_8 will divide between Q_1 and Q_2.

The emitters of Q_1 and Q_2 are coupled to the emitters of Q_3 and Q_4. The *pnp* transistors have poor high-frequency-response characteristics because of their construction. Q_3 and Q_4 are operated common-base (no base signal) to minimize their effect on the overall frequency response as will be shown. The collectors of Q_3 and Q_4 operate into the high-resistance current-source load of Q_5 and Q_6. A further advantage of this loading can be seen from Fig. 5-21. With the input terminals 2 and 3 at the same potential (balanced condition), $I_3 = I_4 = I_6$, assuming matched transistors as usual. If the input signal changes to increase I_4, then $I_3 = I_6$ decreases by the same amount. The output current is $I_4 - I_6 = I_4 - I_3$, or the single output is proportional to the full differential current gain of the stage. Another advantage for the circuit of Fig. 5-21 is that voltage level shifting can also be accomplished in this first stage. Other advantages are: (1) higher ac input impedance because of the two extra base-emitter junctions of Q_3 and Q_4, (2) the output impedance of the first stage is large, (3) the loading on the first stage is only the input impedance of the second stage so that large voltage gain is achieved, (4) the output is single-ended so that a simple common-emitter amplifier can be used for the second stage.

5-4.1 SECOND STAGE

Figure 5-22 displays a simplified form of the second stage. Transistors Q_{16} and Q_{17} form a Darlington pair operated as a common-emitter amplifier. Transistor

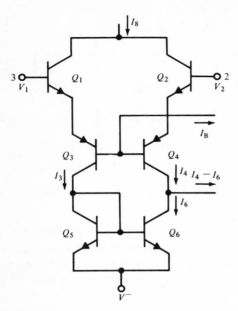

FIG. 5-21   The input stage of Fig. 5-19 simplified for explanation of the double-ended-input single-ended-output feature.

Q_{18} is operated as a diode for correct voltage levels at the output stage. Transistor Q_{22} has been omitted because it normally is inoperative, its purpose being to limit the input voltage to Q_{16}. A large signal at the emitter of Q_{17} turns on Q_{22}, and shunts the input to Q_{16} through the low collector-to-emitter resistance of Q_{22}.

The second-stage load is again a constant-current source involving Q_{12} and Q_{13} for a large load resistance on the $Q_{16} - Q_{17}$ combination. Capacitor C_1 is added, as shown, because the impedance level at the input of Q_{16} is high, and a small value of capacitance can be used to cause the low-frequency breakpoint to occur at about 7 Hz.

5-4.2 OUTPUT STAGE

The output stage appears to be similar to that of Fig. 5-10, but the operation is not the same. The stage is described as a complementary symmetry push-pull amplifier, and it is redrawn for further discussion in Fig. 5-23a. With $v_I = 0$, both Q_{14} and Q_{20} are off (no collector current) because the base-emitter junctions are not forward-biased. When v_I is positive, no current flows until v_I reaches the cutin voltage of Q_{14}. When v_I is greater than the cutin voltage of Q_{14}, current flows in Q_{14} and through the load. If v_I is negative, the same argument applies except current now flows in Q_{20} and through R_L in the reverse direction. Waveforms for sinusoidal input are shown in Fig. 5-23b. The output is a somewhat distorted replica of the input (cross-over distortion), and since each transistor is operating in an emitter-follower configuration, the voltage gain is near unity, but current gain does occur.

To eliminate the distortion in the output, the effect of the cutin voltage for Q_{14} and Q_{20} is balanced by placing biased diodes in the circuit as in Fig. 5-24. Transistor Q_{18} is biased to balance the base emitter voltages of Q_{14} and Q_{20}. Actually, these base-emitter voltages are slightly forward-biased to eliminate the cutin voltage effect, yet operation is such that only one transistor is effectively operating at any time when signal is present. Since very little current flows in Q_{14} and Q_{20} without signal, less standby current and power are required leading to better efficiency in this type of stage. In Fig. 5-14 for the MC 1530, about the same current flows at all times in the output stage whether a signal is present or not. This leads to a continual power drain while supply voltage power is required in Fig. 5-23 only when signals are being amplified.

Assuming that the cutin voltage effects are nullified as in Fig. 5-24, the analysis follows from Fig. 5-25. The equivalent circuit of Fig. 4-6 has been applied with V_{BE} omitted and all emitter resistances included in R_{E1} and R_{E2}. When v_I has the polarity indicated in Fig. 5-25,

$$v_O = (\beta_{14} + 1)i_{B1}R_L \tag{5-37}$$

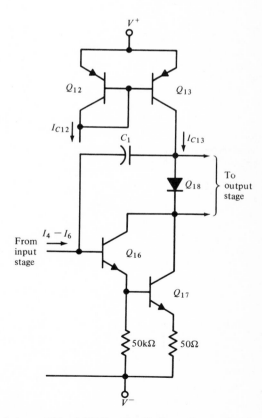

FIG. 5-22   A simplified form of the second stage of Fig. 5-19. For matched transistors, $I_{C12} = I_{C13}$.

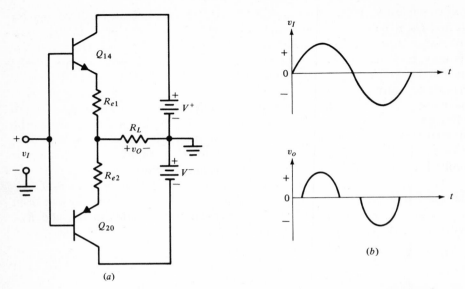

FIG. 5-23 (a) The complementary-symmetry, push-pull stage; (b) the input and output waveforms of (a).

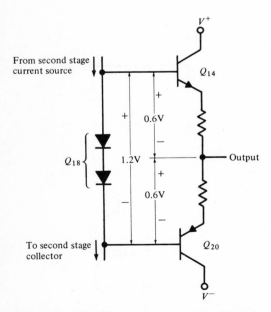

FIG. 5-24 The biasing arrangement for the output stage of Fig. 5-19 to compensate for the cutin voltages of Q_{14} and Q_{20}.

and

$$v_I = (\beta_{14} + 1)i_{B1}(R_{E1} + R_L) \tag{5-38}$$

From Eq. (5-38)

$$i_{B1} = \frac{v_I}{(\beta_{14} + 1)(R_{E1} + R_L)} \tag{5-39}$$

Applying Eq. (5-39) to Eq. (5-37),

$$v_O = \frac{(\beta_{14} + 1)R_L v_I}{(\beta_{14} + 1)(R_{E1} + R_L)} = \left. \frac{R_L v_I}{R_{E1} + R_L} \right|_{v_I > 0} \tag{5-40}$$

Similarly, when v_I is opposite to the polarity shown in Fig. 5-25,

$$v_O = - \left. \frac{R_L v_I}{R_{E2} + R_L} \right|_{v_I < 0} \tag{5-41}$$

assuming $\beta_{14} = \beta_{20}$, and $R_{E1} = R_{E2}$. The voltage gain is near unity for R_E small,

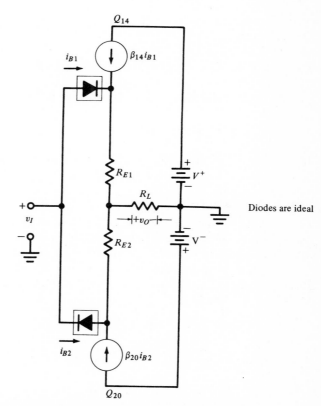

FIG. 5-25 The equivalent circuit of Fig. 5-23 when biased to balance out the cutin voltages.

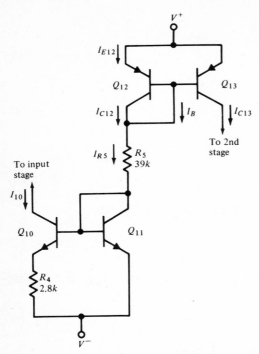

FIG. 5-26 The bias determining elements of Fig. 5-19.

and the input impedance of the output stage is, from Eq. (5-38),

$$\frac{v_I}{i_B} = (\beta + 1)(R_E + R_L) \tag{5-42}$$

The resistance given by Eq. (5-42) is also the effective load on the second stage in Fig. 5-19.

5-4.3 DC BIAS CURRENTS

Determining the bias currents of the first and second stage of Fig. 5-19 will allow calculation of the gain. The bias currents are slightly more difficult to calculate for this IC than for that of Fig. 5-10.

The bias currents are determined by the elements shown in Fig. 5-26. Resistor R_5 has a reference diode formed with Q_{11} and Q_{12} at each end connected to V^+ and V^-. Thus,

$$I_{R_5} = \frac{V^+ + V^- - V_{EB12} - V_{BE11}}{R_5} \tag{5-43}$$

For $V^+ = 15$ V and $V^- = 15$ V, Eq. (5-43) yields

$$I_{R_5} = \frac{15 + 15 - 0.6 - 0.6}{39(10^3)} = 740 \ \mu A$$

To obtain I_{C13}, we must know I_{C12}. Normally for *npn* transistors, the base current would be much less than the collector current. However, because of the type of construction, the *pnp* transistor β is about 5. Thus, $I_B + I_{C12} = I_{R5}$, and $I_B = I_{B12} + I_{B13} = 2I_{B12}$ for matched transistors. Now, $I_{C12} = \beta I_{B12}$, $I_{R5} = 2I_{B12} + I_{C12} = (\beta + 2)I_{B12}$, and

$$I_{B12} = I_{B13} = \frac{I_{R5}}{\beta + 2}$$

Since Q_{12} and Q_{13} are matched,

$$I_{C12} = I_{C13} = \beta I_{B12} = \frac{\beta I_{R5}}{\beta + 2} \tag{5-44}$$

For $\beta = 5$, $I_{C12} = 740(10^{-6})(0.714) = 528 \ \mu A$.

We will employ a graphical procedure to obtain I_{10} of Fig. 5-26. The circuit designer did not have this problem since he knew the desired value of I_{10} and determined R_4 to obtain it. The current I_{10} is less than I_{R5} because of R_4 in the emitter circuit of Q_{10}.

For the base-emitter-junction of an *npn* transistor with forward bias,

$$I_E = I_{ES}[e^{\theta V_{BE}} - 1] \simeq I_{ES}e^{\theta V_{BE}}$$

where I_{ES} is the reverse saturation current. Solving for V_{BE}, $V_{BE} = (1/\theta) \ln(I_E/I_{ES})$. If I_E is increased a decade to $10I_E$, $V'_{BE} = (1/\theta) \ln(10I_E/I_{ES})$, and the change in base-emitter voltage for this current change is

$$\Delta V_{BE} = V'_{BE} - V_{BE} = \frac{1}{\theta}\left[\ln\left(\frac{10I_E}{I_{ES}}\right) - \ln\left(\frac{I_E}{I_{ES}}\right)\right]$$

$$= \frac{1}{\theta}\ln(10) = (0.026)(2.3)$$

or

$$\Delta V_{BE} \simeq 60 \ mV/decade \ change \ in \ I_E \tag{5-45}$$

Equation (5-45) is useful in obtaining I_{10}.

To find I_{10}, refer to Fig. 5-27. On three-cycle semilog graph paper, we plot ΔV_{BE} from 0 to 200 mV on the linear axis and I_E from 1 to 1,000 μA along the log axis (which assumes $I_S = 1 \ \mu A$). A line representing ΔV_{BE} versus I_E for Q_{11} is drawn between the point 0 mV, 1 μA, to point 180 mV, 1,000 μA or 60 mV/decade. Since this ΔV_{BE} is across the base-emitter of Q_{10} and R_4, a second curve of $(\Delta V_{BE10} + R_4I_E)$ versus I_E is plotted. As an example for this second curve, at $I_E = 1 \ \mu A$, plot $\Delta V_{BE} = \Delta V_{BE10} + R_4I_E = 0 + (2.8)(10^3)(10^{-6}) = 2.8$ mV. Similarly, at $I_E = 10 \ \mu A$, we plot $\Delta V_{BE} = \Delta V_{BE10} + I_ER_4 = 60(10^{-3}) + (2.8)(10^3)(10^{-5}) = 88$ mV. Repeating this type of calculation for several points, a curve can be drawn joining the points. We construct a line along the $I_E = I_{R5} = 740$-μA line until it meets the Q_{11} straight line (at about 172 mV). At this latter intersection, which is the voltage across V_{BE11}, we draw another line perpendicular to the 740-μA line until it reaches the $(\Delta V_{BE} + I_ER_4)$ curve for Q_{10}. This last intersection yields I_{10} of about 27 μA.

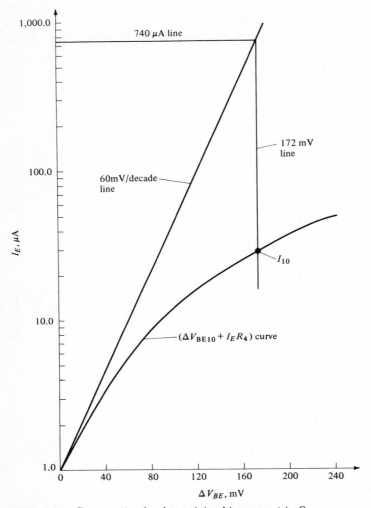

FIG. 5-27 Construction for determining bias current in Q_{10}.

It is tedious to relate I_{10} to the collector currents of Q_1 and Q_2, but we can do so by referring to Fig. 5-28. With $\beta_3 = \beta_4 = \beta_8 = \beta_9 = \beta$ because of identical fabrication, and also with equal emitter currents in Q_8 and Q_9 since $V_{BE8} = V_{BE9}$,

$$I_{C1} + I_{C2} = I_{E8} + \frac{I_9}{\beta} = I_{E9} + \frac{I_9}{\beta} \qquad (5\text{-}46)$$

Also $I_{E9} = I_9 + I_9/\beta$ so that $I_{C1} + I_{C2} = I_9(1 + 2/\beta)$, and with $I_{C1} = I_{C2}$,

$$I_{C1} = I_{C2} = \frac{I_9}{2}\frac{\beta + 2}{\beta} \qquad (5\text{-}47)$$

or

$$I_9 = 2I_{C1}\frac{\beta}{\beta + 2} \tag{5-48}$$

Because Q_1 and Q_2 have high gain, their base currents are neglected and

$$I_{C1} = I_{E1} = I_{C2} = I_{E2} \tag{5-49}$$

The base current of Q_3 and Q_4 are

$$I_{B3} = I_{B4} = \frac{I_{E1}}{\beta + 1} \tag{5-50}$$

and combining Eqs. (5-47), (5-49), and (5-50)

$$I_{B3} + I_{B4} = \frac{2I_{C1}}{\beta + 1} = \frac{I_9}{\beta + 1}\frac{\beta + 2}{\beta}$$

so that

$$I_{10} = I_9 + I_{B3} + I_{B4} = I_9\frac{\beta + 2 + 2/\beta}{\beta + 1} \tag{5-51}$$

Applying Eq. (5-48) in Eq. (5-51),

$$I_{10} = \frac{2\beta I_{C1}}{\beta + 2}\frac{\beta + 2 + 2/\beta}{\beta + 1} \tag{5-52}$$

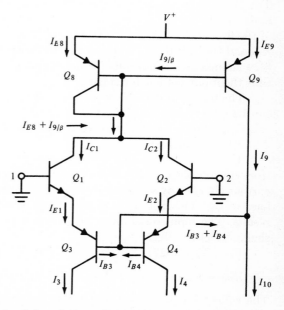

FIG. 5-28 The circuit for relating collector currents of Q_1 and Q_2 to I_{10}.

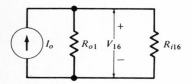

FIG. 5-29 The circuit for obtaining V_{16}. The resistors R_{01} and R_{i16} represent the output resistance of the first stage and the input resistance of the second stage of Fig. 5-19, respectively.

For $\beta = 5$, $I_{10} = 1.76 I_{C1}$, and since I_{10} was found to be 27 μA, $I_{C1} = I_{C2} = (27)(10^{-6})/1.76 = 15 \ \mu$A.

5-4.4 INPUT-STAGE GAIN

In the discussion concerning the input stage, it was shown that the single-ended output current $I_4 - I_6$ of Fig. 5-21 has the same value as if the stage current gain were that of a differential output stage. From Eq. (5-13) the differential output current would be

$$I_o = \frac{V_{02} - V_{01}}{R_L'} = \frac{h_{fe}}{h_{ie}}(V_1 - V_2) \tag{5-53}$$

However, in Fig. 5-21, $(V_1 - V_2)$ appears across four base-emitter junctions rather than two as in Fig. 5-8 so that h_{ie} is twice the value of Eq. (5-53), and

$$I_o = \frac{h_{fe}}{2h_{ie}}(V_1 - V_2) \tag{5-54}$$

The g_m for Q_1 and Q_2 is $38.5 I_{C1} = (38.5)(15)(10^{-6}) = 528 \ \mu$mhos or $h_{fe}/h_{ie} = 528$ μmho where $I_{C1} = I_{C2}$ were found from Eq. (5-52). To calculate the voltage gain of the first stage, it is necessary to know the input resistance of the second stage R_{i16} and the output resistance of the first stage R_{01}. The equivalent circuit of Fig. 5-29 summarizes conditions for determining the output voltage of stage 1 and thus the input voltage for stage 2.

The output resistance of stage 1 is essentially $r_{ce4} \parallel r_{ce6}$ (for small signals) as can be seen from Fig. 5-21 where V_2 and V^- are at ground for ac signals. For the transistors made in ICs, a reasonable value for $r_{ce} = 5 \times 10^6 \ \Omega$ so that $R_{01} = 2.5$ MΩ. See Reference [1].

The second stage has a Darlington pair of transistors with a resistor R_{12} from emitter of Q_{16} to V^-. This resistance is added to reduce the effect of leakage-current changes by shunting some of this current so that it does not go into the base of Q_{12}. The drift of output voltage is reduced, and a more stable system results. The input resistance of Q_{16} is only slightly changed due to R_{12}, and we shall neglect its effect.

From Eq. (5-27),

$$R_{i16} = h_{ie16} + (h_{fe16} + 1)h_{ie17} \tag{5-55}$$

The collector current of Q_{17} is about the same as that of Q_{13} or $I_{C17} = I_{C13} = 528 \ \mu$A from Eq. (5-44). Assuming $\beta_{16} = \beta_{17} = h_{fe17} = 100$, $I_{C16} = I_{C17}/\beta =$

$528(10^{-6})/(100) = 5.28$ μA. From these currents and assuming $h_{fe16} = 50$ because of its reduced current, $h_{ie16} \cong h_{fe16}/g_{m16} = 50/38.5(5.28)(10^{-6}) = 2.96(10^5)$ Ω, and $h_{ie17} \cong 100/(38.5)(528)(10^{-6}) = 4{,}920$ Ω. From Eq. (5-55),

$$R_{i16} = 2.96(10^5) + (101)(4{,}920) = 7.93(10^5) \ \Omega$$

From Eq. (5-54) and Fig. 5-29,

$$V_{16} = I_o(R_{o1} \parallel R_{i16})$$

$$= \frac{h_{fe}}{2h_{ie}} (V_1 - V_2)(R_{o1} \parallel R_{i16})$$

$$= \frac{528(10^{-6})}{2} (V_1 - V_2)[2.5(10^6) \parallel 7.93(10^5)]$$

$$= 159(V_1 - V_2)$$

and the gain for the first stage is

$$A_{v1} = \frac{V_{16}}{V_1 - V_2} = 159 \tag{5-56}$$

5-4.5 SECOND-STAGE GAIN

The effective load on the second stage R_{L2} is the input resistance of the output stage as given by Eq. (5-42). If the total load at the output is $R_E + R_L = 2{,}000$ Ω, $R_{L2} = (\beta + 1)(R_E + R_L) = (101)(2{,}000) = 2.02(10^5)$ Ω. The composite h_{fec} for Q_{16} and Q_{17} is found from Eq. (5-30), and the gain of the second stage can be determined from the equivalent circuit of Fig. 5-30.

For $r_{ce17} = r_{ce13} = 5(10^6)$ Ω, the total load on the second stage is $R_{LT} = r_{ce13} \parallel r_{ce17} \parallel R_{L2} = 1.87(10^5)$ Ω. From the circuit,

$$|V_o| = 5{,}150 \frac{V_{16}}{7.93(10^5)} 1.87(10^5)$$

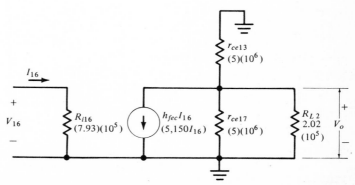

FIG. 5-30   The equivalent circuit for determining the gain of the second stage of Fig. 5-19.

or the gain of the second stage is

$$A_{v2} = \frac{|V_o|}{|V_{16}|} = 1{,}214$$

The total gain of the system is the product of the gains A_{v1} and A_{v2} since each transistor of the push-pull stage operates as an emitter follower and yields a gain near unity. Thus,

$$A_{vtotal} = A_{v1}A_{v2} = (159)(1{,}214) = 1.93(10^5)$$

Comparing this gain with the value in the data sheet of Fig. 2-6 (2×10^5), our result is in good agreement.

We see that the large gain of the 741 is obtained in two stages. The output stage allows for large voltage swings ($\pm 13\ V$) about zero volts. The use of current-source loads not only yields high resistances but also conserves area on the IC chip. The bias circuit is designed so that currents change only slightly with temperature variations.

5-4.6   FREQUENCY RESPONSE

Frequency compensation is effected internally in the 741 by capacitor C_1 in the second stage. Since the breakpoint or rolloff frequency is low, a calculation of the break frequency is relatively easy to perform. We first review Miller's theorem to simplify the calculation.

5-4.7   MILLER'S THEOREM

A simple three-terminal network will be considered to verify Miller's theorem [4] which applies to general networks. In Fig. 5-31a the network terminals are labeled 1, 2, and 3 with terminal 3 as reference or ground. By some independent means, the voltage transfer function $K = V_2/V_1$ for the network must be determined. With impedance Z connected between terminals 1 and 2, the current I_1

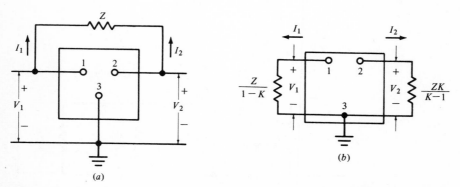

FIG. 5-31   (a) A general three-terminal network; (b) the equivalent of (a) by Miller's theorem.

is

$$I_1 = \frac{V_1 - V_2}{Z} = \frac{V_1(1 - K)}{Z}$$

$$= \frac{V_1}{Z/(1 - K)} \tag{5-58}$$

Thus, if an impedance $Z/(1 - K)$ were connected from terminal 1 to ground outside the network as in Fig. 5-31b, the current I_1 would be the same as in Fig. 5-31a. Similarly,

$$I_2 = \frac{V_2 - V_1}{Z} = \frac{V_2\left(1 - \dfrac{1}{K}\right)}{Z}$$

$$= \frac{V_2}{KZ/(K - 1)} \tag{5-59}$$

If an impedance $KZ/(K - 1)$ were connected between terminal 2 and ground, as in Fig. 5-31b, the current I_2 would be the-same as in Fig. 5-31a. Therefore, if K can be determined for the network, the circuit of Fig. 5-31b is equivalent to Fig. 5-31a.

Example 5-4   Apply Miller's theorem to the hybrid-π circuit for a bipolar transistor.

The hybrid-π circuit has been redrawn in Fig. 5-32. The equivalent shunting input impedance between terminal 1 and ground is found from Eq. (5-58), where $Z = 1/j\omega C_{b'c}$, as

$$\frac{Z}{1 - K} = \frac{1}{j\omega C_{b'c}(1 - K)} = \frac{1}{j\omega C_1}$$

Thus, a capacitor

$$C_1 = C_{b'c}(1 - K) \tag{5-60}$$

placed from terminal 1 to ground makes the circuit equivalent to Fig. 5-32a for input effects.

Similarly, for the output, from Eq. (5-59),

$$\frac{KZ}{K - 1} = \frac{K}{(j\omega C_{b'c})(K - 1)} \cong \frac{1}{j\omega C_{b'c}} \tag{5-61}$$

where the approximation is valid since $\mid K \mid \gg 1$. Thus a capacitor $C_2 = C_{b'c}$ shunting the output from terminal 2 to ground will make the circuits of Fig. 5-32 equivalent. The circuit of Fig. 5-32b shows the added capacitors. The

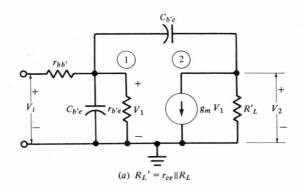

(a) $R_L' = r_{ce} \| R_L$

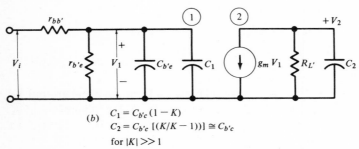

(b) $C_1 = C_{b'c}(1 - K)$
       $C_2 = C_{b'c}[(K/K - 1))] \cong C_{b'c}$
       for $|K| \gg 1$

FIG. 5-32 (a) The hybrid-π circuit for application of Miller's theorem; (b) the equivalent circuit for (a).

latter circuit is simpler to apply because the coupling between the input and output has been removed.

The gain may be calculated from Fig. 5-32b as

$$K = \frac{V_2}{V_1} = \frac{-g_m R_L}{1 + j\omega R_L C_2} \cong -g_m R_L \tag{5-62}$$

where the approximation is valid if $\omega R_L C_2 \ll 1$ which is the usual case since $C_2 = C_{b'c}$. It is also important to note that the capacitor C_1 adds further shunting capacitance in parallel with $C_{b'e}$ and C_1 can be comparable to $C_{b'e}$ even though $C_{b'c}$ may be small.

5-4.8 741—BREAK FREQUENCY

In Fig. 5-33a, the first-stage equivalent-output Norton generator of I_o and R_{o1} is driving the second stage, and the compensating capacitor C_1 is included. The internal capacitors of the transistors of the second stage have been omitted for simplicity. These capacitors affect the response; but our calculation is approximate, and the error is not significant. Invoking the Miller theorem, the values from previous calculations and this theorem (for Fig. 5-33b) yield:

$$R_{o1} = 2.5 \times 10^6 \, \Omega$$

$$R_{i16} = 7.93(10^5)$$

$$h_{fec} = 5{,}150$$

$$R_{LT} = 1.87 \times 10^5 \, \Omega$$

$$A_{v2} = K = -1{,}214$$

From Eq. (5-60) and Eq. (5-61),

$$C_{m1} = C_1(1 + |\,K\,|)$$

$$= 30(10^{-12})(1215) = 3.65(10^{-8}) \, \text{F}$$

$$C_{m2} \cong C_1 = 30 \text{ pF}$$

From the input circuit of Fig. 5-33b, the break frequency at which $V_1 = 0.707$ of the zero frequency value is

$$f_1 = \frac{1}{2\pi(C_{m1})(R_{o1} \,||\, R_{i16})} = \frac{1}{2\pi(3.65)(10^{-8})(6.02)(10^5)} = 7.24 \text{ Hz}$$

From Fig. 2-6, the break frequency is about 7 Hz so that our calculation is reasonable. It can be shown that the effect of C_{m2} in Fig. 5-33b yields a break frequency well beyond the operating range of the IC (see Prob. 5-12).

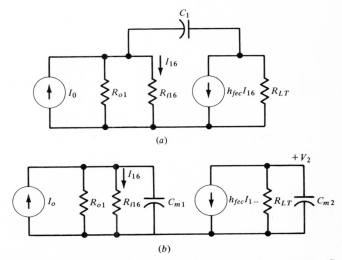

(a)

(b)

FIG. 5-33 (a) The equivalent of stage two from Fig. 5-30 with compensating capacitor C_1 added; (b) the equivalent of (a) with Miller capacitances added.

5-5 COMMON-BASE AMPLIFIER

The first stage of the 741 employed *pnp* transistors in the common-base connection so that the frequency response would be satisfactory. Our purpose here will be to compare the high-frequency response of a transistor-amplifier-operated common base as in Fig. 5-34*a*, and common emitter as in Fig. 5-32*a*.

The basic improvement in frequency response can be shown by use of the hybrid-π circuit at high frequencies. Let us assume $r_{bb}' = 0$ and omit r_{ce} to simplify the analysis and illustrate the features of the *CB* (common-base) stage. Figure 5-34*b* becomes what is shown in Fig. 5-35. The $g_m V_1$ generator has been reconnected so that the effect on the input and output terminals will be the same as for the connection in Fig. 5-34*b*, and this procedure is called substitution. Comparing Fig. 5-35 with a CE (common-emitter) stage such as in Fig. 5-32*b*, we see that the circuits are similar. However, the numerical values are not the same. Generator $g_m V_1$ depends on V_1 so the voltage $V_1(f)$ is the important factor to compare between the two figures. In Fig. 5-32*b*, V_1 appears across $r_{b'e} \parallel C_{b'e} \parallel (C_{b'c}(1 + g_m R_L)$, whereas in Fig. 5-35, V_1 is across $1/g_m \parallel r_{b'e} \parallel C_{b'e}$, and the common-base stage has essentially eliminated the effect of $C_{b'c}$ on the input. With less effective shunting capacitance on the input of the common-base stage, it will operate at higher frequencies before the gain begins to decrease. However, the voltage gain and the current gain of a *CB* stage are less than for a *CE* stage.

At low frequencies such that the capacitors have high reactances, the voltage and current gains may be determined from Fig. 5-35 as follows. At the input node, with $G_E = 1/R_E$ and $g_{b'e} = 1/r_{b'e}$,

$$V_1 = \frac{-V_e G_E}{G_E + g_{b'e} + g_m}$$

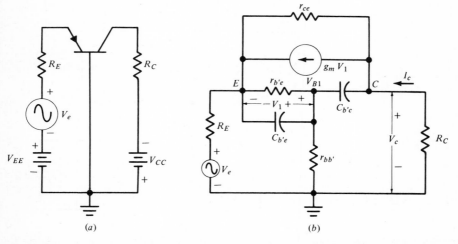

FIG. 5-34   (*a*) A common-base transistor circuit; (*b*) the equivalent of (*a*) for small rms signals.

The output voltage is

$$V_c = -g_m V_1 R_C$$

$$= \frac{g_m R_C V_e G_E}{G_E + g_{b'e} + g_m} \tag{5-63}$$

or

$$A_{vCB} = \frac{V_c}{V_e} = \frac{g_m R_C G_E}{G_E + g_{b'e} + g_m} \tag{5-64}$$

Also, for an input current $= V_e G_E$, the current gain is

$$A_{iCB} = \frac{-g_m}{G_E + g_{b'e} + g_m} \tag{5-65}$$

For the common-emitter stage of Fig. 5-32b driven by the same source as in Fig. 5-35 but including $r_{bb'}$ in R_E,

$$A_{vCE} = \frac{-g_m G_L R'_L}{g_{b'e} + G_E} \tag{5-66}$$

and

$$A_{iCE} = \frac{+g_m}{g_{b'e} + G_E} \tag{5-67}$$

A comparison of Eqs. (5-64) through (5-67) shows that the current and voltage gains are less for the CB stage than for the CE stage for comparable circuit values. A comparison of these same equations also shows that the phase relations are interchanged; i.e., for the CE, voltages are 180° out of phase, currents are in phase and vice versa for the CB.

Finally, the input resistance at low frequencies for the CB stage is lower than for the CE stage. For the CE amplifier, $R_i|_{CE} \simeq r_{b'e}$, and from Fig. 5-35

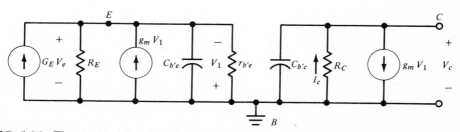

FIG. 5-35 The circuit of Fig. 5-34(b) with $r_{ce} \to \infty$, $r_{bb'} = 0$, and $g_m V_1$ relocated. The input generator has been converted to a Norton equivalent circuit.

with the driving source removed,

$$R_i\,|_{CB} = \frac{1}{g_m + g_{b'e}} = \frac{r_{b'e}}{1 + g_m r_{b'e}} = \frac{r_{b'e}}{1 + h_{fe}} \tag{5-68}$$

Thus, $R_i\,|_{CB} < R_i\,|_{CE}$. It can be shown that cascading CB stages lead to no useful power gain because of the impedance mismatch. In the 741 op-amp, the pnp CB amplifier was applied to ensure that the break frequency was controlled by the compensating capacitor C_1 and not by the poor frequency characteristics of a transistor.

5-6 LINEAR MOSFET INTEGRATED CIRCUITS

An introduction to simple MOSFET linear ICs follows from the circuit of Fig. 5-36. The transistors in this circuit are made simultaneously, but the construction of the load transistor is not the same as the amplifying transistor; the load transistor has a smaller channel width-to-length ratio. The use of an FET for a load saves area on the IC chip and yields large resistance values. The nonlinearities of the two devices compensate for each other, leading to more linear amplification characteristics.

In Fig. 5-36, the drain of the load FET is connected to the gate so that $v_{DSL} = v_{GSL}$. In some cases, the gate lead of the load FET is available for application of a separate voltage so that the Q point can be changed. Note that the gate current is nearly zero so that direct connection is possible as shown. The key to the analysis of the circuit lies in the fact that the drain currents of Q_A and Q_L must be the same. We shall derive the transfer characteristics, v_O versus v_I.

When the input voltage is below the threshold value V_T the drain current is nearly zero. Since $v_{DSL} = v_{GSL}$, the load transistor will always operate in its constant-current region. For a small leakage current ($i_D \approx 0$) in Q_L, the minimum voltage drop must be $v_{DSL}(\text{min}) = V_T$ from Eq. (4-54) where $v_{GSL} = v_{DSL}$.

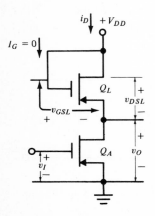

FIG. 5-36 A sample MOSFET p-channel amplifier Q_A with MOSFET p-channel load Q_L.

The output voltage is

$$v_O = V_{DD} - v_{DSL} = V_{DD} - V_T \qquad v_I < V_T \qquad (5\text{-}69)$$

We shall assume that $r_{ds} \to \infty$ for both Q_A and Q_L. For the load transistor, the circuit of Fig. 4-29 becomes what is shown in Fig. 5-37 at low frequencies.

The v_O versus v_I characteristic is determined by use of Eq. (4-54) as follows:

$$i_{DL} = \frac{K_L}{2} (v_{GSL} - V_T)^2 = \frac{K_L}{2} (v_{DSL} - V_T)^2$$

$$i_{DA} = \frac{K_A}{2} (v_{GSA} - V_T)^2 = \frac{K_A}{2} (v_I - V_T)^2$$

and

$$i_{DA} = i_{DL}$$

Thus,

$$\sqrt{\frac{K_A}{2}} (v_I - V_T) = \sqrt{\frac{K_L}{2}} (v_{DSL} - V_T)$$

or

$$v_{DSL} = \sqrt{\frac{K_A}{K_L}} (v_I - V_T) + V_T \qquad (5\text{-}70)$$

The output voltage follows from Eq. (5-70) as

$$v_O = V_{DD} - v_{DSL} = V_{DD} - \sqrt{\frac{K_A}{K_L}} (v_I - V_T) - V_T \qquad \text{for } v_I \geq V_T$$

or

$$v_O = -\sqrt{K_A/K_L}(v_I) + V_{DD} - V_T(1 - \sqrt{K_A/K_L}) \qquad (5\text{-}71)$$

We see that a linear relation exists between v_O and v_I. The signal gain is obtained from Eq. (5-71) as

$$A_v = dv_O/dv_I = -\sqrt{K_A/K_L} \qquad (5\text{-}72)$$

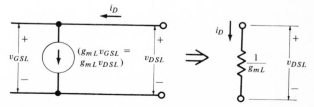

FIG. 5-37 The equivalent circuit of the load transistor in Fig. 5-36 is shown to be equivalent to a resistor $(1/g_{mL})$.

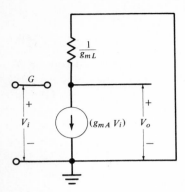

FIG. 5-38   The equivalent circuit of Fig. 5-36 at low frequencies when $V_i \geq V_T$.

The ratio of K_A/K_L can be made as large as about 10, but a typical value is 4. Thus a voltage gain of 2 or 3 can be obtained with one stage.

The load transistor may be replaced by a resistance of value

$$R_L = \frac{1}{g_{mL}} = \frac{1}{\sqrt{2K_L I_D}} \tag{5-73}$$

where Eq. (4-62) has been applied. When the input voltage exceeds the threshold value V_T the signal equivalent circuit becomes that of Fig. 5-38.

Equation (5-72) will no longer be valid when the amplifying transistor enters the ohmic region. Upon entering the ohmic region, we define the drain-source voltage as V_{DSO}. The amplifying transistor enters the ohmic region when $v_{DS} = V_{DSO}$ and from $v_{GS} - V_T = v_{DS}$ at the boundary, $v_I = v_{GS} = V_{DSO} + V_T$. At this point, the conditions on the load transistor are $v_{GSL} = v_{DSL} = V_{DD} - V_{DSO}$. After entering the ohmic region, the operation becomes nonlinear, and we shall consider that V_{DSO} ends the linear transfer curve. The value of V_{DSO} can be found from Eq. (5-71) as

$$V_{DSO} = V_{DD} + A_v(V_{DSO} + V_T - V_T) + V_T \tag{5-74}$$

or

$$V_{DSO} = \frac{V_{DD} - V_T}{1 - A_v} \qquad \text{for } v_I = V_{DSO} + V_T \tag{5-75}$$

The actual transfer characteristic will extend further than V_{DSO}, but it reaches a limit within 2 or 3 V of Eq. (5-75). A practical value for the limit is found by equating i_{DL} to i_{DA} for the load FET in its saturation region and the amplifying FET in its ohmic region.

We assume that the amplifier will be driven by a similar stage so that the voltage limit on the input will be $v_I = v_{GS} = V_{DD} - V_T$ from Eq. (5-69). For this input, the output will be defined as V_O, and $v_{GSL} = V_{DD} - V_O$. Applying

Eqs. (4-54) and (4-53),

$$i_{DL} = \frac{K_L}{2}(V_{DD} - V_O - V_T)^2 \tag{5-76}$$

$$i_{DA} = \frac{K_A}{2}[2(V_{DD} - V_T - V_T)V_O - V_O^2] \tag{5-77}$$

Equating Eqs. (5-76) and (5-77),

$$V_O \simeq \frac{(V_{DD} - V_T)^2}{2A_v{}^2(V_{DD} - 2V_T)} \tag{5-78}$$

Example 5-5 For the amplifier of Fig. 5-36, $V_{DD} = -20$ V, the transistors are made with $K_A/K_L = 4$ and $V_T = -5$ V. Calculate and plot the voltage transfer characteristics.

Solution The transistors are p channel so that v_I will be negative. From Eq. (5-69),

$$v_o = V_{DD} - V_T = -20 - (-5) = -15 \text{ V} \qquad \text{for} \quad -v_I < -5$$

From Eqs. (5-71) and (5-72), for the linear region,

$$v_o = V_{DD} + A_v(V_I - V_T) - V_T$$
$$= -20 - \sqrt{4}(v_I - [-5]) - (-5)$$
$$= -15 - 2(v_I + 5) \qquad \text{for} \quad (-v_I) \geq -5 \text{ V}$$
$$\text{or} \quad v_I \leq -5 \text{ V}$$

The linear region ends when v_o is given by Eq. (5-75). Thus,

$$V_{DSO} = \frac{V_{DD} - V_T}{1 - A_v} = \frac{-20 - (-5)}{1 - (-2)} = -5 \text{ V}$$

$$\text{for} \quad v_I = V_{DSO} + V_T = -5 - 5 = -10 \text{ V}$$

The practical limit on the output is given by Eq. (5-78) and is, for $v_{GS} = (V_{DD} - V_T)$,

$$V_O \simeq \frac{(V_{DD} - V_T)^2}{2A_v{}^2(V_{DD} - 2V_T)} = \frac{(-20 + 5)^2}{2(2)^2(-20 + 10)} = -2.8 \text{ V}$$

The plot of v_O versus v_I is given in Fig. 5-39.

5-7 CMOS AMPLIFIERS

A push-pull type of amplifier is fabricated in integrated-circuit form using n- and p-channel MOSFETs and designated as CMOS (or COS/MOS), the acronym

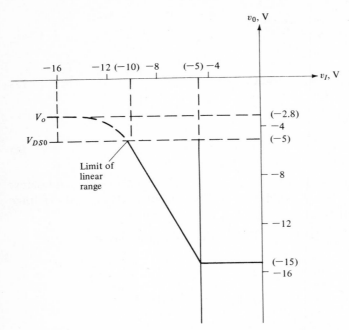

FIG. 5-39 A plot of v_0 versus v_I for Example 5-5. The amplifier FET is in the ohmic region for v_0 between V_{DSO} and V_O.

for a complementary-symmetry MOSFET system. The CMOS ICs can be applied as linear amplifiers or as switching devices (see Chap. 8), and we shall consider two biasing circuits and a small-signal equivalent circuit for amplifier applications in this section.

5-7.1 CMOS TRANSFER CHARACTERISTICS

A single CMOS inverter circuit is sketched in Fig. 5-40a. A particular IC package will have several of these units connected, for example, as a dual complementary pair plus inverter. For inverter or amplifier purposes, the gates and drains of the p-channel and n-channel units are common as in Fig. 5-40.

The transfer characteristics for one value of supply voltage are drawn in Fig. 5-40b. For a single supply voltage, we expect a range of values as shown by the minimum and maximum curves due to manufacturing tolerances in the threshold voltage. To understand the operation of the circuit, consider the maximum curve of Fig. 5-40b with $V_{DD} = 15$ V and the gate-voltage reference polarities of Fig. 5-40a. Starting with $v_I = v_{GSN} = 0$, $v_{GSP} = -15$ V, and the p unit is turned on while the n unit is off; the output is connected to V_{DD} through the low resistance of the p unit. As v_I increases, no change occurs until $v_I = V_{TN} = 5$ V, the threshold voltage of the n unit, and the n unit begins to conduct drain current while the p unit begins to turn off. The curve drops as shown until the

p unit is turned off when $v_{GSP} = V_{TP} = -2$ V, or $v_I = 13$ V, and the n unit is fully on, connecting the output to ground.

The region of the curves in Fig. 5-40b useful for amplifier purposes is the segment having a near-vertical slope. Because a particular transistor pair may have characteristics anywhere between the maximum and minimum curves, the bias problem is that of keeping operation in the vertical-slope area to avoid distortion.

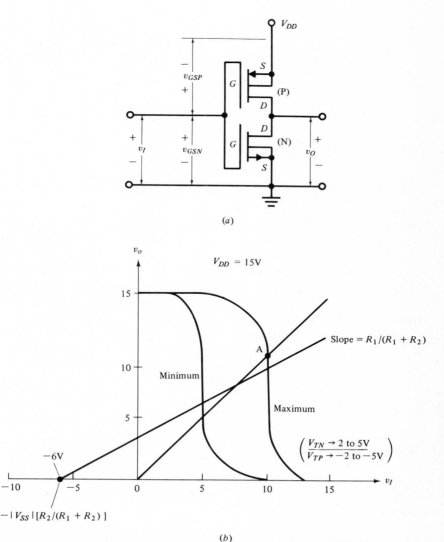

(a)

(b)

FIG. 5-40 (a) A CMOS inverter or amplifier stage; (b) the transfer characteristics of (a) showing tolerances in threshold voltages due to manufacturing techniques. The dc operating lines are for amplifier design.

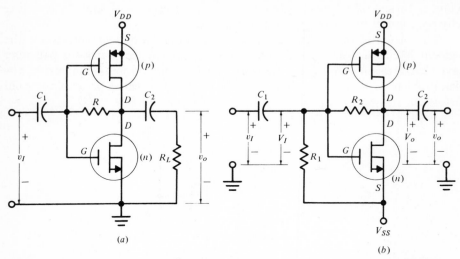

FIG. 5-41 (a) A CMOS amplifier with a single resistor bias; (b) the same amplifier with a two-resistor and second voltage-supply bias arrangement.

5-7.2 CMOS AMPLIFIER BIASING

A simple CMOS amplifier bias circuit is drawn in Fig. 5-41a. The resistor R is large (tens of megohms), fixing the dc voltage of the input and output at the same value (remember that gate current is negligible) while isolating the direct connection of the ac signal between input and output. The dc load line for this bias circuit is indicated in Fig. 5-40b for $v_I = v_O$. If we are operating on the maximum transfer curve, we see that operation may enter the curved region at point A resulting in signal distortion. It may be desirable to have a bias circuit that will allow us to avoid this curved region.

A bias circuit, ensuring that amplifier operation will occur in the vertical-slope region of Fig. 5-40b, is shown in Fig. 5-41b. The circuit requires two dc supply voltages which may be a disadvantage in some applications. Applying the superposition theorem to Fig. 5-41b,

$$V_I = \frac{R_1}{R_1 + R_2} V_o + \frac{R_2}{R_1 + R_2} V_{ss} \tag{5-79}$$

Equation (5-79) is the equation of a straight line having a slope of $R_1/(R_1 + R_2)$ and an intercept at $V_{ss}[R_2/(R_1 + R_2)]$ where $v_o = 0$. If V_{ss} is negative, the operating line is as shown in Fig. 5-40b. By adjusting R_1 and R_2, we can more easily stay in the vertical-slope region, regardless of the tolerances of the threshold voltages.

5-7.3 CMOS SIGNAL GAIN

Having the bias set by the circuit of Fig. 5-41a, from Fig. 4-28, the small-signal equivalent circuit at constant-gain-region frequencies with R very large is shown in Fig. 5-42. We see that the gain is given by

$$\frac{V_o}{V_i} = -(g_{mn} + g_{mp})(R_L \| r_{ds})$$ (5-80)

Example 5-6 For the p-channel MOSFET of Example 4-6, $g_{mp} = 6.5$ mmho and $r_{dsp} = 4,000\ \Omega$. In CMOS ICs, the p-channel unit is normally a higher g_m device. Therefore, assume $g_{mn} = 3$ mmho, and $r_{dsn} = 20,000\ \Omega$, both g_m and r_{ds} values determined at $|V_{DS}|$ between 6 and 9 V. Design a circuit similar to Fig. 5-41b so that the operating line will fall in the region between $V_{DS} = 6$ V and $V_{DS} = 9$ V in Fig. 5-40b if $V_{DD} = -V_{SS} = 15$ V, and $R_1 = 22$-MΩ (megohm). For a load resistance of 2,000 Ω, determine the small-signal gain.

From Fig. 5-40b, the slope of the operating line is $(9 - 6)/(10 - 5) = \frac{3}{5} = R_1/(R_1 + R_2)$. If $R_1 = 22$ MΩ, from the slope, $R_2 = \frac{2}{3}R_1$, or $R_2 = 14.7$ MΩ, and $-V_{SS}[R_2/(R_1 + R_2)] = -15[14.7/36.7] = -6$ V. The dc operating line is drawn in Fig. 5-40b.

The small-signal circuit is shown in Fig. 5-42 where $r_{ds} = r_{dsp} \| r_{dsn} = (4,000) \| (20,000) = 3,333\ \Omega$ and $g_{mp} + g_{mn} = (3 + 6.5) = 9.5$ mmho. From Eq. (5-80),

$$V_o/V_i = -(9.5)(3,333 \| 2,000)(10^{-3})$$

$$= -(9.5)(1,250)(10^{-3}) = -12$$

The voltage gain is not large, but the current and power gain are quite large. The circuit is better than an emitter follower for impedance-matching purposes within the frequency limitations of the devices.

5-8 SUMMARY

Linear integrated circuits have been analyzed in some detail. We have seen how input and output resistances, gain, and output voltage swing can be calculated. It should also be emphasized that we have analyzed the subcircuits of the total systems. Any of these subcircuits can be constructed from discrete components, and

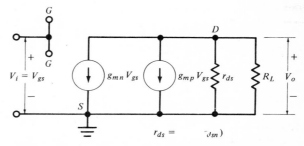

FIG. 5-42 The small-signal equivalent circuit for the circuit of Fig. 5-41(a) with the bias resistor R assumed to be very large.

the operation will be similar. The greatest difference in discrete circuits compared to ICs lies in the fact that discrete transistors will not have matched characteristics. Thus, the bias techniques cannot employ paralleling of base-emitter junctions in discrete circuits. If paralleling is attempted, one transistor will conduct more heavily than the other. Bias circuits, such as shown in Fig. 4-19, should be applied to discrete transistors.

REFERENCES

1. Eimbinder, J., (ed.), "Application Considerations for Linear Integrated Circuits," John Wiley & Sons, Inc., New York, 1970.
2. Fitchen, F. C., "Transistor Circuit Analysis and Design," 2d ed. D. Van Nostrand Company, Inc., Princeton, N. J., 1966.
3. Giles, J. N., "Fairchild Semiconductor Linear Integrated Circuits Applications Handbook," Fairchild Semiconductor, Mountain View, Calif., 1967.
4. Millman, J. and C. C. Halkias, "Integrated Electronics: Analog and Digital Circuits," McGraw-Hill Book Company, New York, 1972.
5. Radio Corporation of America, RCA Linear Integrated Circuits, *Technical Series IC-41*, Harrison, N. J., 1967.
6. RCA Corporation, RCA COS/MOS Integrated Circuits Manual, Somerville, N. J., 1971.
7. Strauss, L., "Wave Generation and Shaping," McGraw-Hill Book Company, New York, 1970.
8. The MC1530, MC1531 Integrated Operational Amplifiers, *Application Note AN-204A*, Motorola Semiconductor Products, Inc., Phoenix, Ariz., 1972.

PROBLEMS

5-1. Assume $h_{fe} = \beta = 150$ for the circuit of Fig. 5-3a, for each transistor. Determine the input and output resistance. Compare your results with the table of Fig. 5-3. Does the voltage gain change appreciably?

5-2. Plot transfer characteristics similar to Fig. 5-9 from Eqs. (5-22) or (5-23) if $\eta = 2$. For comparison, $\alpha = 1$, $I_o = 0.8$ mA, $R_L = 10^4 \,\Omega$, $(kT/q) = 0.025$ V. Also compare the results with Fig. 5-3b.

5-3. Starting with Eq. 5-23, determine a differential voltage-gain expression, v_O/v_I. Evaluate the expression for $\alpha = 1$, $I_o = 0.8$ mA, $R_L = 10^4 \,\Omega$, $(q/\eta kT) = 1/(0.025)$ and $v_I = 0$. Does this gain value agree with the text value of 145?

5-4. To increase the input resistance and the differential voltage range over which linear gain can occur, a resistor is placed in series with each emitter lead of a differential amplifier as shown in Fig. P5-4a. The equivalent circuit is shown in Fig. P5-4b. Noting that $I_1 = -I_2$, show that $R_i = (V_1 - V_2)/I_1 = 2h_{ie} + 2(h_{fe} + 1)(50)$. Evaluate this R_i for $h_{ie} = 6,500 \,\Omega$, $h_{fe} = 100$ and compare with R_i of the circuit of Fig. 5-3a. Also find $V_{o1}/(V_1 - V_2) = -(V_{o2})/(V_1 - V_2)$ and $(V_{o1} - V_{o2})/(V_1 - V_2)$. Since the gain of the diff-amp of Fig. 5-3 was about 145, we see that the voltage gain is decreased by adding the emitter resistances to obtain larger R_i and wider linear range.

5-5. Transistors Q_1 and Q_2 in Fig. 5-3a are replaced by n-channel JFETs having $I_{DSS} = 1.6$ mA, $V_p = -2$ V. The drain, gate, and source are connected to the points where the collector, base, and emitter are connected, respectively. Determine the necessary dc voltage at input terminals 1 and 2 so that the drain currents are 0.4 mA (each) with $V_{DS} = 4$ V. Also determine g_m. Apply the low-frequency equivalent small-signal circuit with r_{ds} infinite, and determine the voltage gain. Compare with the value found for the case of the bipolar transistors. Repeat if $I_{DSS} = 10$ mA, and $V_p = -8$ V. Note that R_i is infinite in both cases.

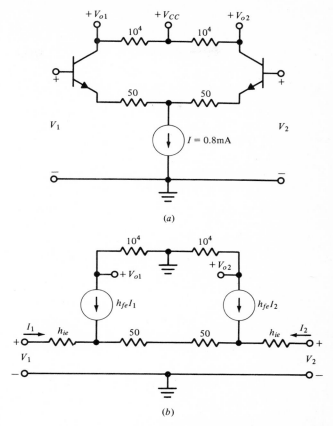

FIG. P5-4   The circuits for Prob. 5-4.

5-6. Apply the composite transistor values found in Example 5-1 for the composite transistors formed by Q_1, Q_2, and Q_4, Q_5, of Fig. 5-11. Determine the input resistance and overall gain (voltage) of the amplifier.

5-7. Consider the push-pull class-A equivalent circuit of Fig. P5-7. Determine V_o in terms of V_i.

5-8. The typical input resistance of the IC in Fig. 5-10 is stated to be 20,000 Ω. What should be the value of h_{fe} for Q_2 and Q_4 to obtain this value of input resistance? If all transistors in the IC have this new value of β, will the overall voltage-gain change?

5-9. Refer to Fig. P5-9. For identical Q_1 and Q_2, $V_{BE\text{active}} = 0.7$ V, $V_{CE\text{sat}} = 0.2$ V, $r_{ce} = 1/h_{oe}$ is assumed to be infinite, $I_C = I_E$, and piecewise-linear models are to be used with $R_E = 0$.

(a) Determine the dc collector currents for Q_1 and Q_2 when $v_1 = v_2 = 0$.

(b) For small signals, determine the differential voltage gain $v_{o2}/(v_1 - v_2)$.

(c) Determine the input common-mode voltage range with $v_1 = v_2$ over which the circuit acts as a differential amplifier, i.e., between cutoff and saturation of Q_1 and Q_2.

(d) Determine the common-mode gain defined as the change in v_{o2}, $\Delta v_{o2} = V_{CC} - v_{o2\text{sat}}$ (where $v_{o2\text{sat}}$ is the output voltage when Q_2 is saturated) to the input common-mode voltage range as found in part (c).

(e) Determine the CMRR in decibels from your answers in parts (b) and (d).

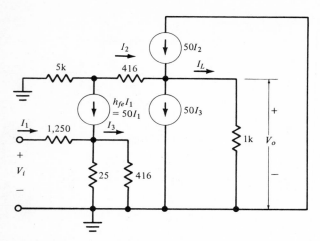

FIG. P5-7   The circuit for Prob. 5-7.

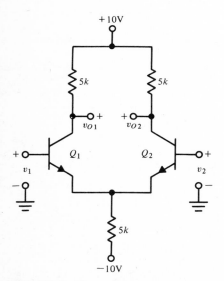

FIG. P5-9   The circuit for Prob. 5-9.

5-10. Fig. P5-10 represents an equivalent circuit of a complementary symmetry push-pull amplifier for dc and low frequencies, where the actual circuit is indicated in Fig. 5-23a. ($R_{E1} = R_{E2} = 0$). Calculate and plot v_O versus v_I if v_I ranges from -5 to $+5$ V. The diodes in Fig. P5-10 are ideal. Sketch v_O versus ωt if $v_I = 5 \sin \omega t$.

5-11. Calculations from Eq. (5-45) lead to the fact that the change in base-emitter voltage, $\Delta V_{BE} = 60$ mV/decade change in I_E. In the diode equation leading to this result, η was assumed to be unity. If $\eta = 2$ for silicon, $\Delta V_{BE} = 120$ mV/decade change in I_E. Repeat the calculation similar to that of Fig. 5-27, and determine the bias current for Q_{10}. Is there a significant change from the value of the text? Repeat if $\Delta V_{BE} = 30$ mV/decade.

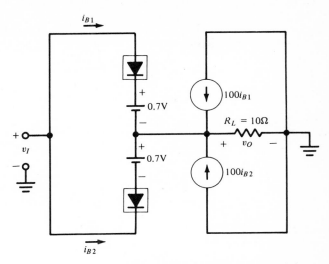

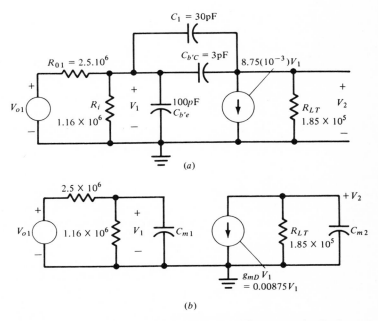

FIG. P5-10 The circuit for Prob. 5-10.

5-12. The circuit of Fig. 5-33a has been modified in Fig. P5-12a to include transistor capacitors Determine the circuit elements in Fig. P5-12b, and write the response function for V_2/V_{o1}. Calculate the dominant frequency break point, and show that the pole due to R_{LT} and C_{m2} is much higher in frequency.

FIG. P5-12 The circuits for Prob. 5-12.

5-13. Fig. P5-13 shows the high-frequency equivalent circuit of Fig. 5-36a with r_{ds} infinite. Simplify this circuit and show that

$$A_v = \frac{V_o}{V_i} = \frac{(-g_{mA})(1 - sC_{gdA}/g_{mA})/(g_{mL})(1 + (sC_o + sC_{gdA})}{1/g_{mL}}$$

where $C_o = C_{gsA} + C_{gsL} + C_{dsL}$. If all Cs are 3 pF, $g_{mA} = 2g_{mL} = 10^{-3}$ mho, determine the upper half-power or break frequency. Using the above equation for A_v, show that

$$Y_i = \frac{I_i}{V_i} = s(C_{gsA} + C_{gdA}) + \frac{(sC_{gdA})(g_{mA} - sC_{gdA})}{[g_{mL} + s(C_o + C_{gdA})]}$$

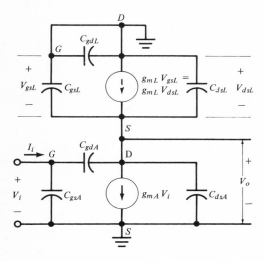

5-14. A 20,000-Ω resistor and a 0.01-μF capacitor are placed in parallel between the Hi-Z output terminals in Fig. 5-3a. Using numerical values found in the text, draw an equivalent circuit for the diff-amp consisting only of a voltage-controlled current generator given by $(h_{fe}/2h_{ie})(V_1 - V_2)$ and a shunt resistance equal to the output resistance at the Hi-Z terminals.

(a) Determine the low-frequency break point or 3 dB down frequency.

(b) If the resistor and capacitor are placed between one Hi-Z output and ground, modify the equivalent circuit of (a) and repeat the calculation. (This is called lag compensation.)

5-15. Lead compensation can be added to the op-amps of Figs. 5-10 and 5-11 at terminals 7 and 8, as shown in Fig. P5-15. Determine the response function of this system, and sketch a Bode plot for $C = 0.01$ μF. Will the system be stable if the uncompensated amplifier has a break frequency at 1 MHz and a rolloff of 6 dB/octave?

5-16. From the numerical values determined for the op-amp of Fig. 5-10, employ an equivalent circuit as shown in Fig. P5-7, and determine the output resistance with R_L removed. Represent the complete amplifier by a voltage-controlled generator and input and output resistance, as in Fig. 2-1.

5-17. Applying the same assumptions of the text as for the MC1530, analyze the op-amp of Fig. P5-17. Determine the gain and input resistance.

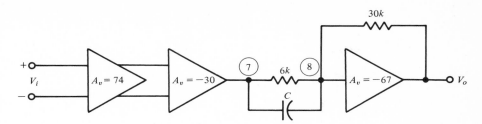

FIG. P5-15 The block diagram for Prob. 5-15.

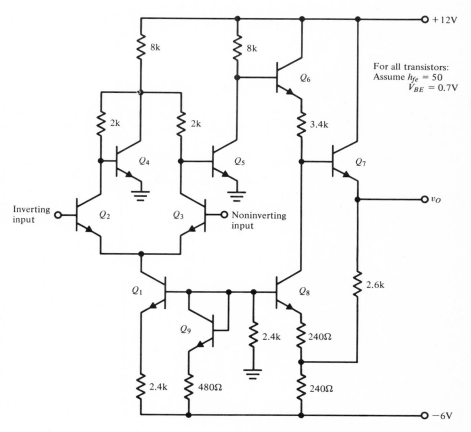

FIG. P5-17 The circuit for Prob. 5-17.

5-18. A three-stage MOSFET amplifier, with each stage made of amplifiers as shown in Fig. 5-36, is drawn in Fig. P5-18. Why can the stages be direct-coupled? What purpose does the 20-MΩ external resistor have? If each stage is made as in Example 5-5, determine the overall small-signal gain at frequencies where the gain is constant. ($V_{DD} = -20$ V).

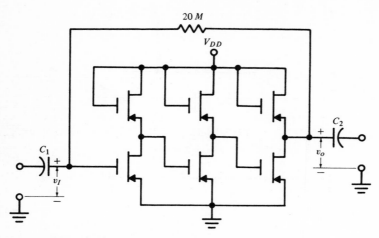

FIG. P5-18   The circuit for Prob. 5-18.

5-19. The gain of the third stage of the MC1530, Fig. 5-10, was determined to be approximately (-5). Determine the error in this calculation by accounting for the open-loop gain having a low value of (-67).

5-20. Sketch the equivalent circuit applicable at high frequencies for the circuit of Fig. 5-41b including bias resistors. Use the numerical values of Example 5-6 and assume that both n- and p-channel FETs have the same capacitance values found in Example 4-6. Determine the high break frequency for a load resistance of 2,000 Ω.

6

Computer-Assisted Analysis

INTRODUCTION

The digital computer is being applied to nearly all phases of electronic circuit design. Early applications of the computer concerned circuit analysis, but later the computer programs were modified to include analysis with various tolerances on the components in order to predict the yield of adjustment free circuits. The design of printed-circuit and integrated-circuit artwork is a more recent addition to the growing field of computer-aided design.

The purpose of this chapter is to examine the computer as a tool in the analysis of electronic circuits. The advantages of the computer become obvious if we consider the computer as a laboratory simulator. The circuit can be described to the computer in a conversational manner by numbering the nodes and listing the components by node connection, element type (R, L, C), and element value. The desired output is specified and an analysis performed. The designer examines the results, makes changes in the circuit and reanalyzes. Through iterative analysis, a suitable design is obtained in a manner similar to that of a breadboard design. If the iterative analysis is to be effective, then obviously an on-line computer facility or a rapid turn-around-time batch facility is required.

Optimization and iterative techniques will be applied in Chap. 9. In this chapter we will analyze two circuits which were presented in Chaps. 2 and 5. No effort will be made to change the components or circuit configuration.

The question that comes to mind when we have decided to apply the digital computer to network design or any general class of problems is: "How shall the user obtain a solution? What kind of program should be used?" We could write a new program for each circuit to be analyzed. Another approach would be to write a main program for each new circuit utilizing a large library of circuit subprograms.

A third and more widely accepted approach is to apply a user-oriented digital computer program in which only the data is supplied. In this method, a variety of commands permit us to select the type of analysis to be performed. Such a program can analyze a wide variety of networks: filters, amplifiers, clipping circuits, switching circuits, etc. The need for more than one such program is dictated more by the type of analysis (linear or nonlinear, steady-state or transient) rather than the type of circuit. Of course, the type of circuit often dictates the analysis required.

A few of the programs that are available are listed in Table 6-1. The type of analysis, network formulation, models available, output options, and type of operations are indicated for each program.

LINCAD and OSUCAD are on-line computer-aided network analysis programs. The term *on-line* implies that we are interacting with the computer, and the results are returned almost immediately. The complete Fortran listings for entering LINCAD and OSUCAD on the computer are given in Appendices C and D, respectively. LINCAD analyzes linear networks, and OSUCAD analyzes nonlinear transistor networks. In OSUCAD, transistors and diodes are specified by model number and node connections. On command, the programs compute the network equations and solve them.

The programs were written with the following properties in mind:

1. The programs are to be on-line to: (*a*) permit ease of inserting the circuit by way of the conversational mode, and (*b*) allow design by iterative analysis.

2. The programs are to be economically feasible; that is, they should not simply demonstrate the principles of on-line operation, but they must be practical from the cost standpoint.

3. The on-line operation is to follow closely the breadboard method of design. We must be able to add new components, change components, delete components, and add new nodes at any point in the analysis.

4. The programs are to require a minimum of instruction. The conversational mode of on-line operation permits the computer to lead the user through the analysis by requests for data.

A short demonstration and a review of an example is sufficient to permit a user to solve circuits unassisted. The procedure for solving a problem follows.

1. The circuit is sketched and component values selected.

2. The circuit nodes are numbered with consecutive integers starting with ground as the zero node.

3. The circuit description is inserted. The components are described to the computer by typing R for resistance, C for capacitance, L for inductance, E for voltage sources, I for current source, T for transistors, and D for diodes. In OSUCAD, a name is assigned to each element and typed after R, L, C, I, E, T, or D. The computer replies with a request for the node connections and value. For transistors and diodes the request includes the model number of the device.

4. A solution is obtained by typing an appropriate command.

The order of entering the node numbers of the possible components (R, L, C, E, and I) is immaterial in either LINCAD or OSUCAD. The following rules apply to polarity and current flow: (1) The voltage across an element is positive on the higher-numbered node; (2) A positive voltage source has the higher-numbered node more positive than the lower-numbered node; (3) A positive current generator has current flow in a direction from the higher- to lower-numbered node.

The insertion of the same type of element (R, L, C, E, I) having the same node numbers as one previously entered simply replaces the old value with the new one.

The user can add, delete, or change components, add new nodes, request a solution, etc., at any time.

TABLE 6-1

	ECAP	LINCAD	POTTLE	CIRCUS	SCEPTRE	OSUCAD
Linear analysis	×	×	×		×	×
Nonlinear analysis	×			×	×	×
Network formulation:						
Nodal	×	×				
State-variable			×	×	×	×
Built-in models				×	×	×
Output options:						
Pole-zero location			×			
Frequency response	×	×	×			
Time response	×		×	×	×	×
Parameter plot		×				
Optimization		×				
Worse-case analysis	×					
DC sensitivities	×	×				
AC sensitivities		×				
Operation:						
Batch	×		×	×	×	
Remote-batch	×			×	×	
On-line	×	×				×

6-1 LINCAD INSTRUCTIONS

The LINCAD program is written in Fortran IV and is listed in Appendix C, and a complete set of instructions is in Appendix A. After inserting the circuit, we can obtain all voltages in the circuit at one frequency, plot a frequency response, plot the node voltages as a function of a component value, tabulate the sensitivity of the circuit, or optimize the circuit. We may select various node voltages or differences of node voltages to be plotted, tabulated, or optimized. The output can be in terms of the absolute value of the voltage, the phase of the voltage, or be expressed in decibels.

A few rules and conventions must be followed to obtain meaningful results:

Only one resistor, one inductor, one capacitor, or one independent current source is permitted between a pair of nodes. Similar parallel components (R, L, C, and I) must be combined before use in the program unless additional nodes connected by negligibly small resistors are used. The same rule applies for voltage sources in series. This rule results in several advantages:

1. The node equations generated are compact, thus saving storage space and computation time.

2. Component modification is rapid in that retyping a component of the same type and node connections replaces the old component. The restriction of one R, L, C, I, and E per node pair is not serious in a linear circuit.

The voltage source specified for a node pair is placed in series with the parallel combination of the R, L, and C specified for the same node pair. Figure 6-1 shows the connection which results when an R, L, C, E, and I are given node numbers of J and K. Note that if a voltage source is specified between nodes J and K, at least one R, L, or C must also be specified between the same nodes. The independent current source can be placed between the node pair without assigning an R, L, or C.

In LINCAD, a component is removed by typing the component with zero mhos, zero farads, zero volts, zero amperes, and zero henries. Zero henries is in reality a short circuit, but LINCAD interprets the zero as the desire to omit the inductor.

When the computer prints the symbol "$> =$", the program is expecting an instruction to insert a component, list the circuit components, select nodes for plotting, select output type, or select one of the many analysis options.

The element types are specified by:

R	Resistance	⎫ all address
G	Conductance	⎬ the same
YR	Real part of the admittance; same as G	⎭ resistor
L	Inductance	
C	Capacitance	
YI	Imaginary part of the admittance	
E	Independent voltage source (Magnitude and Phase)	⎫
ER	Real part of an independent voltage source	⎬ all address the same voltage source
EI	Imaginary part of an independent voltage source	⎭
I	Independent current source (Magnitude and Phase)	⎫
IR	Real part of an independent current source	⎬ all address the same current source
II	Imaginary part of an independent current source	⎭
GM	Voltage-controlled current generator (Frequency Dependent)	
YM	Voltage-controlled current generator (Real and Imaginary Part; Admittance Form)	

The independent voltage sources are addressed by either E to express them in magnitude and phase or by ER and EI to express them in real and imaginary parts. The same convention applies for independent current sources.

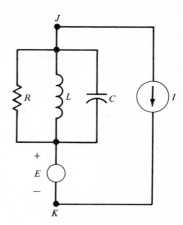

FIG. 6-1 The component placement when all R, L, C, E, and I are given node numbers J, K.

The node connections of the controlled voltage or a dependent current generator are requested immediately after specifying the generator. In the case of GM, the cutoff frequency is also requested. See Eq. (6-1). If there is no cutoff frequency, entering Ø (zero) causes transfer around the frequency calculations of the GM term with a savings in computation time.

The sign of the transconductance G_m may be found by the following convention. If the controlled generator current is in a direction from the higher- to the lower-numbered node, and the controlling voltage is positive on its higher-numbered node, then the sign of G_m is positive. If one of these directions is changed, the sign is negative. If both are changed, the sign is positive.

The last voltage-controlled current generator specified for the same set of node pairs is used. Thus we may correct or change the value by retyping the information. A controlled generator can be removed by retyping all the information with zero (Ø) mho.

The circuit can be modified, that is, elements changed in magnitude, new elements added, new nodes added, and elements deleted at any time.

Specific node voltages or differences between node voltages can be selected by typing NODES for use in PLOT X, VTVM and SENSITIVITIES routines. The node numbers are typed after the computer statement:

NODES TO BE PLOTTED(FILL TO 8 NO. WITH ZEROS)
?

Zeros are used to fill to eight numbers for two purposes:
1. To satisfy the program input requirement of eight numbers.
2. To allow the program to compute the number of curves.

Node-voltage differences can be plotted or printed. The following example will show the method.

Example 6-1 We desire the voltage of node 3 and the difference between nodes 4 and 2. The reply to the NODES TO BE PLOTTED statement is:

3 4 −2 Ø Ø Ø Ø Ø

The "3" indicates that we desire the node voltage of node 3. The "4 −2" indicates that we wish the voltage of node 4 minus the voltage of node 2. Eight numbers have been entered, and blanks must separate each number. The minus sign is a message to the program to subtract that voltage from the previous one.

All node voltages at one frequency are computed by typing ALL for a linear circuit. The node voltages and node-voltage differences previously specified by NODES may be printed at one frequency by typing VTVM. The frequency in hertz is typed after the following computer statement:

FREQ IN HZ

The computed solution at that frequency is then printed on the teletype.

If NODES had not been typed prior to this option, the node voltages of 1, 2, 3, 4, and 5 would be printed.

The selected node voltages or node-voltage differences can be plotted as a function of the magnitude of a component (R, G, L, C) or of frequency on a log or linear scale by the PLOT option. If NODES was not selected prior to this option, then the node voltages of 1, 2, 3, 4, and 5 will be plotted. The element to be varied is typed after the word plot; R for resistance, C for capacitance, and L for inductance; for example, PLOT R.

The node numbers of an element to be varied are typed on a request from the computer. The scale dimensions and number of intervals are typed after the request:

INITIAL AND FINAL(UNITS), MAX AND MIN VOLTAGE, NO. OF INTERVALS
?
where,

INITIAL AND FINAL(UNITS) refer to the scale values to be plotted down the teletype page, MAX AND MIN VOLTAGE refer to the scale for the node voltage to be plotted across the teletype page, and NO. OF INTERVALS refers to the number of intervals or points to be plotted.

To plot frequency on a linear scale, we type PLOT F; and to plot frequency on a log scale, we type PLOT LOG. Since the latter plot is semilog, it is convenient to specify the initial and final hertz in multiples of 1Ø. For a neat plot, the number of points should be an integral multiple of the number of decades.

Decibels or phase can be plotted by typing the output selections DB or PHASE, respectively, prior to any of the analysis options. The output selection remains the same until changed. To return to a magnitude plot, we type MAGNITUDE.

The complete instructions for LINCAD are summarized in Appendix A, and examples of the optimization routine are given in Chap. 9.

6-2 LINCAD EXAMPLE

Consider Prob. 2-7 of Chap. 2 as shown in Fig. 6-2. The circuit represents a differentiator, but for sinusoidal signals, it is also a bandpass filter.

The specifications of the amplifier are as follows:

$$A_v = \frac{v_o}{v_i} = -100,000$$

R_o = output resistance = 1 Ω
R_{in} = input resistance = 1 MΩ
The circuit component values are:
R_1 = 0.1591 kΩ
R_2 = 1.591 kΩ
C_1 = 1 μF
C_2 = 0.1 μF
R_3 = 1 kΩ

The problem is to find the frequency and phase response of the circuit using LINCAD.

LINCAD does not permit voltage-dependent voltage generators. Since voltage-controlled current generators are the only dependent sources allowed, a conversion is required for other types. In linear circuits, any dependent generator can be expressed as a voltage-controlled current generator by application of Norton's theorem and Ohm's law.

The open-loop amplifier as described by the specifications is shown in Fig. 6-3a with the desired form shown in Fig. 6-3b. For the two circuits of

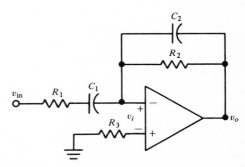

FIG. 6-2 The circuit of Prob. 2-7. The amplifier is represented as in Fig. 2-1.

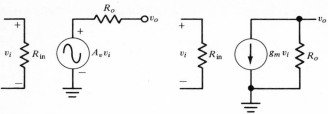

(a) Dependent voltage generator   (b) Dependent current generator

FIG. 6-3   The equivalent linear amplifiers if $g_m = -A_v/R_0$. (a) Dependent voltage generator; (b) dependent current generator.

Fig. 6-3 to be equivalent, the open-circuit voltage of each must be equal. The open-circuit voltages are:

$$-g_m v_i R_o = A_v v_i$$

or

$$g_m = -\frac{A_v}{R_o} = -\frac{-100{,}000}{1} = 100{,}000 \text{ mho}$$

The circuit to be analyzed by LINCAD is shown in Fig. 6-4. The nodes have been numbered with consecutive integers from 1 to 4 with the ground or common node always as zero. All node voltages are computed with respect to the node labeled zero. The voltage source E is selected to be 1 V

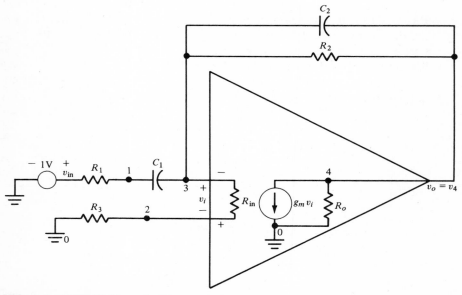

FIG. 6-4   LINCAD Example 6-2.

so that a plot of the node voltage at node 4 will be the closed-loop-amplifier gain $v_o/v_{in} = (v_o/1\text{ V}) = v_o$.

The circuit is now ready to be entered into the computer. We turn on the time-sharing terminal, dial the appropriate telephone number, and make communication with the computer. Figure 6-5 shows an example sheet given to LINCAD users which describes the procedures to establish the terminal hookup. These instructions vary at each installation and can be conveniently adjusted to our needs when we have the program entered into the computer.

When the keyboard is free (the keys do not function unless the computer is ready to receive data) we type our user identification number and the password. Upon verification of the user identification, the typewriter prints the logon date, time and the word READY. We then type a line to load or call LINCAD for execution. When the program begins execution, the following message is printed:

LINEAR COMPUTER-AIDED CIRCUIT DESIGN
$> =$

where the $> =$ symbol pair indicates that the program is awaiting a command to insert a component, make an option selection, choose an analysis type, etc.

The procedure outlined above for turning on the terminal (or computer) and the method of loading the program varies depending on the manufacturer and sometimes on the particular installation. In any case, once the

IBM 370 Time-sharing System (TSO)

---

10 A.M.–10 P.M. MTuesWF
12 A.M.–10 P.M. Thurs

---

1. Turn ON/OFF switch to ON.
2. Make sure the COM/LCL switch is set to COM.
3. Remove handset from telephone (data set).
4. Press TALK button on telephone.
5. Dial extension given on handset.
6. Wait for a high-pitched tone. When you hear this tone you are in contact with the computer. If you get a busy signal or no answer, hang up and repeat from instruction 3 for other extensions.
7. Push the DATA button on the telephone. If the DATA-button light goes off at any point during session, repeat from instruction 3.
8. Replace handset on the cradle.
9. Enter LOGON command.

(*Note:* Please turn ON/OFF switch to OFF after you enter LOGOFF.)

FIG. 6-5 TSO access instructions.

program is loaded there is little difference between various machines. The primary difference concerns how errors are corrected when data is entered. On the IBM time-sharing terminal, a backspace is provided to permit correction of mistyped data, but other computers may erase a character with RUBOUT. In all machines once the carriage return is depressed, little can be done to correct errors. The remainder of this example will be performed on the IBM time-sharing terminal.

Having entered the program, the "> =" symbol pair was printed. Obviously the circuit must be inserted first. The components may be entered in any order. To insert the voltage source an e is typed and the computer replies with a request for the node connections and value. The voltage source is considered a part of the 1-0 node pair. Thus the node connections are 1 and 0 and the value 1.0 V and 0.0 degrees. The sequence of typing and printing to insert the voltage sources is:

> = _e_
NODES, VOLTS, DEG
?
_0 1 1. 0._

where the underlined quantities were typed by the user. Underlining of typed quantities will be employed in all examples. Upper-case and lowercase letters are permissible at most installations.

The resistor connected in series with the voltage source is entered with the same node pair. The command and numerical data are:

> = _r_
NODES, KILOHMS
?
_0 1 .1591_

The dependent current generator is inserted by the command *gm*. The node connections of the generator, and the value in mhos are typed on request. Following that, the node connections of the controlling voltage and the cutoff frequency (f_{co}), in megahertz, are requested. The transconductance in the program is given by

$$g_m(f) = \frac{G_m}{1 + jf/f_{co}} \tag{6-1}$$

where
G_m = value given in first request
f_{co} = half power or break frequency, in megahertz
This expression amounts to a 6 dB/octave falloff at high frequencies or $|g_m(f_{co})| = 0.707 G_m$. If a zero is typed for f_{co}, the program interprets the

desire to have an infinite cutoff frequency or

$$g_m(f) = G_m$$

The series of commands and numerical input are:

> =
gm
NODES, MHOS
?
0 4 100000
CONTROLLING NODES,FCO IN MHZ
?
2 3 0

A zero was entered for f_{co} so that the open-loop-amplifier response would be independent of frequency. The remaining components are entered as follows:

> =
c
NODES,MICROFARAD
?
1 3 1
> =
r
NODES,KILOHMS
?
2 3 1000
> =
r
NODES,KILOHMS
?
3 4 1.591
> =
c
NODES,MICROFARAD
?
3 4 .1
> =
r

NODES,KILOHMS
?
<u>0 2 1</u>
<u>> =</u>
r

NODES,KILOHMS
?
<u>0 4 .001</u>

The output of the circuit is v_4. In order to print only that node voltage, the following selection and reply is typed by the user:

<u>> =</u>
nodes
NODES TO BE PLOTTED(FILL TO 8 NOS. WITH ZEROS)
?
<u>4 0 0 0 0 0 0 0</u>

Eight numbers were typed as requested of which seven were zeros. Had only the 4 been typed, the program would have printed another question mark until eight integers were typed. The seven zeros are a flag to indicate that only one voltage v_4 is desired. Had this selection not been used, the node voltage v_1, v_2, v_3, v_4, and v_5 would have been printed or plotted.

Prior to plotting, the user should examine the response of the circuit at a few frequencies. The command VTVM prints the node voltage selected above at one frequency. Below are shown the results when we request the value of v_4 at 10, 100 and 1,000 Hz, respectively.

<u>> =</u>
vtvm
FREQ IN HZ
?
<u>10</u>
   4— 0 .99954D-01 V, −91.15 DEG
<u>> =</u>
vtvm
FREQ IN HZ
?
<u>100</u>
   4— 0 .98975 V, −101.42 DEG
vtvm
FREQ IN HZ
?
<u>1000</u>
   4— 0 4.9997 V, −179.98 DEG

At this point, we observe that v_4 is increasing with frequency. We then use VTVM once more at 10,000 Hz, as

> =
vtvm
FREQ IN HZ
?
10000
   4— 0 .99042 V, 101.43 DEG

Some knowledge of the circuit and the above computations indicate that a peak of v_4 occurs between 100 and 10,000 Hz. Since a plot of v_4 in decibels will eventually be desired, we change the output selection to decibels by the following command:

> =
db

Having confidence, we proceed to find the frequency at which v_4 is a maximum. We first type "tune f" followed by "maximum." The program requests the node-voltage difference for which the maximum is to be found. The sequence and results are:

> =
tune f
ESTIMATED FREQ IN HZ
?
10
> =
maximize
NODE DIFF TO MAXIMIZE
?
4   0
$F =$ 1000.3 HZ
   4— 0 13.979 DECIBELS

The peak of v_4 occurs close to 1,000 Hz and $v_4 = 13.979$ dB. The value in decibels is actually $20 \log v_4$, but since the input is 1 V, the voltage at node 4 is also the gain of the network in decibels.

At this point we are ready for a frequency response, and we type the following:

> =
<u>plot log</u>
INITIAL AND FINAL HERTZ   , MAX AND MIN DECIBELS,
NO. OF INTERVALS
?
<u>100 10000 15 0 20</u>

The 100 and 10,000 are the scale limits to be plotted down the type-written page on a log scale, 15 and 0 the scale of decibels across the page, and 20 the number of intervals. The results are shown in Fig. 6-6. Note that a choice of 20 intervals (10 intervals/decade) produces a neatly labeled frequency scale. The solution is extremely close to that calculated in Prob. 2-7 of Chap. 2.

The phase response is obtained by a similar command, but the output selection must be made first. Unless a change is typed, the program assumes that decibels are still desired. The command is:

> =
<u>phase</u>

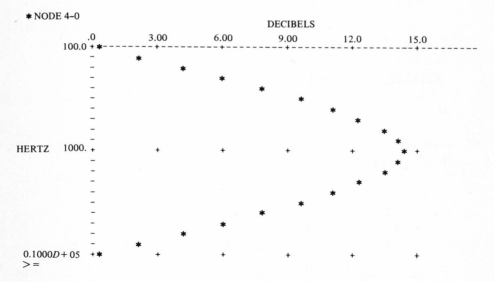

FIG. 6-6 Frequency response of the amplifier of Fig. 6-4.

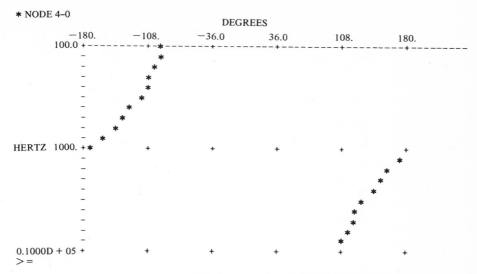

FIG. 6-7   Phase response of the amplifier of Fig. 6-4.

The phase response is plotted after the following command and data:

> =
plot log

INITIAL AND FINAL HERTZ   , MAX AND MIN DEGREES,
NO. OF INTERVALS
?
100 10000 180 −180 20

The scale across the typewritten page was requested and typed in degrees. The results are shown in Fig. 6-7.

6-3   OSUCAD INSTRUCTIONS

As stated previously, OSUCAD analyzes nonlinear transistor circuits. Diodes, bipolar transistors, and FETs are the only nonlinear elements permitted since models are built in for these devices. Diodes are modeled by the exponential diode equation and a nonlinear capacitance, and bipolar transistors are modeled by the Ebers-Moll equations. Bipolar-transistor dynamics are simulated by nonlinear emitter- and collector-junction capacitances. The FET model is similar to that of Ebers-Moll but differs in that the equations are quadratic instead of exponential (see Chap. 4). The models and nonlinear equations are shown in Fig. 6-8. Two FET models are used where the FET model in Fig. 6-8d contains an

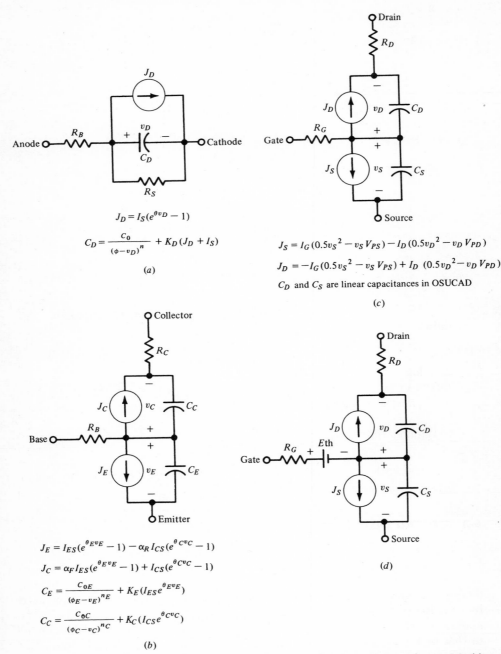

$$J_D = I_S(e^{\theta v_D} - 1)$$

$$C_D = \frac{C_0}{(\phi - v_D)^n} + K_D(J_D + I_S)$$

(a)

$$J_S = I_G(0.5v_S{}^2 - v_S V_{PS}) - I_D(0.5v_D{}^2 - v_D V_{PD})$$

$$J_D = -I_G(0.5v_S{}^2 - v_S V_{PS}) + I_D(0.5v_D{}^2 - v_D V_{PD})$$

C_D and C_S are linear capacitances in OSUCAD

(c)

$$J_E = I_{ES}(e^{\theta_E v_E} - 1) - \alpha_R I_{CS}(e^{\theta_C v_C} - 1)$$

$$J_C = \alpha_F I_{ES}(e^{\theta_E v_E} - 1) + I_{CS}(e^{\theta_C v_C} - 1)$$

$$C_E = \frac{C_{0E}}{(\phi_E - v_E)^{n_E}} + K_E(I_{ES}e^{\theta_E v_E})$$

$$C_C = \frac{C_{0C}}{(\phi_C - v_C)^{n_C}} + K_C(I_{CS}e^{\theta_C v_C})$$

(b)

(d)

FIG. 6-8 OSUCAD device models. (a) Diode model; (b) *npn* bipolar transistor model; (c) nMOS model; (d) enhancement mode nMOS model.

independent voltage source in series with the input to simulate the enhancement-mode threshold voltage.

From Fig. 6-8 notice that the diode is represented by eight parameters, the bipolar transistor by sixteen parameters, and the FET by fourteen parameters. The program has an internal library of parameters for representative devices which are automatically implemented when specified by the user. We can insert our own set of parameters if desired by a procedure to be described in Appendix B. The diode and bipolar transistor models are identical to that used in SCEPTRE, a large network-analysis program available from the Department of the Air Force, Kirkland Air Force Base, New Mexico.

The SCEPTRE documentation includes a Model Library reference which lists the measured parameters of many diodes and bipolar transistors. This reference is an excellent source of device parameters. A sample listing was given in Table 4-2, Chap. 4.

The use of OSUCAD is very similar to LINCAD. Many of the commands are the same; that is, E, C, R, L, I, LIST, SAVE, OLD, and NODES. The main difference between the two programs is in the type of analysis performed.

Components may be given individual names in OSUCAD. The name is typed after the component type—E, C, R, L, I, T, D, or MF; for example, E ONE. The name consists of up to five characters. Any five characters are permitted as selected by the user. One space must be typed after E, C, R, L, I, T, and D followed by the name. A component name is necessary in order to permit:

1. Parallel resistances
2. Removal of a component by the REMOVE command
3. Identification of a transistor or diode when a device, voltage, or current is to be plotted
4. Identification of transistors and diodes in printouts

A diode, bipolar transistor, or FET is added to a circuit by the D, T, or MF command. The name of the device is typed after the D or T or MF as described earlier; for example, D ONE. The program then requests the model number by typing MODEL NO.? The user types the model number (1N4003, 2N2368, etc.). The program searches its device library for the number. If it is one of the permanently stored devices or one previously added by the user, the program requests the pin connections. Otherwise, an error message is printed followed by the device library contents.

The pin connection requests are

NODES, (B, C, E)?

for the base, collector and emitter nodes, respectively, of a bipolar transistor;

NODES, (G,D,S)?

for the gate, drain and source nodes, respectively of a FET; and

NODES, (C,A)?

for the cathode and anode nodes, respectively, of a diode. The device model number or connections can be changed by retyping all information beginning with T *name*, D *name*, or MF *name*.

The equations generated by OSUCAD are the state-variable equations. These are differential equations in which the unknown values are the capacitor voltages and inductor currents. A second set of equations, called the output equations, computes the node voltages from the capacitor voltages, inductor currents, and independent sources. The state equations are a minimal set; that is, a circuit cannot be represented by fewer equations.

The primary advantage of the state-variable equations is that the solution time is also a minimum since solution time is dependent on the number of equations. The solution time increases as the square of the number of equations. For example, consider a circuit with two capacitors, two inductors, and ten nodes. There would be ten node equations (LINCAD generates node equations) each of which is a second-order differential equation. There would only be four state-variable equations. These state equations are first-order differential equations, and for this example, the node equations consist of no less than 10×11 or 110 terms; whereas the state-variable equations have a minimum of 4×5 or 20 terms. Of course, to obtain the node voltages, another set of equations is required by the state method. However, the calculation of the node voltage need only be done when a printout is required. In practice, nonlinear differential equations are solved many times between the printing of results.

The penalty for such a compact set of equations is the complexity that exists in generating the state equations. Many matrix multiplications as well as matrix inversions are required to compute the state equations, whereas node equations require little computational time. For this reason OSUCAD has three modes of operation: insertion of components and service mode, the DC mode, and the DYNAMIC mode. These modes are distinguished by the symbol printed by the computer program when in that mode; "=" for the insertion of components and the service mode, "+" for the DC mode, and ". ." for the DYNAMIC mode. Components may be added, deleted, etc., in the "=" mode. Once the user requests a DC or DYNAMIC solution, the state equations and output equations are generated, and the computer types the "+" or ". .", respectively. In the DC or DYNAMIC modes, the equations cannot be altered but may be solved repeatedly. The value of independent sources and some transistor parameters can be changed in the latter two modes. The technique will be shown by examples in the next section.

The following list contains the service options. These are the only replies to "=" other than the insertion of components.

OFF   To initially start a transistor in the cutoff state, the reply must follow immediately after the base, collector, and emitter nodes are typed

SAVE	To save a circuit on the disk
OLD	To retrieve a previously saved circuit
	(To implement the SAVE and OLD features, your own computer consultants will instruct you as to how data is saved)
NEW	To begin a new circuit. (This is automatic on entering OSUCAD)
LIST	To list the circuit components
DEVICES	To list the device library model numbers
PARAMETERS	To print the device parameters
DC	To obtain the dc operating point
DCPLOT	To go directly to the dc mode for plotting
DYNAMIC	To obtain the dynamic response (the Ebers-Moll charge-control model is used for transistors)
EQUATIONS	To print the state-variable equations
REMOVE X *name*	To remove an element where X is E, I, R, L, C, T, D, or MF and *name* is the name given under ' = '
NODES	To select specific nodes or node voltage differences to be printed or plotted later
D MODEL	To insert the parameters of a new device or to change the parameters of a library device
END	To exit gracefully from OSUCAD

The selection of NODES must be done under " = " so that OSUCAD can compute the required output equations. Device voltages and currents are selected in the DC and DYNAMIC modes.

6-4 OSUCAD EXAMPLE

The first OSUCAD example will be a computer analysis of the μA730 differential amplifier shown in Fig. 5-1 and repeated in Fig. 6-9. The dc operating point as well as most of the typical electrical characteristics will be computed in the simulation. We shall find that the Chap. 5 calculations of the dc voltages and currents were quite good. No design attempts will be made in this chapter as these design variations are reserved for Chap. 9. This example will also extend the analysis of the differential amplifier so as to compute and compare typical electrical characteristics not covered by the simple analysis of Chap. 5.

The nodes have been numbered consecutively beginning with zero as the ground node. The components have been assigned names even though the names are not necessary since all elements have different node connections. Sources have been included at the inputs which correspond to V_{CM} in Chap. 5.

The logon procedure and program call is similar to that of LINCAD. Once loaded, the program types:

OSUCAD—OSU COMPUTER-AIDED DESIGN

=

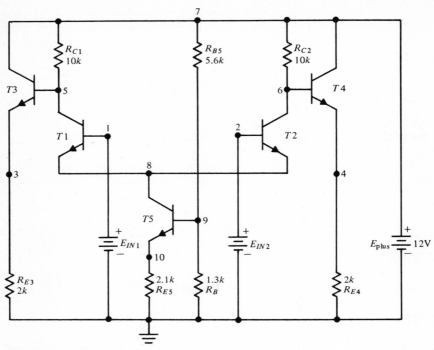

FIG. 6-9 μA730 single-stage differential amplifier.

The "$=$" sign indicates that the program is awaiting the insertion of a component, the selection of one of the service routines, or one of the analysis options. The following responses insert the input sources of 3.5 V and transistor $T1$.

<u>e in1</u>
NODES, VOLTS$(T=0)$
?
<u>0 1 3.5</u>
=
<u>e in2</u>
NODES, VOLTS$(T=0)$
?
<u>0 2 3.5</u>
=
<u>t 1</u>
MODEL NO.?
<u>2n2369</u>

NODES, (B,C,E)
?
1 5 8
=

The 2N2369 is one of the built-in devices. The node connections were requested after the program found the 2N2369 in its library.

A shortcut entry exists in OSUCAD to speed the insertion of transistors.
If a nonblank set of characters are present beginning in space 10 of the line
which specifies a T, D, or MF, the characters are assumed to be the model
number. The insertion of *T2*, *T3* and *T4* are thus:

t 2   2n2369
NODES,(B,C,E)
?
2 6 8
=

t 3   2n2369
NODES,(B,C,E)
?
5 7 3
=

t 4   2n2369
NODES,(B,C,E)
?
6 7 4
=

The remaining transistor and components are entered as shown below.

t 5
MODEL NO.?
2n2369
NODES,(B,C,E)
?
9 8 10
=

r *e*3
NODES,KILOHMS
?
0 3 2
=

r *e*5

NODES,KILOHMS
?
<u>0 10 2.1</u>
<u>=</u>
<u>r b</u>
NODES,KILOHMS
?
<u>0 9 1.3</u>
<u>=</u>
<u>r e4</u>
NODES,KILOHMS
?
<u>0 4 2</u>
<u>=</u>
<u>r c1</u>
NODES,KILOHMS
?
<u>5 7 10</u>
<u>=</u>
<u>r b5</u>
NODES,KILOHMS
?
<u>9 7 5.6</u>
<u>=</u>
<u>r c2</u>
NODES,KILOHMS
?
<u>6 7 10</u>
<u>=</u>
<u>e plus</u>
NODES,VOLTS$(T=0)$
?
<u>0 7 12</u>
<u>=</u>

A title, the nodes selection, and file save are entered by the following commands:

<u>title</u>
<u>a730 single-stage differential amplifier</u>
<u>=</u>
<u>nodes</u>

NODES(8)
?
1 -2 6 -5 6 9 3 10
=
save
=

The title was typed by the user after the command TITLE. The entry of nodes is the same as in LINCAD. We requested the node difference between nodes 1 and 2 (the differential input), the node difference between nodes 6 and 5 (the differential output), and the voltage at node 6 (the single-ended output) and the voltages at nodes 9, 3, and 10.

The command SAVE stored the circuit, title, and node selection in the file A730 which was allocated prior to the OSUCAD call.

The next two commands typed in response to " = " were LIST and DC. The printed page is shown in Fig. 6-10. The LIST command causes a listing of the circuit including the title to be printed. The DC command results in the generation of the state equations, the dc solution of those equations, and a print of the dc operating point. Also shown in Fig. 6-10 at the bottom is the " + " sign indicating that the user is in the dc portion of the program.

It is of interest to compare the measured typical characteristics of the μA730 (Fig. 6-1) with the simulated results of OSUCAD and the simple calculations of Chap. 5.

The collector current of $T5$ in Fig. 6-10 is 0.7503 mA compared with 0.79 mA by the hand calculation. The difference is attributed to the approximation of $V_{BE} = 0.6$ V and to the neglect of base current in the simple hand calculation. The total supply current is given by

$$I_S = I_{C3} + I_{C4} + I_{C1} + I_{C2} + I_{B3} + I_{B4} + \frac{V_+ - V_9}{5.6 \text{ k}} = 9.589 \text{ mA}$$

These and other dc voltages and current are compared in Table 6-2. The favorable comparison illustrates the accuracy of the simple calculations of Chap. 5.

The low-frequency small-signal analysis can be made in the dc portion of OSUCAD by utilizing the source variation options. A table of operating points can be printed as a function of any independent source. The procedure to obtain such a table is to enter the source and its name as typed under " = ". The minimum, maximum, and incremental values of that source will be requested. The table is printed following the user response. Other quantities can be varied as shown in Appendix B and later examples.

In response to the " + " sign, the user types the following in order to obtain a table of operating points as E_{in1} is varied:

e in1

list
A730 SINGLE-STAGE DIFFERENTIAL AMPLIFIER

E IN1	1	0	3.50000	VOLTS(T=0)
E IN2	2	0	3.50000	VOLTS(T=0)
E PLUS	7	0	12.0000	VOLTS(T=0)
R E3	3	0	2.00000	KILOHMS
R E5	10	0	2.10000	KILOHMS
R B	9	0	1.30000	KILOHMS
R E4	4	0	2.00000	KILOHMS
R C1	7	5	10.0000	KILOHMS
R B5	9	7	5.60000	KILOHMS
R C2	7	6	10.0000	KILOHMS
T 5	2N2369	(B,C,E) =	9 8 10	
T 4	2N2369	(B,C,E) =	6 7 4	
T 3	2N2369	(B,C,E) =	5 7 3	
T 2	2N2369	(B,C,E) =	2 6 8	
T 1	2N2369	(B,C,E) =	1 5 8	

=
dc

A730 SINGLE-STAGE DIFFERENTIAL AMPLIFIER

DEVICE	VCE.VD OR VDS (VOLTS)	IC OR ID (MA)	VBE OR VGS (VOLTS)	IB OR IG (MICROAMPERE)
T 1	4.9330	.36983	.63330	5.2892
T 2	4.9330	.36983	.63330	5.2892
T 3	4.8920	3.5038	.69184	50.110
T 4	4.8920	3.5038	.69184	50.110
T 5	1.2684	.75030	.65124	10.731

$V\ 1 - V\ 2 = .0$ $V\ 6 - V\ 5 = -.25513D\text{-}12$

$V\ 6 - V\ 0 = 7.7998$ $V\ 9 - V\ 0 = 2.2495$

$V\ 3 - V\ 0 = 7.1080$ $V10 - V\ 0 = 1.5983$

E IN1 = 3.5000 VOLTS
E IN2 = 3.5000 VOLTS
E PLUS = 12.000 VOLTS
+

FIG. 6-10 Listing and dc operating point of the circuit in Fig. 6-9.

TABLE 6-2

	OSUCAD Simulation		Calculations of Chap. 5
	$\alpha = .9858$	$\alpha = .99$	
$V_{B5} = V_9$	2.2495 V	2.2528 V	2.26 V
$V_{E5} = V_{10}$	1.5983 V	1.6017 V	1.66 V
I_{C5}	0.7503 mA	0.75502 mA	0.79 mA
$V_{B4} = V_6$	7.7998 V	7.9014 V	8 V
$V_{E3} = V_3$	7.108 V	7.21 V	7.4 V
I_{C3}	3.5038 mA	3.5689 mA	3.7 mA
I_{B1}	5.2892 μA	3.7748 μA	4 μA

MIN, MAX, INCREMENT IN VOLTS

?

3.49 3.51 .02

Typing "$e\, in1$" selected that particular source to vary. The minimum, maximum, and incremental value for the table were specified after the appropriate request. The results are shown in Fig. 6-11. From Fig. 6-11,

$$A_v = \frac{V_6 - V_5}{V_1 - V_2} = \frac{-1.3487}{-0.01} = 134.87$$

The OSUCAD gain of 134.87 compares with the Chap. 5 value of 146 and the Fig. 5-3 value of 145.

The input resistance of $T1$ (or $T2$) with respect to ground is given by the incremental change in the base voltage of $T1$ (or $T2$) divided by the incremental change in the base current of $T1$ (or $T2$). The input resistance is:

$$R_{ig} = \frac{3.51 - 3.49}{(6.3233 - 4.2557)(10^{-6})} = 9,673 \ \Omega$$

The differential input resistance between the two inputs is the sum of the

A730 SINGLE-STAGE DIFFERENTIAL AMPLIFIER

DEVICE	VCE.VD OR VDS (VOLTS)	IC OR ID (MA)	VBE OR VGS (VOLTS)	IB OR IG (MICROAMPERE)
E IN1 = 3.4900	VOLTS			
T 1	5.6119	.29757	.62782	4.2557
T 2	4.2632	.44209	.63781	6.3226
T 3	4.2202	3.8348	.69435	54.844
T 4	5.5637	3.1729	.68910	45.378
T 5	1.2639	.75030	.65124	10.731
V 1 − V 2 = .10000D-01			V 6 − V 5 = −1.3487	
V 6 − V 0 = 7.1254			V 9 − V 0 = 2.2495	
V 3 − V 0 = 7.7798			V10 − V 0 = 1.5983	
E IN1 = 3.5100	VOLTS			
T 1	4.2532	.44214	.63781	6.3233
T 2	5.6019	.29752	.62782	4.2550
T 3	5.5637	3.1727	.68910	45.375
T 4	4.2202	3.8350	.69435	54.847
T 5	1.2739	.75030	.65124	10.731
V 1 − V 2 = .10000D-01			V 6 − V 5 = 1.3487	
V 6 − V 0 = 8.4741			V 9 − V 0 = 2.2495	
V 3 − V 0 = 6.4363			V10 − V 0 = 1.5983	
E IN1 = 3.5000	VOLTS			
E IN2 = 3.5000	VOLTS			
E PLUS = 12.000	VOLTS			

FIG. 6-11   Table of operating points for $E_{in1} = 3.49$ and 3.51 V.

two input resistances with respect to ground or $2R_{ig}$, as

$$R_i = 2R_{ig} = 2 \times 9{,}673 = 19{,}346 \text{ k}$$

This compares with 13.0 k from the approximations of Chap. 5 and 20 k from Fig. 5-3.

There are various laboratory techniques for measuring the output resistance. A convenient method is to measure the open-circuit voltage at the terminals and then to load the terminals with a known resistance. The output resistance can be calculated from the loaded and unloaded output voltages. Consider the amplifier output represented by a Thévenin equivalent circuit as shown in Fig. 6-12a where the open-circuit voltage is V_{oc}. The Thévenin equivalent circuit is loaded with a known resistance, R_{load}, with the voltage across R_{load} denoted as V. The unknown output resistance is

$$R_o = \frac{V_{oc} - V}{V} R_{\text{load}}$$

The resistance can be inserted only in the "$=$" mode. A blank response to "$+$" or carriage return, returns control to the "$=$" mode. The following

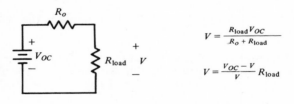

$$V = \frac{R_{\text{load}} V_{OC}}{R_o + R_{\text{load}}}$$

$$V = \frac{V_{OC} - V}{V} R_{\text{load}}$$

(a)

A730 SINGLE-STAGE DIFFERENTIAL AMPLIFIER

DEVICE	VCE.VD OR VDS (VOLTS)	IC OR ID (MA)	VBE OR VGS (VOLTS)	IB OR IG (MICROAMPERE)
T 1	3.4645	.36977	.63329	5.2883
T 2	4.9325	.36989	.63330	5.2900
T 3	6.4028	13.795	.73411	197.30
T 4	4.8926	3.5035	.69183	50.106
T 5	1.2684	.75030	.65124	10.731

$$
\begin{array}{llll}
\text{V } 1 - \text{V } 2 = & .0 & \text{V } 6 \ - \text{V } 5 = & 1.4680 \\
\text{V } 6 - \text{V } 0 = & 7.7992 & \text{V } 9 \ - \text{V } 0 = & 2.2495 \\
\text{V } 3 - \text{V } 0 = & 5.5971 & \text{V}10 - \text{V } 0 = & 1.5983
\end{array}
$$

E IN1	=	3.5000	VOLTS
E IN2	=	3.5000	VOLTS
E PLUS	=	12.000	VOLTS

(b)

FIG. 6-12 Measurement of the output resistance. (a) Method of measuring the output resistance by open circuit and fixed load R_{load}; (b) OSUCAD output with $R_{\text{load}} = 500 \, \Omega$.

commands insert a 500-Ω resistance between node 3 and ground, and select the dc solution:

$+$

_____ (blank or simply a carriage return)

$=$

r load

NODES, KILOHMS

?

0 3 .5

$=$

dc

From Fig. 6-10 and Fig. 6-12b

$$R_o = \frac{7.1080 - 5.5971}{5.5971} \, 500 = 135 \, \Omega$$

This value is higher than the 70 Ω listed in Fig. 5-3 and the 106 Ω calculated in Chap. 5. The difference can be attributed to the value selected for β. From Fig. 6-12b the β of transistor T_3 is

$$\beta_3 = \frac{13.795}{0.1973} = 70$$

The value of h_{ie3} is

$$h_{ie3} = \frac{h_{fe3}}{g_{m3}} = \frac{h_{fe3}}{40 I_{C3}}$$

The collector current of T_3 is 3.5 mA in Fig. 6-10 and 13.8 mA in Fig. 6-12. Thus,

$$127 < h_{ie} < 500 \, \Omega$$

Choosing the average $h_{ie} = 314 \, \Omega$, the calculation of Chapter 5 yields

$$R_o = \frac{10,314}{70} = 147 \, \Omega$$

Thus the OSUCAD and the calculated value compare reasonably well. No doubt the μA730 has transistors with β higher than 100 as indicated in the discussion of Chap. 5.

The common-mode rejection is computed by increasing both inputs by the same amount and computing the gain with reference to Fig. 6-10. If the incremental value of the source is typed as a zero, OSUCAD interprets the user's desire to change the source value for subsequent analysis. On return to "=", the old values will be reinserted. The responses shown below change the value of both input sources to 3.51 V., and select a dc analysis.

+
e in1
MIN, MAX, INCREMENT IN VOLTS
?
3.51 0 0
+
e in2
MIN, MAX, INCREMENT IN VOLTS
?
3.51 0 0
+
dc

The results are shown in Fig. 6-13. No change is detected in V_6 to four decimal places. None should be expected since the simulated circuit is perfectly symmetrical. The simulation will show deviation in V_6 when the transistors reach cutoff or saturation. However, the common-mode rejection of Fig. 5-3 does not refer to that case. Knowledge of the unsymmetrical nature of the μA730 is necessary to verify the common-mode rejection.

The various small-signal characteristics as listed in Fig. 5-3, calculated in Chap. 5, and computed with OSUCAD are compared in Table 6-3.

The transfer characteristic can be plotted in the dc mode of OSUCAD. We desire to plot $(V_6 - V_5)$, and to avoid plotting the other voltages, a new "nodes" selection must be made. A return to the "=" mode is neces-

A 730 SINGLE-STAGE DIFFERENTIAL AMPLIFIER

DEVICE	VCE.VD OR VDS (VOLTS)	IC OR ID (MA)	VBE OR VGS (VOLTS)	IB OR IG (MICROAMPERE)
T 1	4.9230	.36983	.63330	5.2892
T 2	4.9230	.36983	.63330	5.2892
T 3	4.8920	3.5038	.69184	50.110
T 4	4.8920	3.5038	.69184	50.110
T 5	1.2784	.75030	.65124	10.731

$$V\ 1\ -\ V\ 2\ =\ \ .0 \qquad\qquad V\ 6\ -\ V\ 5\ =\ -.82601D\text{-}13$$
$$V\ 6\ -\ V\ 0\ =\ \ 7.7998 \qquad\qquad V\ 9\ -\ V\ 0\ =\ \ 2.2495$$
$$V\ 3\ -\ V\ 0\ =\ \ 7.1080 \qquad\qquad V10\ -\ V\ 0\ =\ \ 1.5983$$

E IN1 =   3.5100   VOLTS
E IN2 =   3.5100   VOLTS
E PLUS =   12.000   VOLTS

FIG. 6-13 Operating Point to be compared with Fig. 6-10 for the common mode rejection.

TABLE 6-3

	Typical electrical characteristic (*Fig.* 5-3)	OSUCAD simulation		Calculations of *Chap.* 5
		$\alpha = 0.9858$	$\alpha = 0.99$	
I_S	9.5 mA	9.589 mA	9.698 mA	9.9 mA
Power consumption	114 mw	115 mw	116.4 mw	119 mw
Input bias current	3.5 µA	5.28 µA	3.7748 µA	4 µA
Differential voltage gain	145	134.87	139.37	146
Input resistance	20 k	19.3 k	27 k	13.0 k
Output resistance	70 Ω	135 Ω		106 Ω
Bandwidth	1.5 MHz	1.77 MHz (2N2369)		
		1.0 MHz (2N2368)		

sary. Therefore, the user's reply to the "+" is a carriage return. The following responses were typed by the user:

<u>+</u>

_____ (a carriage return)

<u>=</u>

<u>nodes</u>

NODES (8)

?

<u>6 −5 0 0 0 0 0 0</u>

<u>=</u>

<u>dcplot</u>

+

The command DCPLOT skips the dc operating point printout. The user is now in the dc mode as indicated by the "+". Since a plot was requested, the next variation of an independent source or parameter will result in a plot. Figure 6-14 was printed after the following responses.

<u>e in1</u>

MIN, MAX, INCREMENT IN VOLTS

?

<u>3.4 3.6 .02</u>

MIN. AND MAX VOLTS

?

<u>−10 5</u>

A730 SINGLE-STAGE DIFFERENTIAL AMPLIFIER

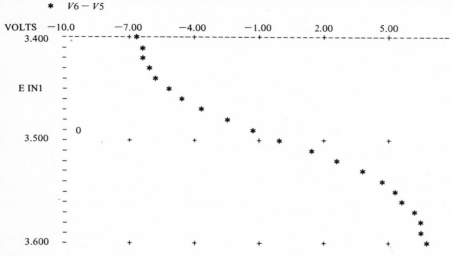

FIG. 6-14   Transfer characteristic of the μA730.

The MIN AND MAX VOLTS refer to the scale of voltage across the printed page.

The transfer characteristics compare reasonably well with Fig. 5-3 and the calculations of Chap. 5, and vice-versa.

If another plot were desired, we must answer the "+" with the word PLOT prior to the variation of a source or parameter. Otherwise a table of operating points will be printed. The word PLOT must be typed prior to every plot except when the command is DCPLOT.

Since the OSUCAD program solves nonlinear circuits, the solution of problems with sinusoidal sources must be in the time domain. Future plans call for an option in OSUCAD in which small-signal equivalent circuits are generated for a given dc operating point. Then steady-state sinusoidal analysis will be possible.

However, it is not necessary to generate small-signal equivalent circuits, or have sinusoidal sources in order to observe and compute the upper cutoff frequency of the μA730. Square-wave techniques, that have been used in the laboratory for decades, can be employed.

The upper half-power frequency can be calculated from the output response of a circuit to a square-wave input. Actually a step function is sufficient. The output response is assumed to be the result of a single pole; or another way of stating this assumption is to say that the circuit can be represented by an ideal amplifier with an RC series input, as shown in Fig. 6-15. The output of the circuit of Fig. 6-15a to a step input, as shown in Fig. 6-15b, is

$$v_o = Av_{in} = AV(1 - e^{-t/\tau})$$

where $\tau = RC$. The time constant τ is the time it takes for the output to reach 63.2 percent of its final value.

The response of the circuit of Fig. 6-15a to a sinusoidal signal is

$$v_o = Av_s = A \; \frac{\dfrac{1}{j\omega C}}{R + \dfrac{1}{j\omega C}} \; v_s$$

The circuit voltage gain is

$$\frac{v_o}{v_s} = A \frac{1}{1 + j\omega RC}$$

The upper half-power frequency occurs when ωRC is equal to one. Thus

$$\omega_2 = \frac{1}{RC}$$

or

$$f_2 = \frac{1}{2\pi RC} = \frac{1}{2\pi\tau} \tag{6-2}$$

Thus the half-power frequency of an amplifier can be found by using a step function or square-wave input, observing the time required for the output to change 63.2 percent, and applying Eq. (6-2).

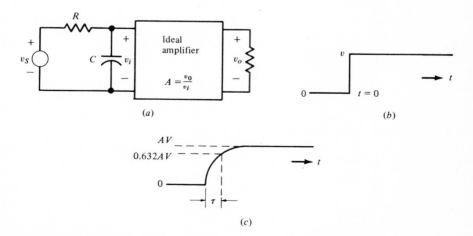

(a)

(b)

(c)

FIG. 6-15   Single-pole equivalent circuit of an amplifier. (a) Single-pole equivalent circuit; (b) step input $v_S(t)$; (c) $v_0(t)$ for the step input of (b).

To obtain the step response of the μA730, we type DYNAMIC in reply to " = ". The program computes the dc operating point which is the initial condition for a transient analysis. When this is completed OSUCAD prints a double period (. .) to indicate that we are in the dynamic mode. Any source can be made a function of time by typing its name as entered under " = ". The program then requests the parameters of the waveform as indicated in Fig. 6-16. To begin a plot of the time-domain response we type PLOT. This is followed by requests for the scale and number of intervals. The procedure is as follows:

=
dynamic
. .
e in1
TDELAY,TWIDTH,TREP,MAX,MIN (NS,VOLTS OR A)
?
25 500 1000 3.51 3.5
. .
plot
NO. OF INTERVALS,TFINAL(NS)
?
40 1000
MIN AND MAX VOLTS
?
0 1.5

The output response is printed as shown in Fig. 6-17. The output changes from a minimum of zero to a maximum of 1.35 V in response to an input change of 0.01 V. This agrees with the gain of 134.87 computed earlier. Now 63.2 percent of 1.35 V is 0.853 V, and from Fig. 6-17, $\tau = 90$ ns. Thus, from Eq. (6-2)

$$f_2 = \frac{1}{2\pi \times 90 \times 10^{-9}} = 1.77 \text{ MHz}$$

This compares favorably with the bandwidth of 1.5 MHz given in the manufacturer's characteristics in Fig. 5-3.

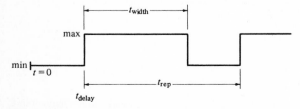

FIG. 6-16 Waveform used in the DYNAMIC solution.

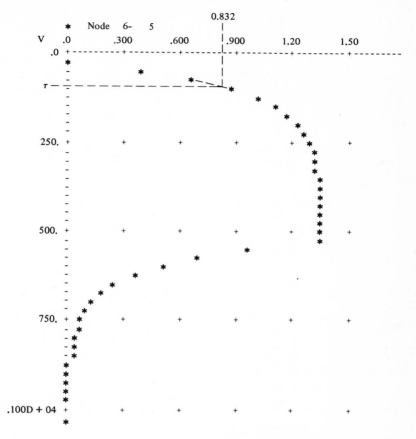

FIG. 6-17 Dynamic characteristics of the μA730.

6-5 SUMMARY

The digital computer has been shown to be a useful tool in the analysis of electronic circuits. The general programs described in this chapter enable the circuit designer to quickly analyze a variety of circuits. No attempt was made to modify the circuits and obtain more desirable responses. The design of circuits by iterative analysis is reserved for Chap. 9 where other features of LINCAD and OSUCAD will be used and discussed. The optimization routines of LINCAD will be shown to be a powerful tool that goes a step beyond analysis.

REFERENCES

1. "Sceptre Support," AFWL-TR-67-124, vol. 1, Contract F29601-67-C-0049, April, 1968.
2. "Transistor and Diode Model Handbook," AFWL-TR-69-44, Contract F29601-68-C-0117, October, 1969.

PROBLEMS

6-1. Find an equivalent circuit for the CA3000 with the active component a controlled current generator of the form

$$g_m = \frac{G_m}{1 + j^f/f_{co}}$$

The gain response of the CA3000 is given in Fig. 1-10.

6-2. The measured gain-vs.-frequency response curve for an amplifier is shown in Figure P1-10. The amplifier has 180° phase shift in the constant-gain region. Find an equivalent circuit of the amplifier suitable for entry into LINCAD. The output resistance is 1 k and the input resistance 200 k.

6-3. Verify the response in Prob. 6-2 by simulation using LINCAD or a suitable CAD program.

6-4. A transistor can be modeled by the circuit shown in Fig. P6-4. Find R_1, R_2 g_{m1} and g_{m2} in terms of the common-emitter h parameters.

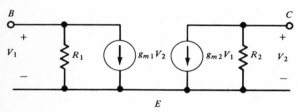

FIG. P6-4

6-5. Solve Example 4-7 using LINCAD or another CAD program.

6-6. Solve Example 5-3 using LINCAD or a suitable CAD program.

6-7. Verify the validity of the Miller theorem by simulating Fig. 5-33a and b. Use the component values given and calculated for the 741 in Sect. 5-4.9. Compare the break frequency of the circuits.

6-8. Find the cutoff frequency in Prob. 5-13 by simulation.

6-9. Using OSUCAD or a similar nonlinear analysis CAD program, plot the transfer function v_O versus v_I of the complementary-symmetry push-pull stage shown in Fig. 5-23. Assume 2N404 and 2N2369 transistors and $V^+ = V^- = 15$ V.

6-10. Find the frequency and phase response of the twin-tee circuit shown in Fig. P6-10.

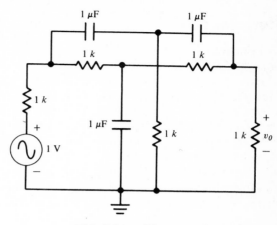

FIG. P6-10   The circuit for Prob. 6-10.

6-11. Modify the circuit of Fig. P6-10 to produce an output (v_O) response with a minimum at 100 Hz.

7

Wave Generation and Switching

INTRODUCTION

In the preceding chapters, emphasis has been placed on electronic circuits which operate with analog signals; i.e., signals which vary continuously with time. However, some of the most precise electronic instrumentation has been implemented by circuits which operate with discrete rather than continuous voltage levels. The digital computer is an example of such a system. The digital computer is a complex system which we cannot discuss completely. We shall consider some of the characteristics of subsystems of digital computers, showing how they operate and how they may be applied.

Generation of electrical voltages having other than sinusoidal waveforms is an important subject. These waveforms find application in TV and radar sweep circuits, digital computers, and digital communication circuits.

7-1 DIGITAL VOLTMETER

It is desirable to introduce waveform generation and digital circuits by use of a system example. In this manner, we learn how subsystems operate, why they are used, and how they combine together to function as a complete system. No attempt will be made to show all of the many possible ways of performing a function. One or two examples will be presented to illustrate methods of implementation. It is always a probability that a simpler, improved, or less expensive method can be developed to perform the same function. The experience and ingenuity of the designer are vital to the development of these simpler, or better, systems.

The digital voltmeter serves as an introductory system since it contains several wave generation subsystems and digital building blocks. This type of meter has a display in numerical form rather than a meter dial reading. The accuracy can be controlled to close limits, the display is unambiguous, the input resistance can be large for minimum circuit loading, and as many significant figures as desired can be included.

Several schemes have been devised to implement digital meters. We shall select one of the simpler systems to illustrate the principles involved. In block-diagram form, Fig. 7-1 indicates the various functions in the system. To aid in understanding the operation, waveforms at several points are also sketched. In general, the input analog voltage is changed to a series of voltage pulses which are counted; the output of the counter is then displayed on a numerical display system.

The input selector switch connects to a calibration current source, reference voltage, or to a voltage range in which measurement is to be made. The table of Fig. 7-1 shows waveforms and operations for one complete cycle of the reference cycle, 60 Hz in this case. We note that the timing generator has zero output for one-half the reference cycle and a constant value of $+V_G$ during the second

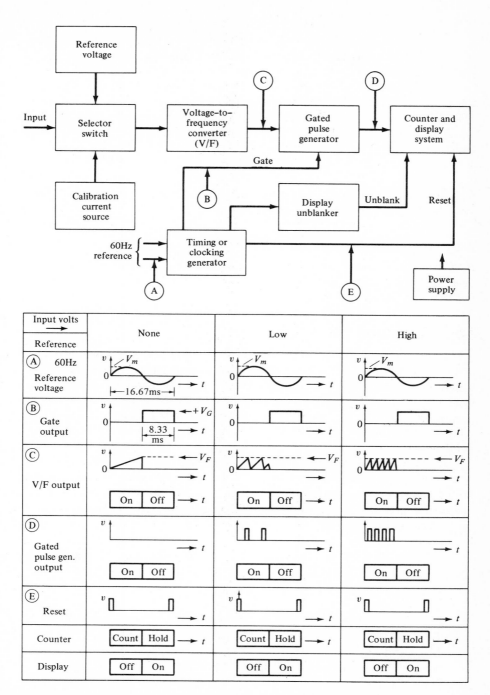

FIG. 7-1 Digital voltmeter system with waveforms and indicated operations. (*Courtesy Popular Electronics, September, 1970. Ziff-Davis Publishing Co.*).

half. The timing generator signal switches the voltage-to-frequency (V/F) converter and the gated pulse generator as indicated. It also generates a reset pulse that erases the counter so that counting can begin again and control the blanking. As the input voltage increases, the triangular output voltage of the V/F converter increases in frequency. The gated pulse generator is controlled by the V/F converter and yields a pulse at the end of each V/F cycle as shown by (D). While the display is *off* and pulse generator is *on*, the counter counts the number of pulses generated. At the end of the counting half cycle, the counter stops and the display is unblanked so as to read the count. The frequency of the V/F converter is linearly related to the input voltage and thus the pulses and count yield a reading of this input. The display is blanked fast enough that flicker is not objectionable to the eye. Other methods of counting and display vary the time the display is *on* or *off*.

7-2   V/F CONVERTER

A voltage-to-frequency converter can be composed of integrated circuits. However, the circuit chosen has discrete devices which introduce another transistor not normally found in ICs.

7-3   UNIJUNCTION TRANSISTOR (UJT)

The unijunction transistor (UJT) is a three-terminal negative-resistance device. We will use the UJT to generate short, high-current pulses. The construction is somewhat like that shown in Figure 7-2a. A *pn* junction and two ohmic contacts are formed on the silicon base material. The contact to the junction is designated as emitter while the two ohmic contacts are designated base 1 ($B1$) and base 2 ($B2$). In the circuit symbol of Fig. 7-2b, the arrow of the emitter slants so as to point toward $B1$.

In normal application, the interbase voltage V_{BB} is maintained constant. The important characteristic of the UJT is that of v_E versus i_E. An expanded sketch of v_E versus i_E with defined regions is indicated in Fig. 7-3. A normal diode characteristic is obtained when $B2$ is disconnected so that $i_{B2} = 0$. For $i_{B2} > 0$, a family of curves will result for increasing values of V_{BB}, only one of which is shown in Fig. 7-3.

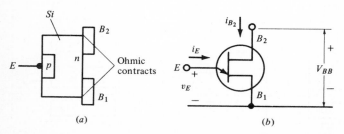

(a)   (b)

FIG. 7-2   (a) A sketch of the unijunction transistor; (b) the UJT circuit symbol and voltage designations.

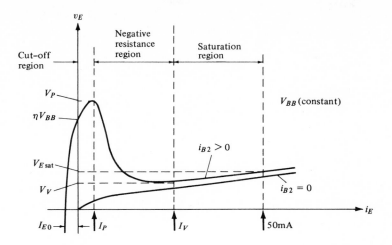

FIG. 7-3 The unijunction transistor emitter-to-base-1 characteristics.

When $i_E = 0$, there is an ohmic resistance R_{BB} between $B1$ and $B2$ having a value in the range of 5,000 to 10,000 Ω. The emitter divides R_{BB} into two parts: R_{B1} (R_{B2}) between $B1$ ($B2$) and the n side of the emitter so that $R_{BB} = R_{B1} + R_{B2}$. The voltage on the n side of the junction is ηV_{BB} at $i_E = 0$, where $\eta \equiv R_{B1}/R_{BB}$ and η is defined as the intrinsic standoff ratio. The value of η is a manufacturer's specification and has a value between 0.5 and 1.

When v_E increases above ηV_{BB}, the emitter-to-base-1 junction becomes forward-biased. A peak voltage V_P is reached at a very small value of current I_P on the order of a few microamperes. The holes injected into the $B1$ region increase the conductivity (decrease the resistance) in this region. Thus, as the emitter current increases, the R_{B1} resistance decreases, and the v_E voltage decreases. Since i_E is increasing as v_E decreases, the device possesses a negative-resistance characteristic.

As i_E increases, eventually $i_E \gg i_{B2}$ and i_{B2} may be considered negligible. Thus, for large i_E, the characteristic approaches the $i_{B2} = 0$ curve as shown in Fig. 7-3. The negative-resistance region stops at $v_E = V_v$ and $i_E = I_v$, the valley point. The resistance is positive for $i_E > I_v$ in the saturation region. Arbitrarily, the saturation voltage, V_{Esat}, is defined as the voltage at $I_E = 50$ mA and V_{Esat} has a value of about 3 V.

The important parameter of UJTs for most applications is the peak voltage V_P given by

$$V_P = \eta V_{BB} + 0.6 \text{ V} \tag{7-1}$$

This voltage is not a sensitive function of temperature and the addition of a resistance in series with $B2$ reduces temperature effects even further. In Eq. (7-1), the 0.6-V factor is the cutin voltage for a Si pn junction, and we see that V_P is directly proportional to V_{BB}. Figure 7-4 shows a family of curves for a UJT. We can estimate η at $i_E \simeq 0$ from these curves by use of Eq. (7-1). The value of

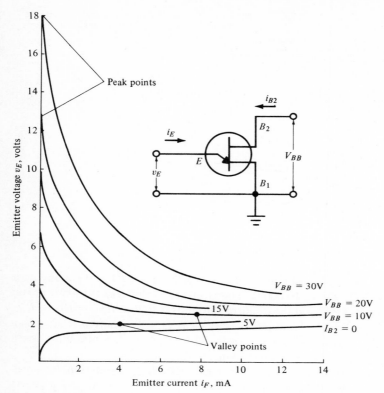

FIG. 7-4 A family of unijunction transistor emitter-input characteristics. (*Courtesy General Electric Semiconductor Products Department, Syracuse, N.Y.*)

$\eta \simeq 0.6$; e.g., for $V_{BB} = 20$ V, $V_P = \eta V_{BB} + 0.6 = 0.6(20.0) + 0.6 = 12.6$ V which is the value of v_E at $i_E \simeq 0$ for this curve.

7-3.1 NEGATIVE-RESISTANCE SWITCHING

We noted that the UJT possessed a negative-resistance region in which i_E increased as v_E decreased. Other devices such as tunnel diodes, *pnpn* diodes, silicon-controlled rectifiers, and Thyristors also have negative-resistance characteristics. Such devices can be used to construct switching or wave-generating circuits. The principles of negative-resistance (NR) switching will be presented and applied to the UJT. Figure 7-5a shows the basic circuit applied with negative-resistance devices for switching or waveform generation. In Fig. 7-5b the NR device characteristic has been idealized by piecewise linear segments. Between points 0 and 1, and between 2 and 3, the device has a positive resistance given by the slope of the lines. Between points 1 and 2, the device displays a negative resistance given by the slope of the line.

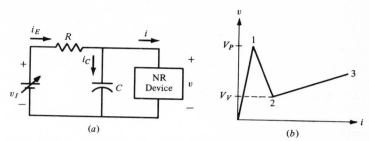

FIG. 7-5   (a) A basic negative-resistance switching circuit; (b) an idealized negative-resistance device characteristic.

The circuit may be generalized to that of Fig. 7-6 where R_{NR} has a value dependent on the region of operation. The time constant for Fig. 7-6 is

$$\tau = \frac{RR_{NR}}{R + R_{NR}} C \tag{7-2}$$

The expression for the voltage v across R_{NR} is

$$v = \frac{v_I R_{NR}}{R + R_{NR}} (1 - e^{-t/\tau}) \tag{7-3}$$

where C is initially uncharged. This solution is easily determined by noting that v must be zero initially and as $t \rightarrow \infty$, $v = v_I R_{NR}/(R + R_{NR})$ by voltage division when the capacitor is charged. The usual exponential response must exist between these limits.

The input voltage and i_R may be expressed in terms of v across the NR device as

$$v_I = i_R R + v$$

or

$$v = v_I - i_R R \tag{7-4}$$

If a final static voltage can be achieved, Eqs. (7-3) and (7-4) must agree. Since we are confronted with a nonlinear problem, several possible conditions must be considered. With a fixed value of R, the circuit operation will be analyzed for

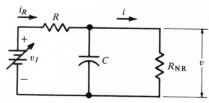

FIG. 7-6   The circuit of Fig. 7-5 with the NR device replaced by an equivalent resistance.

application of unit step voltages having three amplitudes V_1, V_2, and V_3 as in Fig. 7-7. The lines representing Eq. (7-4) having slope $= -R$ are sketched on the NR characteristic coordinates. If a static condition can occur, the intersection of the lines representing Eq. (7-4) and the NR characteristic yields the final voltage and current condition (C is charged).

7-3.2 APPLICATION OF STEP VOLTAGES

(a) *Application of V_1 to the circuit of Fig. 7-5a.* Referring to Fig. 7-7, the voltage v starts at zero and rises to the value at point 1 where the V_1 load line and the NR characteristic intersect. The negative-resistance region is not entered. The solution from Eq. (7-3) is

$$v = \frac{V_1 R_1}{R + R_1}\left(1 - e^{-t/\tau_1}\right) \tag{7-5}$$

where

$$R_1 = \frac{V_P}{I_P} \qquad \text{and} \qquad \tau_1 = \frac{R_1 R}{R_1 + R}\,C$$

The final voltage is $V(\text{point } 1)$ and the current i_R from Eq. 7-4 is

$$I_R(\text{point } 1) = \frac{V - V(\text{point } 1)}{R}$$

The time t_1 to reach $V(\text{point } 1)$ and $I_R(\text{point } 1)$ is found by solving for t_1 in Eq. 7-5 for $v = V(\text{point } 1)$. In this case a final static voltage (and current) is attained.

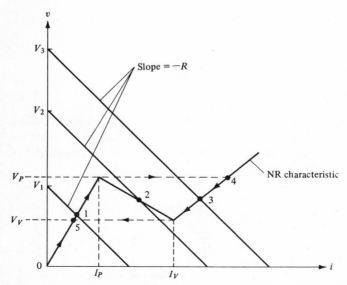

FIG. 7-7 A plot of load lines from Eq. (7-4) on the ideal NR characteristic for three applied voltages: V_1, V_2, and V_3.

(b) *Application of V_3 to the circuit of Fig. 7-5a.* Referring to Fig. 7-7, it appears that a final static condition should be attained at point 3. The operating path in attaining this point is somewhat surprising. The solution is the same as in Sec. (a) until $v = V_P$, the peak voltage. At this point, a discontinuity occurs in which v, which is also across capacitor C, would have to reduce abruptly in value if operation followed the NR characteristic. From circuit theory, we know that the voltage across a capacitor cannot change instantaneously without an impulse (which we do not have). Thus, rather than following the NR characteristic, the operating point jumps instantaneously to point 4. We can further expand on this phenomenon by considering Fig. 7-8 which shows the diode as a negative resistance and the region being considered lies between the voltage points marked as V_P and V_V in Fig. 7-7. From Fig. 7-7 we see that $R > R_n$ in this region. Converting all elements to the left of capacitor C into a Thévenin equivalent of voltage $[V_3(-R_n)/(R - R_n)]$ and a series resistance of $[-RR_n/(R - R_n)]$, the differential equation for the circuit becomes

$$\left[\frac{V_3(-R_n)}{R - R_n} - v\right]\frac{1}{(-RR_n)/(R - R_n)} = C\frac{dv}{dt}$$

or

$$\frac{dv}{dt} = \frac{V_3}{RC} + \frac{v(R - R_n)}{RR_nC} \tag{7-6}$$

From the conditions of Fig. 7-7, we know V_3/RC will always be larger than the second term in Eq. (7-6) because $V_3 > v$. Therefore, the equation states that dv/dt is *positive*. However, we are in the negative-resistance region, and the voltage should decrease if we follow the characteristic curve. To avoid the anomoly, the voltage remains constant at V_P across the diode-capacitor combination, but the current jumps to point 4 in Fig. 7-7 *instantaneously*. If we built the circuit, the operation would be essentially as explained, but the change would not be instantaneous because of the inherent series resistance and inductance. We see that the voltage across the capacitor has remained constant, but current i (and i_C) has increased considerably. The final operating path travels from point 4 to point 3 along the positive-resistance slope of the NR characteristic. The solution consists of two parts: (1) $0 \le v \le V_P$, and (2) V(point 3) $\le$

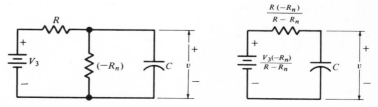

FIG. 7-8 The basic switching circuit converted to its Thévenin equivalent for the discussion of its operation in the negative resistance region.

$v \leq V_P = V$(point 4). Replacing V_1 by V_3, Eq. 7-5 yields the solution until $v = V_P$ at $t = t_2$. After t_2, $R_2 = R_{NR} = (V(\text{point } 4) - V_V)/(I(\text{point } 4) - I_V)$ and, $\tau_2 = CR_2R/(R + R_2)$. To solve for v when $t > t_2$, we must find a new equivalent circuit for the NR characteristic since, if the line segment between V_V and point 4 of Fig. 7-7 is extended to $i = 0$, the extended line does not necessarily pass through $v = 0$. The equation for this line segment is

$$v = R_2i + V_C \tag{7-7}$$

where

$$V_C = \text{constant}$$

Since the point (V_V, I_V) is on the line, the constant in Eq. (7-7) is

$$V_C = V_V - R_2I_V \tag{7-8}$$

With an initial voltage of V_P on C of Fig. 7-5 the new equivalent circuit for $t \geq t_2$ is that of Fig. 7-9. The general solution for v in this circuit is

$$v = [V_{\text{final}} + (V_{\text{initial}} - V_{\text{final}})]e^{-(t-t_2)/\tau_2} \tag{7-9}$$

where

$$V_{\text{final}} = \frac{(V_3 - V_C)}{(R + R_2)}(R_2) + V_C \qquad (C \text{ is charged to final value})$$

$$V_{\text{initial}} = V_P$$

The final static voltage is reached when $v = V$(point 3), Fig. 7-7, at $t = t_3$. The solution from Eq. (7-5) (with V_3 replacing V_1) and Eq. (7-9) is,

$$v = \left[\left(\frac{V_3 - V_C}{R + R_2}\right)R_2 + V_C\right] + \left\{V_P - \left[\frac{V_3 - V_C}{R + R_2}(R_2) + V_C\right]\right\}e^{-(t-t_2)/\tau_2} \quad t_2 \leq t \leq t_3 \tag{7-10}$$

Again, the solution is valid until $t = t_3$ at $v = V$(point 3). Equations (7-5) and (7-10) are more easily visualized by the sketch of Fig. 7-10. The dashed lines indicate the response if circuit conditions did not change.

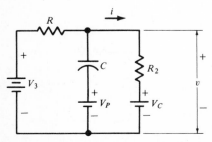

FIG. 7-9 The equivalent circuit of Fig. 7-5 for $t \geq t_2$ in Sec. 7-3.3b.

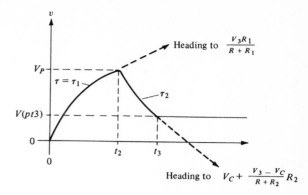

FIG. 7-10 The general plot of Eqs. (7-5) and (7-10).

Shorting Capacitor C in Fig. 7-5a returns the voltage v to zero. Removing the short, the cycle of Fig. 7-10 will repeat. This system operation is called *monostable* since there is one stable operating point at V(point 3) in Fig. 7-7.

(c) *Application of V_2 to the circuit of 7-5.* Referring again to Fig. 7-7, the operation begins as in section (b). The voltage v rises from zero to V_P. The operation jumps instantaneously to point 4 and starts down the characteristic from point 4 toward V_V. To this point, the solution is the same as in Sec. (b). The circuit tries to stabilize at a voltage corresponding to point 2 in the negative-resistance region, but we find that this point is never attained. When operation reaches voltage V_V, the characteristic discontinuity requires an instantaneous increase in V, but since the voltage across the capacitor cannot change instantaneously, the operation jumps immediately to point 5. The path of operation then repeats as above. From the previous solutions, we can sketch the results as in Fig. 7-11. Since the circuit continues to oscillate with the waveform of Fig. 7-11, it is said to be *astable* because there is no static voltage at which operation settles.

We have indicated the ideal path of operation in Fig. 7-7 when V_2 is applied. The actual path depends on the device and inductance which is inherent in the circuit wiring. The operating path may become oval in shape. A computer solution can display various paths rather easily.

7-3.3 BISTABLE OPERATION OF NR DEVICES

One type of operation was not shown in Fig. 7-7. If v_I and R in Fig. 7-5 are altered so that the plot of Eq. 7-4 and the NR characteristics are as shown in Fig. 7-12, another type of operation is possible.

Applying $v_I = V_4$ in Fig. 7-5, the voltage v responds as in Sec. (a) until the voltage at point 1, Fig. 7-12, is attained. If no further changes are made on the circuit, the static voltage $v = V$(point 1) will remain. Now, if the voltage across the capacitor is increased to V_P or beyond by external means, the path of operation will be as in Sec. 7-3.3(b) until point 3 is attained. If no further circuit changes are made, the circuit voltage v will remain at V(point 3). Finally, if the capacitor is shorted momentarily, the voltage v will return to V(point 1). Thus,

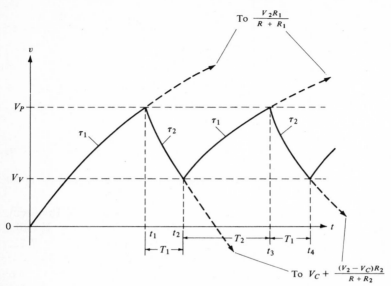

FIG. 7-11 Plot of voltage v for the circuit of Fig. 7-5 when V_2 is applied as in Fig. 7-7. Time constants are the same as in Fig. 7-10.

there are two stable points at which the voltage v may terminate, and the operation is said to be *bistable*.

Comparing Figs. 7-7 and 7-12, we see the conditions that must exist between the circuit load line and the NR characteristic in order to obtain monostable, astable, or bistable conditions.

7-4 VOLTAGE-TO-FREQUENCY CONVERTER

The output of the dc voltage-to-frequency converter in Fig. 7-1 has the appearance of the waveform of Fig. 7-11. Thus, we can use an NR device with a load adjusted for astable operation to construct a V/F converter. The NR device characteristics of Fig. 7-4 will be applied to a circuit similar to Fig. 7-5a.

Consider Fig. 7-13. Resistors R_{2B} and R_3 have been added as compared to Fig. 7-5a. These resistors do not affect circuit operation until the UJT conducts current. We wish to investigate the voltage waveforms at nodes B, C, and D.

The circuit of Fig. 7-13 is to be designed to operate in the astable mode. The load line for the emitter circuit of the UJT must be adjusted to intercept the NR characteristics as in Fig. 7-7 with applied $V_I = V_2$. If $V_{BB} = 10$ V, the input voltage V_I will have to be larger than 6.7 V to enter the NR region according to Fig. 7-4. Assume that V_I is about twice this voltage, or $V_I = 14$ V. For this value of V_I, the minimum value of R may be found by drawing a line between $V_E = 14$ V and the valley point at $I_E = 8$ mA on Fig. 7-4. From Fig. 7-4 this value is $R(\text{min}) \simeq 14/0.01 = 1{,}400$ Ω. So that the intersection of the load line with the

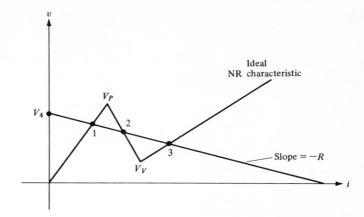

FIG. 7-12 Conditions for bistable operation.

negative-resistance characteristic will occur at a low I_E value, we arbitrarily choose $R = 10^5 \ \Omega$. (Note that R and C control period T_2 in Fig. 7-11). The voltage waveform at point C in Fig. 7-13 will be of the form shown in Fig. 7-11. The time constant τ_1 can be determined and the resulting time intervals calculated from the characteristics of Fig. 7-4 and a selected value of C. We select C to be 1,000 pF for our example.

The time constant τ_1 in Fig. 7-11 is

$$\tau_1 = \frac{RR_1}{R + R_1} C$$

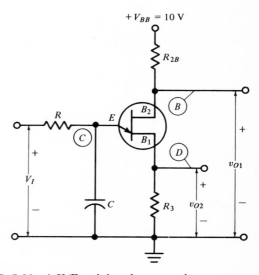

FIG. 7-13 A V/F and dc voltage-to-pulse converter.

where $R_1 = V_P/I_P$. From Fig. 7-4 R_1 is very large $(R_1 > 1\ \text{M}\Omega)$ so that

$$\tau_1 \simeq RC = 10^5(10^{-9}) = 100\ \mu\text{s}$$

The value of τ_2 and T_1 during which C discharges is difficult to determine. An approximate value for τ_2 and T_1 may be determined as follows: After switching at the peak voltage V_P the emitter current becomes large and then decreases along a line having the slope of the $I_B = 0$ curve of Fig. 7-4. The path of operation is followed by arrows in Fig. 7-14. From Figure 7-4, the slope of the $I_B = 0$ curve in the linear region is $R_2 \simeq \Delta v_E/\Delta i_E = 0.4/12(10^{-3}) = 333\ \Omega$. This will be the same value for R_2 in τ_2 since the $V_{BB} = 10$ V and $I_B = 0$ curves tend to be parallel. Thus,

$$\tau_2 = \frac{RR_2}{R + R_2}C = \frac{10^5(333)}{100,333}(10^{-9}) \simeq 0.33\ \mu\text{s}$$

From Eq. (7-8), $V_C = V_V - R_2I_V = 2.5 - (333)(8)(10^{-3}) = 2.5 - 2.67 = -0.17$ V. As indicated in Fig. 7-11, the response is heading to $V_F = V_C + (V_I - V_C)R_2/(R + R_2) = -0.17 + (14 + 0.17)(333)/100,333 = -0.17 + 0.047 = -0.123$ V. Thus the solution from Fig. 7-11 is

$$v = V_F - (V_F - V_P)e^{-(t-t_1)/\tau_2} \tag{7-11}$$

The time T can be found from Eq. (7-11) when $t - t_1 = T_1$ and $v = V_V$. We find

$$2.5 = -0.123 - [-0.123 - 6.7]e^{-T_1/0.33\mu\text{s}}$$

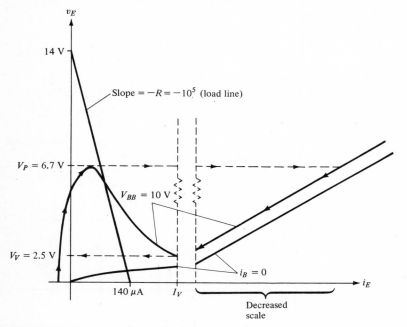

FIG. 7-14   The path of operation for the circuit of Fig. 7-13.

or

$$T_1 = 0.318 \ \mu\text{s}$$

The value of T_2 is found from Fig. 7-11 for $V_I = V_2$, $R_1 \gg R$. We find

$$v = V_I - (V_I - V_V)e^{-(t-t_2)/\tau_1} \tag{7-12}$$

To find T, when $v = V_P$,

$$v = V_P = V_I - (V_I - V_V)e^{-T_2/\tau_1} \tag{7-13}$$

or

$$6.7 = 14 - (14 - 2)e^{-T_2/(10^{-4})}$$

yielding $T_2 = 50 \ \mu\text{s}$. The resulting steady-state waveform is shown in Fig. 7-15a.

Figures 7-15b and c show the voltage waveforms at bases $B1$ and $B2$. We see that these voltages are short pulses such as desired from the gated pulse generator in row D of Fig. 7-1. The amplitude of the pulses varies with the device and external circuit. Manufacturer data sheets should be consulted since much of the information calculated above is given in curve form on these data sheets.

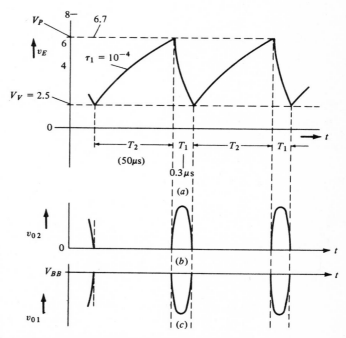

FIG. 7-15 Waveforms developed in the circuit of Fig. 7-13 with $R = 10^5$, $C = 10^{-9}$ F. (a) Voltage at point C; (b) voltage at point D; (c) voltage at point B.

From Eq. (7-13), it will be found that T_2, or the frequency, changes with V_I, the input voltage. Thus, if V_I is proportional to the input voltage in our digital meter, the input voltage has been converted to an oscillating waveform with frequency proportional to V_I. The V/F converter of Fig. 7-13 yields pulse output at base $B1$ so that short pulses are also available which have a frequency proportional to V_I.

7-4.1 LINEARIZING A V/F CONVERTER

The voltage waveform for charging the capacitor in Fig. 7-13 is not linear as shown during period T_2 in Fig. 7-15a. The waveform can be made more linear if the capacitor charging current is nearly constant.

Consider the circuit of Fig. 7-16a. The constant-current generator is charging capacitor C. Since $i_c = C \, dv_c/dt$,

$$v_c = \frac{1}{C} \int_0^t i_c \, dt = \frac{1}{C} \int_0^t I_o \, dt = \frac{I_o}{C} t \tag{7-14}$$

Thus v_c rises linearly with time if a constant-current source charges a capacitor.

A transistor can be applied to approximate a constant-current source and the charging circuit of Fig. 7-16b results. In this circuit, the emitter current I_E is a constant, and the collector current is $I_C \simeq -I_E$. Thus, to the extent that the transistor is nearly ideal, the capacitor charging current is linear. The analysis of the circuit is of interest.

With the switch SW closed in Figure 7-16b, $v_o = -V_{CC}$ and

$$I_E = \frac{V_{EE} - V_{EB}}{R_E} \simeq -I_C \tag{7-15}$$

When the switch opens, I_C remains the same as given by Eq. (7-15), but it now charges the capacitor. The equivalent circuit and output-voltage waveform will

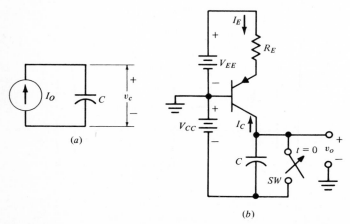

FIG. 7-16 (a) Linear charging circuit; (b) transistor approximation to (a).

be as shown in Fig. 7-17a and b. The charging voltage of C is linear with time
until the cutin voltage of the collector-base junction is attained. Thus,

$$v_o = -V_{CC} + \frac{I_E}{C}t \qquad 0 \le t \le t_1 \tag{7-16}$$

The CB cutin voltage will be approximately 0.5 V and the forward-biased CB
voltage will be about 0.7 V. After t_1 when CB is forward-biased, the equivalent
circuit is as shown in Fig. 7-17c. The solution of v_o for this latter circuit is straightforward.

The linear charging of the capacitor as indicated here may be applied to the
V/F converter of Fig. 7-14.

Before continuing the discussion of our digital meter circuits, two basic
linear-voltage time-base generators encountered in practice will be presented.
These two generators are described generically as the Miller and Bootstrap
sweep circuits.

7-5 MILLER SWEEP

An integrator applied to a linear-voltage sweep generator is designated as a
Miller sweep circuit. The circuit representation is drawn in Fig. 7-18. The input

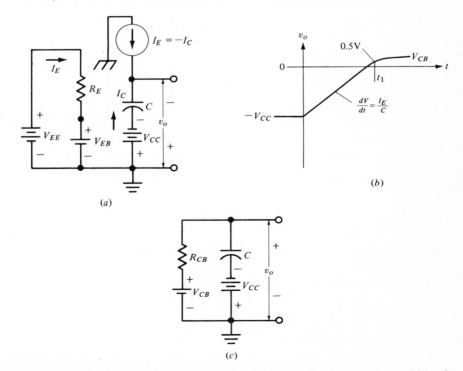

FIG. 7-17 (a) Equivalent circuit of Fig. 7-16(b) until CB junction becomes forward-biased;
(b) output waveform of (a); (c) equivalent circuit of Fig. 7-16(b) when CB is forward-biased.

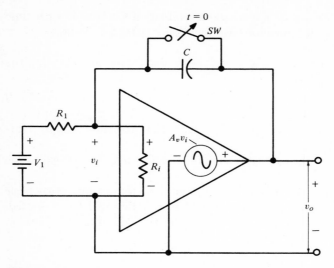

FIG. 7-18   The Miller integrator sweep circuit.

circuit of this figure has been converted to its Thévenin equivalent in Fig. 7-19. We wish to develop an equivalent circuit which is simpler and displays the essential characteristics of the Miller sweep generator. Assume that the output resistance of the op-amp is negligible to further simplify our task.

Before the switch opens at $t = 0$, from Eq. (2-3), $v_o = (-R_F/R_1) = 0$, since $R_F = 0$ and the input v_i is driven to zero in Fig. 7-19. After the switch

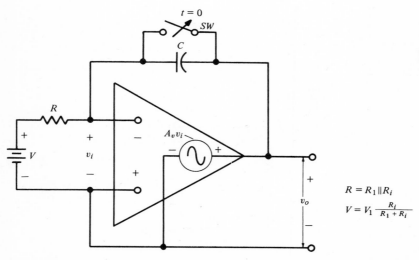

$$R = R_1 \| R_i$$
$$V = V_1 \frac{R_i}{R_1 + R_i}$$

FIG. 7-19   The Miller sweep circuit of Fig. 7-18 with the input converted into its Thévenin equivalent. The switch opens at $t = 0$.

opens, we apply the Miller approximations of Fig. 5-31b with $K = A_v$ to obtain Fig. 7-20. The input shunting capacitance is given by $C\,(1 - K) = C\,(1 - A_v)$, and the output shunting capacitance is given by $KC/(K - 1) \simeq C$.

At $t = 0^+$, the voltage across capacitor C is zero since it cannot change from the value it had at $t = 0^-$ with switch SW closed. Thus, $v_i(0^+) = v_o(0^+) = 0$. The expression for v_i becomes

$$v_i = V(1 - e^{-t/\tau_1}) \tag{7-17}$$

where $\tau_1 = (1 - A_v)RC$. We see that the effective input time constant is $R \times C$ multiplied by the voltage gain of the amplifier and will normally be large. The output voltage is

$$v_o = A_v v_i(1 - e^{-t/\tau_2}) \tag{7-18}$$

where $\tau_2 \simeq R_o C$. The value of τ_2 will be quite small compared to τ_1 so that its effect will die out rapidly. The output voltage will be nearly linear for $t \ll \tau_1$.

The percentage of linearity can be specified. From the series expansion of the exponential (neglecting τ_2 effect), Eqs. (7-17) and (7-18) become,

$$v_o \simeq A_v v_i = A_v V\left(\frac{t}{\tau_1} - \frac{t^2}{2\tau_1^2}\right) = \frac{A_v V t}{\tau_1}\left(1 - \frac{t}{2\tau_1}\right) \tag{7-19}$$

and at the end of the sweep period,

$$v_o(T) = \frac{A_v V T}{\tau_1}\left(1 - \frac{T}{2\tau_1}\right) \tag{7-20}$$

where $T =$ time at end of sweep period. The second term in Eq. (7-20) represents deviation from linearity. If a nonlinearity of 1 percent is desired, then from Eq. (7-20), $T/2\tau_1 = .01$. For a specified period T the necessary minimum time constant can be determined from $\tau_1 = (1 - A_v)RC$. Since $(-A_v)$ can be quite large in the case of an operational amplifier, large values of τ_1 may be obtained and relatively long period sweep voltages can be generated.

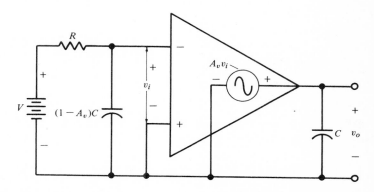

FIG. 7-20   The circuit of Fig. 7-19 when $R_1 \gg R_0$, and $-A_v \gg 1$ for $t > 0$.

Example 7-1   For the circuit of Fig. 7-21a the op-amp has a minimum voltage gain of $(-5,000)$, a very large input impedance Z_i, an output voltage limit of ± 10 V, and the switch is open for 100 ms. Sketch and label the output waveform. Determine the percent of deviation from linearity.

At $t = 0^-$, $v_o = 0$ and at $t = 0^+$, $v_o = 0$, since the capacitor has zero charge. From $\tau_1 = (1 - A_v)RC$,

$$\tau_1(\text{min}) \cong (-A_v)RC = (5,000)(10^4)(10^{-7}) = 5 \text{ s}$$

Applying Eq. (7-20),

$$v_o(0.1) = (-5,000)(5)(0.1)\frac{1 - 0.1/(2)(5)}{5}$$

$$v_o(0.1) \cong -500 \text{ V} \tag{7-21}$$

However, the output voltage can only reach -10 V due to op-amp limits. Again applying Eq. (7-20), the time t_1 to reach -10 V is approximately

$$t_1 = \frac{V_{O}\tau_1}{A_v V} = \frac{V_O A_v RC}{A_v V} = \frac{V_O}{V}RC = \frac{(10)(10^{-4})}{5} = 2 \text{ ms} \tag{7-22}$$

The waveform will be indicated in Fig. 7-21b. For the duration of the sweep, the nonlinearity is within $T/2\tau_1 \times 100 = .002/2(5) \times 100 = 0.02$ percent. A longer sweep time can be obtained by increasing R and C or by decreasing V. From Eq. (7-22), the sweep period is given by $(V_O/V)RC$.

7-6   BOOTSTRAP SWEEP

Another approach to generating a linear sweep voltage can be seen from Fig. 7-22a. We note that

$$i = \frac{V - v_c}{R} \tag{7-23}$$

Thus, if the effect of the nonlinearity in v_c can be removed, the charging current will be constant at $I = V/R$. If V could be replaced by $V + v_c$, then the desired constant current would result. This can be done by amplifying v_c with unity gain and introducing the output in series with V. From Fig. 7-22b, $t \geq 0$,

$$i_c = \frac{V - v_c + A_v v_c}{R} = \frac{V - (1 - A_v)v_c}{R} \tag{7-24}$$

assuming the amplifier input impedance to be very large. From Eq. (7-24) we see that $A_u = 1$ yields a constant charging current and $v_o = A_v I_c t/C$ or

$$v_o = \frac{V}{RC} t \tag{7-25}$$

We note that the input of Fig. 7-22b is lifted (or raised) by its own output or by its "bootstraps," leading to the circuit name.

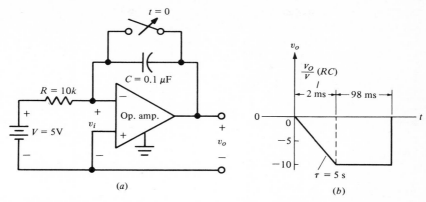

FIG. 7-21 (a) The circuit for Example 7-1; (b) the output of (a) if the switch is open for 100 ms.

Bootstrap Example 7-2 The operational amplifier of Example 7-1 is to be applied to the Bootstrap circuit of Fig. 7-23. Determine R_1 and R_2 for linear voltage sweep and find the time t_1 in Fig. 7-23b for the output to attain the saturation voltage of the amplifier, 10 V.

Solution In our discussion of operational amplifiers, we found that the gain of the circuit in Fig. 7-23 was given by Eq. (2-4) as $|A| = 1 + R_2/R_1$. Since the overall gain should be unity, from Eq. (2-4) and Example 2-2, $R_2 = 0$ and $R_1 \to \infty$, or open circuit. The output voltage is given by Eq. (7-25) as

$$v_o = \frac{10}{10^4(10^{-7})} t = 10^4 t$$

Thus, the time for v_o to reach 10 V is,

$$t_1 = \frac{10}{10^4} = 10^{-3} \text{ s} = 1 \text{ ms}$$

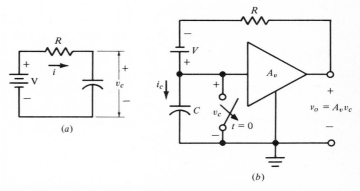

FIG. 7-22 (a) General capacitor-charging circuit; (b) basic bootstrap circuit for removing the effect of the nonlinear capacitor-charging voltage.

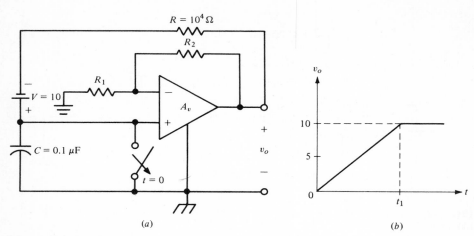

FIG. 7-23 (*a*) Bootstrap circuit; (*b*) output response of (*a*).

7-7 SWITCHING SWEEP CIRCUITS

In the preceding sweep circuits, a knife switch was indicated for short-circuiting the capacitor before $t = 0$ or at some desired time. It is desirable to control the switching electronically so that the sweep voltages may be controlled from some reference information or signal. This switching may be accomplished by use of transistors or IC gates (see Chap. 8).

We know that a bipolar transistor (or FET) has very low resistance between collector and emitter (or drain and source) when operated in the saturation region (or ohmic region). Thus, a transistor switch appears feasible. Switching times may not be negligible, however, as we shall indicate by an example.

Example 7-3 A bipolar transistor's collector characteristics and a circuit for switching across a capacitor is shown in Fig. 7-24*a* and *b*. The collector characteristics have been linearized, and the transistor is made of silicon so that the base-emitter voltage is 0.7 V when forward-biased. The input is a step voltage of 6.7 V value, and the circuit is in a steady-state condition prior to $t = 0$. Sketch and label the capacitor voltage v_c as a function of time.

Solution Prior to $t = 0$, no base or collector current flows in the transistor ($v_i = 0$). The capacitor C is charged to 10 V, having been charged through R_C. We shall neglect the transistor internal capacitance assuming negligible rise and fall times due to these elements. At $t = 0^+$,

$$i_B = \frac{v_i - V_{BE}}{R_B} = \frac{6.7 - 0.7}{10^5} = 60 \ \mu A$$

The base current will remain constant for all time that $t \geqq 0$. The voltage across C cannot change instantaneously so that operation jumps to point A in Fig. 7-24*a* in zero time. With i_B constant at 60 μA, operation continues from point A to point B where the transistor becomes saturated. By writing an equation for the line segment $A - B$, we can obtain an equivalent circuit for the transistor and find the time necessary to traverse this part of the operating path. The extension of line $A - B$ intersects i_C at 12 mA. Thus, the equation for this line is:

$$i_C = \frac{\Delta i_C}{\Delta v_{CE}} v_{CE} + I_{C1}$$

where I_{C1} = intercept at $v_{CE} = 0$. The slope of the $i_B = 60 \ \mu A$ line $(A - B)$ is

$$\frac{\Delta i_C}{\Delta v_{CE}} \simeq \frac{0.001}{10} = 10^{-4}$$

Thus, $i_C = 10^{-4} v_{CE} + 0.012$. The equivalent circuit of Fig. 7-24b during the

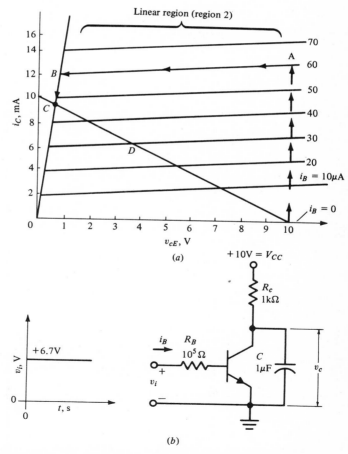

FIG. 7-24   (a) Transistor-linearized collector characteristics; (b) switching circuit.

time to travel from A to B is then as drawn in Fig. 7-25a. A Thévenin equivalent of all circuit elements to the left of the capacitor is given in Fig. 7-25b. From Fig. 7-25b, we may write

$$v_c = -1.8 + 11.8e^{-t/\tau_1}$$

where $\tau_1 = (910)(10^{-6}) = 910 \ \mu s$
The time necessary to attain point B where $v_{CE} = 0.7$ V is then $0.7 = -1.8 + 11.8e^{-t_1/\tau_1}$ or $t_1 = 1410 \ \mu s$. The final part of the operating path is from point B to point C in Fig. 7-24a. Operation stops at point C because the dc load-line and $i_B = 60\mu A$ characteristic curve intersect at this point. The equation for the saturation line from B to C is

$$i_C \cong \frac{v_{CE}}{55}$$

and the equivalent circuit for $t \geq t_1$ is that of Fig. 7-26a. Again, a Thévenin source has been applied to replace all elements to the left of the capacitor resulting in the circuit of Fig. 7-26b. From the latter figure,

$$v_c = 0.7e^{-t/\tau_2}$$

where $\tau_2 = 52.6(10^{-6}) = 52.6 \ \mu s$. The time required to follow the path from point B to point C in Fig. 7-24a is found when $v_o = 0.5$ V. Thus, $0.5 = 0.7e^{t_2/52.6 \ \mu s}$ or $t_2 = 52.6(10^{-6})(0.337) = 17.7 \ \mu s$. A sketch of v_c is plotted in Fig. 7-27. The voltage v_c is nearly at the final value at $t = t_1$. We see that a finite and significant time may be consumed in switching C to a low voltage.

We note that if $v_i = 3.7$ V in Fig. 7-24b, the final operating point would be at point D in Fig. 7-24a. Here, the voltage would not be low enough if we are trying to get near zero volts across capacitor C. We also see that the switching time would be decreased if the slope in region 2 of Fig. 7-24 were increased. The series resistance of Fig. 7-25b would be decreased so that τ_1 and thus t_1 would be smaller.

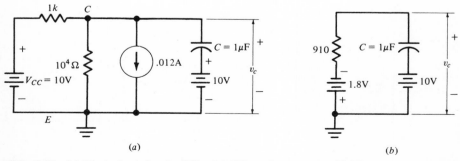

FIG. 7-25 (a) Equivalent circuit of Fig. 7-24(b) until saturation; (b) Thévenin equivalent of (a). The 10-V battery in series with C is the initial condition on C after $t = 0$.

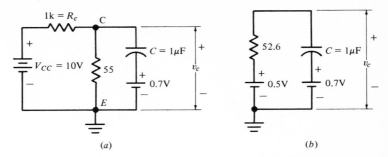

FIG. 7-26   (a) Equivalent circuit for the circuit of Fig. 7-24(b) for operation from point B to point C in Fig. 7-24(a); (b) Thévenin equivalent of (a). The 0.7-V battery in series with C is the initial value for $t > t_1$.

7-8   A V/F CIRCUIT

A gated voltage-to-frequency circuit is drawn in Fig. 7-28. The source for the gating voltage at node B will be considered later. Transistor Q_1 operates as an emitter follower for high input impedance. The current gain β of both Q_1 and Q_2 is greater than 300. Note that when V_{in} increases, the voltage from base to ground and emitter to ground in Q_1 decreases, leading to increased forward bias on Q_2 and, thus, increased collector current in Q_2. We see that Q_2 is acting as a controllable current source charging capacitor C_1. The voltage across C_1 controls the breakdown of UJT Q_5. Thus, Q_2 acts as a linear charging device similar to that of Fig. 7-16b. However, the charging current into C_1 and the voltage across C_1 is linearly proportional to the input voltage. When the voltage across C_1 reaches V_P, the peak voltage—Eq. (7-1)—of the UJT, an output pulse occurs, and C_1 is discharged. The number of pulses generated will be linearly related to

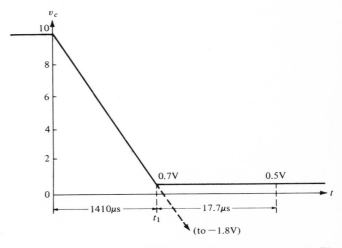

FIG. 7-27   Plot of v_c for Fig. 7-24(b).

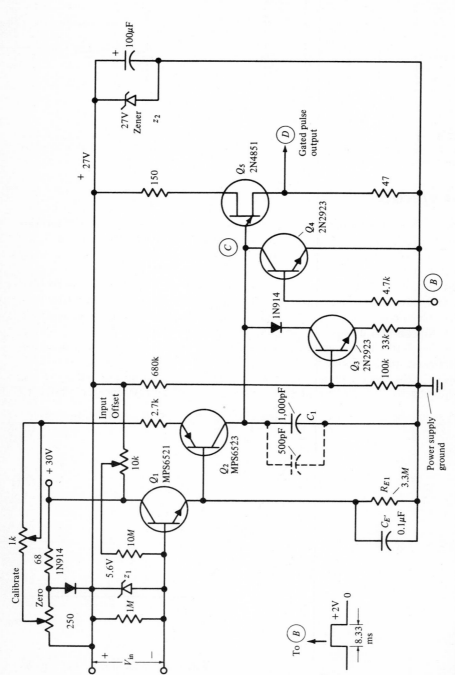

FIG. 7-28 A gated voltage-to-frequency converter. Zener diode z_1 is for input protection. (*Courtesy Popular Electronics, September, 1970.* *Ziff-Davis Publishing Co.*)

the input voltage during a specified time interval. Transistor Q_3 is biased to draw a constant current of about 100 μA from Q_2 to improve the linearity of the charging circuit when the base-emitter voltage is low. Transistor Q_4 is a switching transistor which turns on and off (or gates) the oscillator circuit. This transistor keeps the UJT from firing when on, and the timing wave at node B determines the counting time interval of 8.33 ms in this case.

The circuit of Fig. 7-28 is designed for an input voltage of $0 \rightarrow 2$ V. Switching resistors in the input circuit (and decimal point in the display) will allow input voltages of larger value (20, 200). The output count must be from zero to 200 within the time interval of 8.33 ms (for three significant figures). Thus the oscillator maximum frequency must be $200/8.33 \times 10^3 = 24$ kHz (kilohertz). The calibrating control, which is on the panel of the instrument, is adjusted to set this latter frequency. Capacitor C_1 is adjusted so that the calibrating control is near midrange for source reference voltage. The zero control, also on the panel, sets the frequency at node C so that no pulses occur during the 8.33-ms time interval. The counting and display circuitry (not shown) is used to set the input offset control. With the input shorted or open-circuited, an unchanging reading of 0.01 on the display should be obtained by adjusting the input offset control. The 0.00 output reading is set by the zero panel control. A reference diode having a voltage near midrange can be added to set and check the calibration control.

It will be necessary to become familiar with various gates and logic circuits before we can show the timing and counting circuits. These topics will be presented in the next chapter.

7-8.1   DESIGN OF V/F CONVERTER

Transistor Q_1:   By choosing the emitter resistance to be 3.3 MΩ, the input resistance to the base of Q_1 only is $R_{in} \simeq R\beta_{E1} = 300(3.3)(10^6) = 990$ MΩ. The 10-MΩ resistor from the input offset control also shunts R_{in}. The total shunting resistance at the input is 1 M//10 M//990 M $\simeq$ 1 MΩ. Capacitor C_{E1} smooths out any rapid changes in input.

Transistor Q_3:   The transistor β is specified as being in the range of 100 to 200. Neglecting base current, the base-to-ground voltage is

$$V_{BG3} = \frac{27(100)}{780} = 3.46 \text{ V}$$

The voltage from emitter to ground is:

$$V_{EG3} = V_{BG3} - V_{BE3} = 3.46 - 0.7 = 2.76 \text{ V}$$

Thus, the emitter current which is nearly the same as the collector current is

$$I_{E3} = \frac{V_{EG3}}{33 \ k} = \frac{2.76}{33 \ k} = 84 \ \mu\text{A}$$

The collector current $I_{C3} \simeq I_{E3}$ passes through Q_2 maintaining a minimum current so as to eliminate some nonlinearity that occurs at low collector currents.

Transistor Q_5:   The intrinsic standoff ratio of the 2N4851 is between 0.55 and 0.82. Applying a value of $\eta = 0.7$, the emitter voltage changes state at

$$V_P \simeq \eta V_{BB} + 0.6 = 0.7(27) + 0.6 = 19.5 \text{ V}$$

This value of V_P can lie between 17 and 22 V, depending on the particular UJT so that some external adjustment is necessary. The adjustment is made by trimming C_1 and varying the calibration control. The V_{EB1sat} value is 4 V. Thus, V_{EG} of the UJT varies between about 4 V and 20 V. The resistance of 47 and 150 Ω connected to $B1$ and $B2$ of Q_5 are chosen so that the pulse output voltage is at least 2 V to drive the counting circuit.

Transistor Q_2:   Fig. 7-29 shows the capacitor charging circuit in simplified form. Note that Fig. 7-29 is similar to Fig. 7-16*b*. From Eq. (7-16),

$$\Delta v_o(t) = \frac{I_E}{C} t$$

The maximum frequency for $V_{\text{in}} = 2$ V must be at least 24 kHz (see Sec. 7-8). Thus,

$$I_E \simeq \frac{V_{\text{in}} - V_{BE}}{R_E} \simeq \frac{2 - 0.7}{3 \ k} = 0.425 \text{ mA}$$

assuming R_E set to 3 kΩ. Not all of this current charges C_1, due to transistor Q_3 (84 µA). Thus, the charging current is

$$I_c = I_E - I_{C3} = 0.425 - 0.084 = 0.341 \text{ mA}$$

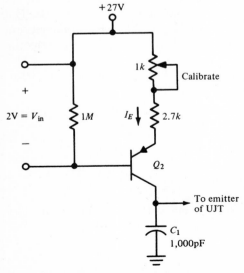

FIG. 7-29 Simplified circuit of the capacitor-charging transistor section of Fig. 7-28. The emitter-follower and zero control have been omitted.

To find the approximate capacitance value,

$$C_1 = \frac{I_c T}{\Delta v_o(t)}$$

The $\Delta v_o(t)$ is $V_{Emax} - V_{Esat} = 20 - 4 = 16$ V to the emitter of the UJT Q_5 and $T = 1/24{,}000$. Thus,

$$C_1 = \frac{0.341(10^{-3}) \left(\dfrac{1}{24{,}000}\right)}{16}$$

$$= \frac{0.341(10^{-3})(4.17)(10^{-5})}{16}$$

$$= 890 \times 10^{-12} = 890 \text{ pF}$$

This value of C_1 indicates the approximate value needed, but some adjustment will be necessary. The design is sufficiently complete at this point for construction of the circuit. It is only necessary that the base resistor of Q_4 be such that Q_4 can be driven into saturation. Since the maximum collector current of Q_2 is approximately 0.4 mA, Q_4 must be saturated with this current. Thus,

$$I_{B4sat} = \frac{0.4}{\beta_4} = \frac{0.4}{90} = 45 \ \mu\text{A} = \frac{V(\text{point } B) - 0.7}{4.7 \text{ k}}$$

or $V(\text{point } B) \simeq 0.9$ V. Since this calculation brings Q_4 to the edge of saturation, a blanking pulse at point B of 2 V should be adequate.

The design of the digital voltmeter will be completed in Chap. 8.

REFERENCES

1. Lancaster, D., Assembling the Popular Electronics Mini-DVM, *Popular Electronics,* September, 1970.
2. Millman, J. and H. Taub, "Pulse, Digital and Switching Waveforms," McGraw-Hill Book Company, New York, 1965.
3. Pettit, J. M., "Electronic Switching, Timing, and Pulse Circuits," McGraw-Hill Book Company, New York, 1959.
4. General Electric Company, "SCR Manual," 3d ed., Syracuse, N.Y., 1964.

PROBLEMS

7-1. Refer to the circuit and device characteristics of Fig. 7-5. For the characteristics:

$V_P \qquad = 10$ V at $I_P = 10$ mA

$V_V \qquad = 2$ V at $I_V = 10$ mA

$V_{Esat} = 3$ V at $I_E = 50$ mA

(a) If $R = 1$ k and $v_I = 5V$ is applied, determine the steady-state voltage across the NR device v.

(b) If $R = 1$ k and $v_I = 20V$, repeat a.

(c) If $R = 200\Omega$ and $v_I = 6V$, repeat a. Are two stable points possible?

(d) If $R = 10^4, C = 100$ pF, and $v_I = 20V$, find v across the NR device; sketch and label. The initial charge on C is zero.

7-2. Assume operation of the circuit in Fig. 7-13 follows the arrows in Fig. 7-14. Ideally the base voltage at $B1$ would be as in Fig. P7-2. If $R_3 = 100$, estimate V_{B1P} and V_{B1M}, assuming the device to have the characteristics of Prob. 7-1. Also estimate T_1.

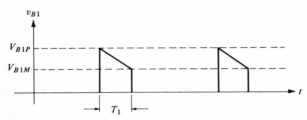

FIG. P7-2   The waveform for Prob. 7-2.

7-3. In the circuit of Fig. 7-16, $V_{EE} = V_{CC} = 10$ V, $R_E = 9.3$ k, and $C = 1$ μF. The forward-biased emitter-base junction can be considered constant at 0.7 V. Sketch and label v_o versus t if the switch opens at $t = 0$.

7-4. Design a circuit similar to Fig. 7-16 to develop the output voltage waveform of Fig. P7-4. Assume $|V_{EE}| = |V_{CC}|$.

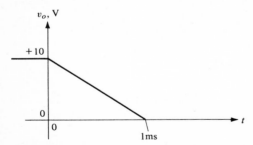

FIG. P7-4   The waveform for Prob. 7-4.

7-5. Consider the circuit of Fig. 7-21, and the op-amp of Example 7-1. The input voltage V is changed to that of Fig. P7-5. Sketch and label $v_o(t)$ if (a) $T = 2$ ms, (b) $T = 3$ ms, (c) $T = 1$ ms. Capacitor C is uncharged at $t = 0$.

7-6. Repeat Prob. 7-5 for the circuit of Fig. 7-23 for $R_2 = 0$, R_1 infinite, and where V is replaced by the square wave of Prob. 7-5.

7-7. Determine V, R, and C in the circuit of Fig. P7-7 for astable operation at a frequency of 1,000 Hz. Neglect the time necessary to discharge the capacitor assuming it to be small compared to the charging time.

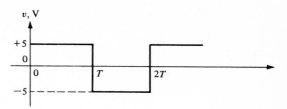

FIG. P7-5   The waveform for Prob. 7-5.

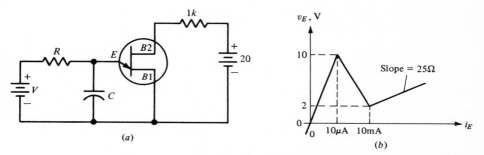

FIG. P7-7   The circuit and UJT characteristics for Prob. 7-7.

7-8. The linearized UJT characteristics shown in Fig. P7-8 include the effect of the 100-Ω resistor in the base 1 circuit. Sketch and label numerically v_E versus t when steady-state operation is achieved. Include time constants and time periods in your sketch.

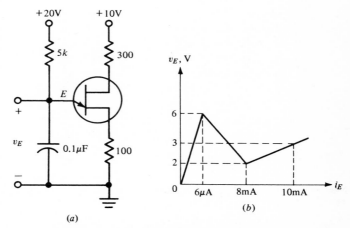

FIG. P7-8   The circuit and UJT characteristics for Prob. 7-8.

7-9. For Example 7-3 and Fig. 7-24, assume steady-state conditions have been reached (point C). Sketch and label $v_e(t)$ if (a) v_i is reduced abruptly to 1.7 V; (b) if v_i is reduced abruptly to zero.

8

Basic Logic Circuits and Applications

INTRODUCTION

Most logic elements operate on a binary or two-state voltage basis. The two states are not perfectly defined but lie within specified limits. Further, the two states are defined as being in the "1" or "0" level depending upon the logic system employed. Figure 8-1 illustrates the two systems in binary logic. Either positive or negative logic or combinations of the two may be used to implement a logic function. The voltage ranges which are defined as the "0" or "1" levels depend upon the type of logic elements applied.

8-1 LOGIC SYMBOLS AND DEFINITIONS

We shall consider positive logic in the following symbols and definitions unless otherwise specified:

AND   The AND gate has two or more input terminals and one output terminal. If all inputs are in the "1" state, the output is in the "1" state. If any one input is in the "0" state, the output is in the "0" state. Figure 8-2 shows the symbol for a two input AND gate and the logic or truth table for this gate. Note that the equation $C = A \cdot B = AB$ is read C equals A and B with the meaning as defined above. Steady-state (or dc) voltage conditions are assumed.

OR   The output of an OR gate is in the "1" state if one or more of the inputs is in the "1" state. Figure 8-3 indicates the logic symbol and truth table for a two input OR gate. The logic equation $C = A + B$ is read as C equals A or B.

NAND   The output of a NAND gate is in the "1" state if any input is in the "0" state (see Fig. 8-4). Stated another way, the output is in the "0" state if all inputs are in the "1" state. Compared to the AND gate, the output is inverted or "negated" in the NAND gate. The symbol of Fig. 8-4 is that for an AND gate with a small circle added on the output. This small circle will always mean negation or inversion. The logic equation $C = \overline{AB}$ is read as C equals not A and B. The overbar in the logic equation indicates negation.

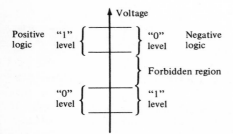

FIG. 8-1 Logic-level definitions.

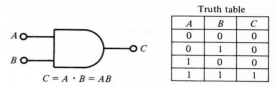

FIG. 8-2   The AND gate with two input terminals and the corresponding truth table.

NOR   The output of a NOR gate is in the "0" state if one or more of the inputs is in the "1" state. Compared to the OR gate, negation of the OR operation is indicated for the NOR gate (see Fig. 8-5). The logic equation $C = \overline{A + B}$ is read as C equals not A or B. Again the symbol is that of the OR gate with the negation symbol on the output.

8-2   WIRED OR DOTTED LOGIC GATES

Certain gates are provided with an "open" collector output to permit the wired-AND (called a wired-OR) function. This is achieved by connecting open collector outputs together and adding an external pullup resistor, as in Fig. 8-6b. The logic equation is

$$C = \overline{A + B} = \bar{A}\bar{B} \tag{8-1}$$

where the second expression is read as C equals not A and not B. Equation (8-1) can be verified from the truth table of Fig. 8-5 and also from the truth table of Fig. 8-2 where each A and B logic level is negated; i.e., each "1" is replaced by "0", and each "0" is replaced by "1" for A and B only. The last two expressions in Eq. 8-1 form an equality known as DeMorgan's theorem. This theorem is useful for converting a circuit with NOR gates to AND gates and vice versa.

An understanding of the basic gates permits us to apply more complex devices since the more complex devices have inputs and outputs identical to the basic gate.

We shall consider commonly employed gates of simple construction. Gate improvements and modifications are continually being made and manufacturers' literature will note these changes. If we understand the characteristics of simple gates, it is not too difficult to follow the changes which a manufacturer introduces.

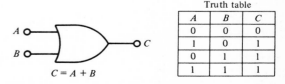

FIG. 8-3   The OR gate symbol, logic equation, and truth table.

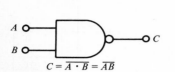

A	B	C
0	0	1
1	0	1
0	1	1
1	1	0

$C = \overline{A \cdot B} = \overline{AB}$

FIG. 8-4   The NAND gate symbol, logic equation, and truth table.

8-3   RESISTOR-TRANSISTOR LOGIC (RTL)

The basic RTL NOR gate in integrated circuit form and its transfer function are shown in Fig. 8-7. With all inputs at ground ($V_A = V_B = V_C = 0$), the transistors' base-emitter junctions are not forward-biased, and the transistors are off. Thus, the output voltage, $V_Y = 3$ V. If any one of the inputs, say V_A, reaches the transistor cutin voltage of 0.5 V, collector current begins to flow and V_Y decreases. For a transistor $\beta = 25$ and $V_{CEsat} = 0.1$, the transistor becomes saturated when the base current is

$$I_{Bsat} = \frac{I_{Csat}}{\beta} = \frac{V_{CC} - V_{CEsat}}{600(\beta)}$$

$$= \frac{3 - 0.1}{(600)(25)} \approx 0.2 \text{ mA}$$

$$V_{Asat} = I_{Bsat}(400) + V_{BEsat}$$

$$= (0.2)(0.4) + 0.7 \cong 0.8 \text{ V}$$

The output voltage is $V_{CEsat} = 0.1$ V. The important points are labeled in Fig. 8-7b. Similar transfer characteristics apply for inputs B and C. Since any one input or all inputs at a voltage greater than 0.8 V causes the output voltage to be low, the RTL gate *performs the NOR function*.

When the output is loaded by a similar gate, the high voltage level changes. Consider Fig. 8-8 in which the RTL gate on the left has its output level high; i.e., its inputs are all low. The equivalent circuit loading the left-hand gate is illustrated in Fig. 8-8b. We see that, for a load of three gates, defined as a fanout of three,

$$V_Y = V_{CC} - 3I_B(600)$$

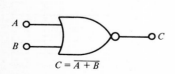

A	B	C
0	0	1
0	1	0
1	0	0
1	1	0

$C = \overline{A + B}$

FIG. 8-5   The NOR gate symbol, logic equation, and truth table.

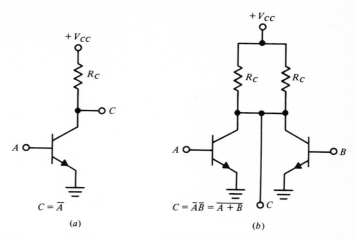

FIG. 8-6 (a) Transistor inverter; (b) the "wired-AND," "dotted collector," or NOR gate.

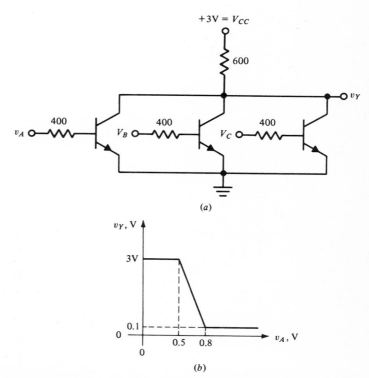

FIG. 8-7 (a) RTL gate circuit with three inputs; (b) transfer characteristics of (a) with $V_B = V_C = 0$.

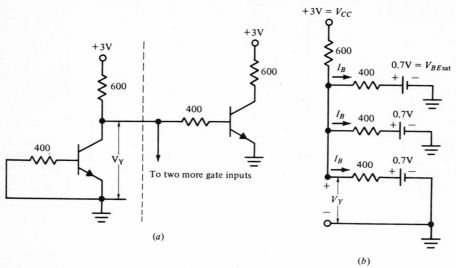

FIG. 8-8   (a) Loading a RTL gate; (b) equivalent circuit of (a).

where $I_B = (V_Y - V_{BEsat})/400$. Thus,

$$V_Y = V_{CC} - (1,800)\left(\frac{V_Y - V_{BEsat}}{400}\right)$$

or

$$V_Y\left(1 + \frac{18}{4}\right) = V_{CC} + \frac{18}{4}V_{BEsat}$$

and

$$V_Y = \frac{3 + 4.5(0.7)}{5.5} = \frac{6.15}{5.5} \simeq 1.1 \text{ V}$$

Comparing this result with the unloaded output of 3 V, we see that the amount of loading is important. According to Fig. 8-7b, if the V_Y voltage drops below 0.8 V the driven gate may not see a logical 1 as desired.

With a fanout of 3, a margin of $1.1 - 0.8 = 0.3$ V exists between V_Y and the critical voltage of 0.8 V. The 0.3 V is called the "noise margin" since a negative spike of 0.3 V at the input of the driven gate can be tolerated. The manufacturer guarantees a noise margin of 0.3 V for these gates. Thus with $V_{CC} = 3$ V, a fanout greater than three will reduce the noise margin below manufacturers' specification.

The logic levels will then be defined as: "0" for voltages between 0 and 0.6 V; "1" for voltages between 1.1 and 3 V. "Fanin," or the number of input terminals to the gate, is also specified in data sheets; e.g., fanin is three for Fig. 8-7a.

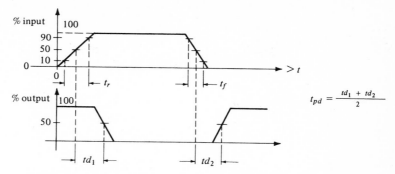

FIG. 8-9 Definitions of rise, fall, and propagation-delay times, t_r, t_f, t_{pd}.

The switching speed and propagation-delay time of a gate are important parameters. If a pulse signal passes through several separate gates and must be combined at some point, it is important that the combined pulses all arrive at the same time. Thus the propagation delay in passing through gates must be known. Figure 8-9 defines rise, fall, and propagation-delay times. The propagation-delay time for an RTL gate is of the order of 50 ns, and this time is considered to be relatively large.

8-3.1 DIODE TRANSISTOR LOGIC (DTL)

The DTL gate is an integrated circuit that functions as a NAND gate. One form of this gate with three inputs is shown in Fig. 8-10a and its transfer function in Fig. 8-10b. We note that if all inputs are high (≈ 5 V), the output is low. If any one input is low, the output is high. Transistor Q_1 appears as a diode as far as

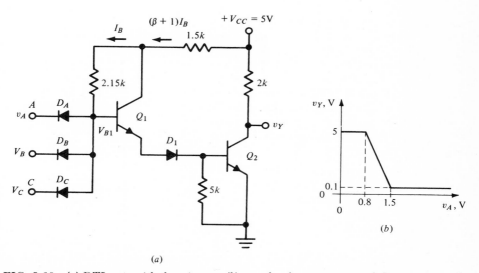

FIG. 8-10 (a) DTL gate with three inputs; (b) transfer characteristics with $V_B = V_C = 5$ V.

schematic (each gate)

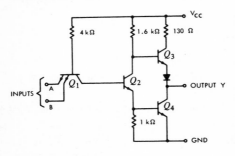

Component values shown are nominal.

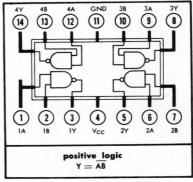

positive logic
$$Y = \overline{AB}$$

recommended operating conditions

Supply Voltage V_{CC} . . . . 4.5 V to 5.5 V
Fan-Out From Each Output, N . . . . 1 to 10

electrical characteristics, $T_A = -55°C$ to $125°C$

PARAMETER		TEST CONDITIONS		MIN	TYP	MAX	UNIT
$V_{in(1)}$	Logical 1 input voltage required at all input terminals to ensure logical 0 level at output	$V_{CC} = 4.5$ V,	$V_{out(0)} \leq 0.4$ V	2			V
$V_{in(0)}$	Logical 0 input voltage required at any input terminal to ensure logical 1 level at output	$V_{CC} = 4.5$ V,	$V_{out(1)} \geq 2.4$ V			0.8	V
$V_{out(1)}$	Logical 1 output voltage	$V_{CC} = 4.5$ V, $I_{load} = -400$ μA	$V_{in} = 0.8$ V,	2.4	3.3‡		V
$V_{out(0)}$	Logical 0 output voltage	$V_{CC} = 4.5$ V, $I_{sink} = 16$ mA	$V_{in} = 2$ V,		0.22‡	0.4	V
$I_{in(0)}$	Logical 0 level input current (each input)	$V_{CC} = 5.5$ V,	$V_{in} = 0.4$ V			−1.6	mA
$I_{in(1)}$	Logical 1 level input current (each input)	$V_{CC} = 5.5$ V,	$V_{in} = 2.4$ V			40	μA
		$V_{CC} = 5.5$ V,	$V_{in} = 5.5$ V			1	mA
I_{OS}	Short-circuit output current†	$V_{CC} = 5.5$ V		−20		−55	mA
$I_{CC(0)}$	Logical 0 level supply current (each gate)	$V_{CC} = 5$ V,	$V_{in} = 5$ V		3		mA
$I_{CC(1)}$	Logical 1 level supply current (each gate)	$V_{CC} = 5$ V,	$V_{in} = 0$		1		mA

switching characteristics, $V_{CC} = 5$ V, $T_A = 25°C$, N = 10

PARAMETER		TEST CONDITIONS	MIN	TYP	MAX	UNIT
t_{pd0}	Propagation delay time to logical 0 level	$C_1 = 15$ pF		8	15	ns
t_{pd1}	Propagation delay time to logical 1 level	$C_1 = 15$ pF		18	29	ns

†Not more than one output should be shorted at a time.
‡These typical values are at $V_{CC} = 5$ V, $T_A = 25°C$.

FIG. 8-11 The typical characteristics of a two-input NAND gate. (*By permission of Texas Instruments Inc., Dallas, Texas.*)

the input is concerned. It also controls the base current to Q_2 when all inputs are high so that greater fanout is possible than for the case where a diode replaces Q_1.

The propagation-delay time for DTL gates is only slightly less than for RTL gates (≈ 30 ns).

The breakpoints of Fig. 8-10b can be found as follows: no current begins to flow in Q_1, D_1, and Q_2 until the base of Q_1 becomes $V_{B1} = V_{BE1cutin} + V_{D1cutin} + V_{BE2cutin}$. Assuming all cutin voltages to be at 0.5 V, $V_{B1} = 1.5$ V for the current to flow. The voltage at point A is one diode drop below V_{B1} or 0.8 V. Diode D_A will not turn off until $v_A = V_{B1} - V_{DAcutin} = V_{BE1} - V_{DAcutin} + V_{D1} + V_{BE2} \cong 0.6 - 0.5 + 0.7 + 0.7 = 1.5$ V, where current stops flowing in D_A at $V_{DAcutin} = 0.5$ V. It can be shown that the current into the base of Q_2 is about 2 mA if all β's $= 30$ and thus, Q_2 will be saturated. For the DTL gate, fanout is about 7 and fanin is 2 or more. Several similar gates are enclosed in one package; e.g., a quadruple two-input NAND gate, which have four gates of the type shown in Fig. 8-11 in one package.

8-3.2 TRANSISTOR-TRANSISTOR LOGIC (TTL OR T²L)

A typical data page for a TTL NAND gate is given in Fig. 8-11. This type of gate switches faster and has lower propagation-delay time than the previous two types. The output of the schematic diagram in Fig. 8-11 is made up of Q_2, Q_3, and Q_4. This particular arrangement is called a "totem-pole" circuit and is designed for faster switching. The transfer characteristic is plotted in Fig. 8-12, and the breakpoints may be found as follows. With input A at ground, the base of Q_1 will be 0.6 V above ground. Thus, voltage at the base of Q_2 is not sufficient to turn on transistors Q_2 and Q_4. When $v_A > 0.9$ V, the collector-base junction of Q_1 becomes forward-biased, and Q_2 and Q_4 turn on. When $v_A \geq 1.5$ V, the output $v_Y = V_{CE4sat} \cong 0.1$ V. At the same time, Q_3 turns off because its base voltage is equal to or lower than its emitter voltage, ($V_{CE2} = V_{CE4} = 0.1$ V; $V_{BE4} = 0.7$ V; $V_D \leq 0.7$ V; then $V_{B3} = 0.8$ V, $V_{E3} \leq 0.8$ V, therefore $V_{BE3} \leq 0$.) Note from

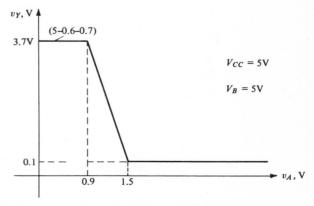

FIG. 8-12 TTL transfer characteristics assuming enough load to cause Q_3 and D_1 of Fig. 8-11 to conduct. Both inputs must be high for output to be low.

Fig. 8-11 that fanout can be as large as 10 and propagation-delay time as small as 8 ns, an average being 13 ns. The transistor symbol having two emitters indicates that transistor Q_1 is constructed with two emitters and one collector, separated by a common base area. More emitters per transistor are possible.

8-3.3 EMITTER-COUPLED LOGIC (ECL OR CCL)

The ECL gate sometimes called current-controlled logic (CCL) can be used as either an OR or a NOR gate. None of the transistors operate under saturation conditions so that a small propagation-delay time of 1 to 4 ns is possible. Figure 8-13a and b display the circuit and transfer characteristics, respectively. Note that the input circuit is similar to a differential amplifier except that four transistors are in parallel on the left side of Fig. 8-13a instead of the usual single transistor. The base voltage of Q_5 is fixed at $V_{BB} = -1.175$ V by transistor Q_6 and diodes D_1 and D_2. This voltage V_{BB} remains constant. The transfer characteristics of Fig. 8-13b show that switching takes place when any one input is near this voltage of V_{BB}. The input and output high levels ("1") are indicated by

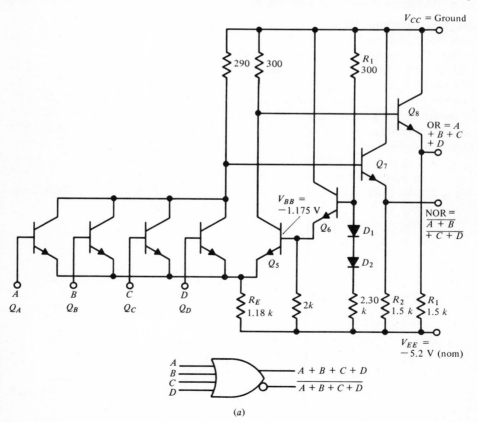

FIG. 8-13 (a) The basic ECL gate; and (b) its transfer characteristics. (By permission Motorola Semiconductor Products, Inc., Phoenix, Ariz.)

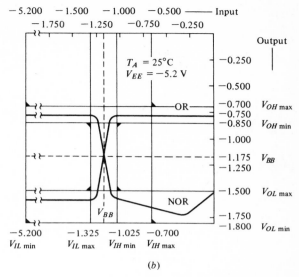

$$-5.200 \quad -1.500 \quad -1.000 \quad -0.500 \quad\text{——— Input}$$
$$-1.750 \quad -1.250 \quad -0.750 \quad -0.250$$

Output

$T_A = 25°C$
$V_{EE} = -5.2$ V

-0.250

-0.500

OR
$-0.700 \quad V_{OH\,max}$
-0.750
$-0.850 \quad V_{OH\,min}$

-1.000

$-1.175 \quad V_{BB}$
-1.250

$-1.500 \quad V_{OL\,max}$

NOR

V_{BB}
-1.750
$-1.800 \quad V_{OL\,min}$

$$-5.200 \qquad -1.325 \quad -1.025 \quad -0.700$$
$$V_{IL\,min} \qquad V_{IL\,max} \quad V_{IH\,min} \quad V_{IH\,max}$$

(b)

FIG.8-13 Continued.

V_{IH} and V_{OH}, and the input and output low levels are indicated by V_{IL} and V_{OL}. The maximum and minimum (or tolerance) due to the manufacturing process are also specified. Note that the nominal logic "1" level is -0.75 V and the nominal logic "0" level is -1.55 V. Both these voltages are negative or below ground, but the "1" level is larger than the "0" level. The nominal transfer characteristics can be calculated, but this calculation will be left to the problems (see Prob. 8-4).

Other gates are available, such as the complementary-symmetry-metal-oxide semi-conductor (CMOS) series of gates. These gates have several advantages but have a relatively long propagation-delay time (≈ 30 ns), (see Sec. 8-13).

8-4 APPLICATIONS OF GATES

Some of the applications of IC gates which we shall consider are made use of in computers. However, gates have many applications which are not employed in computers. We shall attempt to show the versatility of gates in several types of circuits so that, with some ingenuity, other applications can be devised.

8-4.1 SINE-WAVE-TO-SQUARE-WAVE CIRCUITS

We consider an inexpensive RTL gate as a circuit to transform a voltage sine wave to a positive-going square-wave voltage. In Fig. 8-14a, we use only one gate to perform this function. In Fig. 8-14b, the input and output waveforms are indicated. Transistor Q_1 does not conduct current until the input voltage reaches $v_{BE\text{cutin}}$. For v_A between the cutin and saturation levels, the transistor Q_1 acts as a linear amplifier. When v_A reaches the saturation level of Q_1, the output remains

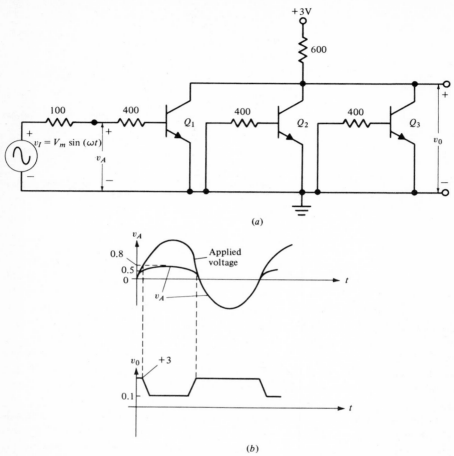

FIG. 8-14 (a) RTL circuit for converting a sine wave to a positive going square wave; (b) input and output waveforms of (a).

constant at about v_{CEsat}. The amount of slope on the edges of the output square wave is a function of the input voltage. If $V_m \simeq 7V_{BE}$, but less than the reverse breakdown voltage of the base-emitter junction, the rise and fall times of the output wave are satisfactory for some applications.

A better square wave can be obtained by using two gates as indicated in Fig. 8-15a. The transfer characteristic of Fig. 8-7b has been applied to obtain Fig. 8-15b. The construction between Fig. 8-15b and c shows that the output voltage v_Z of gate 2 changes levels with much less voltage difference for v_A (660 to 720 mV.) The output of the first gate is loaded so that the "1" level is 1.6 V since $I_{Bsat} = (V_{CC} - V_{BEsat})/(600 + 400) = 2.3$ mA and $v_Y ("1") = V_{CC} - I_{Bsat}(600) = 3 - 2.3(0.6) \simeq 1.6$ V. The points at which gate 2 switches levels are found from the slope of Fig. 8-15b and from the fact that the input switching levels of gate 2 are also 0.5 and 0.8 V. We may use the characteristics of Fig. 8-15c to see how to obtain a better square wave.

Figure 8-16 indicates an improved two-gate circuit which develops a positive-going square-wave output with a sine-wave input. Actually, the input does not need to be a good sine wave. The sine-wave input starts the switching, but the connection of one gate output to the other gate input improves the switching as indicated by Fig. 8-15. The output is a good square wave that does not depend on the amplitude of the input sine wave as long as its amplitude is large enough to cause a switching change. A steep slope of v_i at switching times is desirable. The reader should follow through a cycle of operation to verify the switching levels. The circuit of Fig. 8-16 will be used in the digital meter of Fig. 7-1.

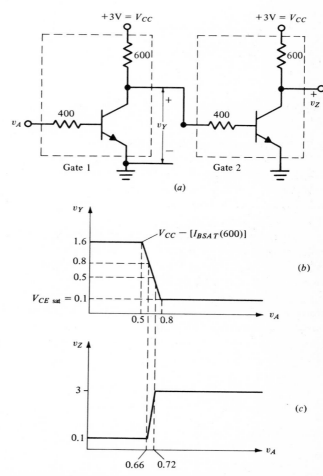

FIG. 8-15 (a) A two-gate RTL circuit for sharper transfer characteristics; (b) gate 1 transfer characteristic with a load of the second gate; (c) output of second gate vs. input of first gate.

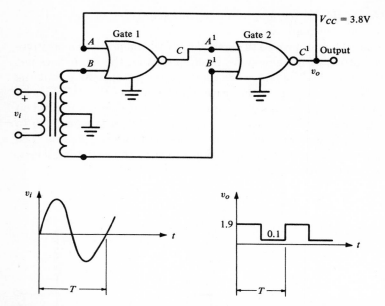

FIG. 8-16 A two-gate circuit for obtaining a square wave. Note that $V_{CC} = 3.8$ V.

8-5 BISTABLE OR MEMORY CIRCUIT

The circuit of Fig. 8-16 is a gated (or driven) bistable multivibrator. It can be shown that the rise and fall times of the output are approximately **17 ns**. See Reference [4]. The circuit is bistable because there are two stable output levels, 2 V("1") and ≈ 0.1 V("0").

　　　The so-called SR Multivibrator is usually labelled as shown in Fig. 8-17a and b. We see that this circuit is the same as Fig. 8-16 without the sinusoidal drive. The operation of the SR bistable circuit is more easily followed from the circuit diagram of Fig. 8-18. Assume Q at logic "1" which means that Q_R and Q_1 are off. Due to the cross-coupling from Q to the base of Q_2, $\bar{Q}$ is at logic "0". If we now apply a logic "1" pulse to R, Q_R saturates, and Q changes to logic "0" while the cross-coupling changes $\bar{Q}$ to logic "1". When S is set at a logic "1" level, the $\bar{Q}$ output becomes logic "0", and Q goes to logic "1". Thus, a set input, $S = 1$, yields Q at logic "1", and a reset input, $R = $ "1", yields Q at logic "0".

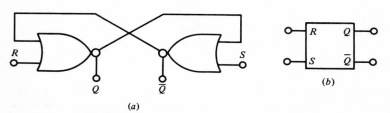

FIG. 8-17 SR NOR-input multivibrator designations.

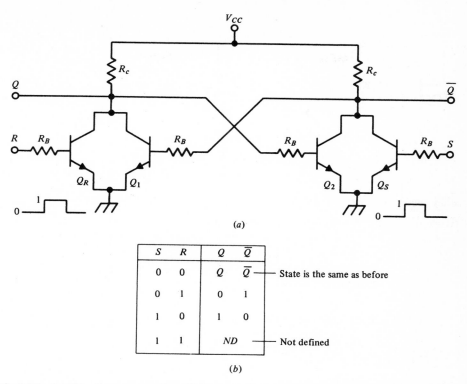

FIG. 8-18 (a) Circuit representation of Fig. 8-17; (b) truth table for (a). Note that S and R pulses go from zero to one and return.

For $S = $ "0" and $R = $ "0", the bistable circuit remains in the previously existing state. When $R = $ "1" and $S = $ "1" both outputs are forced to "0", and when these R and S excitations are removed, the output is not determinate because of random unbalance in the circuit and which signal, R or S, is removed first. Digital systems containing RS bistable units are designed to avoid the $R = $ "1", $S = $ "1" conditions. A truth table can be prepared assuming pulses to the R and S terminals so that the pulse returns to zero after each pulse. Of course, a "0" is the lack of a pulse (see Fig. 8-18b). A faster switching RS bistable circuit may be made of cross-coupled TTL-NAND gates. The circuit and truth table are shown in Fig. 8-19. The reader should refer to Figs. 8-11 and 8-12 to verify the circuit operation.

Some manufacturers use the terms set and clear instead of set and reset. The Q and $\bar{Q}$ terminals are sometimes interchanged, but a check of the truth table on the data sheets will clarify the operation.

8-6 PULSE WAVEFORMS

Short time-length pulses are necessary for many applications. An example is the reset pulse indicated for the digital voltmeter of Fig. 7-1, row E. We shall introduce some methods of obtaining these pulses.

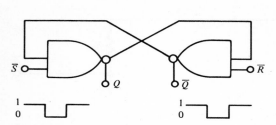

$\overline{S}$	$\overline{R}$	Q	$\overline{Q}$	
0	0	*ND*		Not defined
0	1	1	0	
1	0	0	1	
1	1	Q	$\overline{Q}$	Same as before

FIG. 8-19 Circuit and truth table for the TTL-NAND gate bistable circuit. Note that S and R are negated with respect to Fig. 8-17, and that the pulses go from one to zero and return.

8-6.1 PULSES BY DIFFERENTIATION

If we differentiate a square waveform, a series of short pulses can be obtained. A passive circuit which can differentiate an input signal is drawn in Fig. 8-20. Writing the expression for $v_O(t)$, we obtain

$$v_O(t) = Ri = v_I(t) - \frac{1}{C} \int i \, dt \tag{8-2}$$

If the peak value of $v_O(t) \ll v_I(t)$,

$$v_I(t) \cong \frac{1}{C} \int i \, dt \tag{8-3}$$

or

$$i \cong C \frac{dv_I(t)}{dt} \tag{8-4}$$

Thus,

$$v_O(t) = Ri = RC \frac{dv_I(t)}{dt} \tag{8-5}$$

It is necessary that most of the voltage drop occur across C, and this can occur when the circuit time constant is small compared to the time interval of interest. If the input voltage v_I has discontinuities, the circuit of Fig. 8-20 will not give a

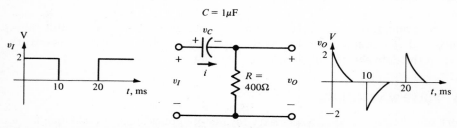

FIG. 8-20 Differentiator with example input and output waveforms.

true differentiation since the differential of a discontinuity yields an infinitely high pulse. However, the approximation for practical cases is sufficient so that the circuit approximates differentiation even at these discontinuities.

An example input and output waveform is also sketched in Fig. 8-20. For $v_I = +2$ V between 0 and 10 ms, the capacitor voltage is

$$v_C = 2(1 - e^{-t/\tau}) = 2(1 - e^{-t/4(10^{-4})}) \qquad (8\text{-}6)$$

Thus,

$$v_O = v_I - v_C = 2 - 2(1 - e^{-t/4(10^{-4})})$$
$$= 2e^{-t/4\times10^{-4}} \text{ V} \qquad (8\text{-}7)$$

After about 3τ, v_O has settled to zero volts. By placing a negative sign in front of Eqs. (8-6) and (8-7) and replacing t by $t - 10$ ms, the output between 10 and 20 ms is obtained. We note that when $v_I = 0$ at $t = 10^+$ ms, the capacitor is charged to $v_C = +2$ V and the effective input due to this charge is (-2) V (initial condition at $t = 10^+$ ms).

An output load will affect the pulse duration for the circuit of Fig. 8-20. As long as the load is about $10R$, the circuit operates as shown.

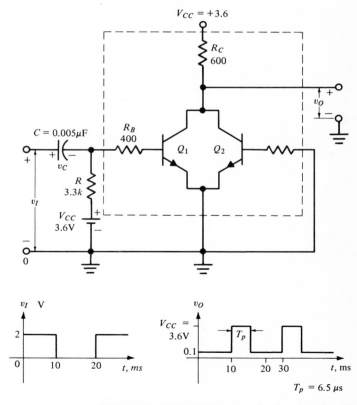

FIG. 8-21   RTL NOR gate pulse-forming circuit.

8-6.2 IC PULSE CIRCUITS

It is sometimes desirable to have a faster pulse fall time than is possible with the circuit of Fig. 8-20. We shall consider two possible methods of obtaining short rectangular pulses.

8-6.3 NOR-GATE PULSE-FORMING CIRCUIT

A modification of the passive differentiator is sketched in Fig. 8-21. An RTL NOR gate is added to shorten the pulse fall time. Assuming settled conditions prior to $t = 0$, Q_1 is on and saturated since $I_B = (3.6 - 0.7)/3.7$ k $= 0.78$ mA and $I_{C1} = \beta I_{B1} \simeq 50\,(0.78)\,(10^{-3}) = 39$ mA $\geq (V_{CC} - V_{CEsat})/R_C \simeq 5.8$ mA. With $v_I = 0$, $v_C = -[I_B R_B + V_{BEsat}] = -[0.78\,(0.4) + 0.7] = -1.0$ V. At $t = 0^+$, $v_I = 2$ V and $v_C = 1.0 - 2e^{-t/\tau}$ where the time constant τ is approximately $R_B C = 2 \times 10^{-6}$ s. At $t = 10^-$ ms, $v_C = 1 - e^{-5000} = 1$ V. Transistor Q_1 remains saturated so that the output voltage does not change.

When v_I returns to zero at $t = 10^+$ ms, the equivalent circuit at the input is that of Fig. 8-22. Since the input to the base is negative due to v_C, Q_1 is turned off and v_{B1} rises toward $V_{CC} = 3.6$ V. When $v_{B1} = 0.5$ V, the gate will switch rapidly back to $v_O \approx 0.1$ V. From Fig. 8-22,

$$v_{B1} = 3.6 - 4.6e^{(t-10\ \text{ms})/\tau_2} \tag{8-8}$$

where $\tau_2 = RC = 3.3(10^3)\,(.005)\,(10^{-6}) = 16.5 \times 10^{-6}$ s. Setting $v_{B1} = 0.5$ V in Eq. (8-8) and solving for t, we obtain $t - 10$ ms $= T_p \cong 6.5$ μs. The output waveform is indicated in Fig. 8-21. The pulse rise and fall time depend upon the gate characteristics and can be much shorter than for the differentiator circuit of Fig. 8-20. The pulses are also unidirectional. The circuit of Fig. 8-21 will be applied in the final design of the digital voltmeter.

8-6.4 MONOSTABLE PULSE CIRCUIT

A monostable multivibrator has one stable state but can be forced into a second state for a short period of time. Such a circuit is useful for generating pulses of a specified length of time.

Figure 8-23 indicates an IC arrangement which operates as a monostable

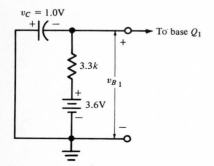

FIG. 8-22   Input circuit of Fig. 8-21 at $t = 10^+$ ms.

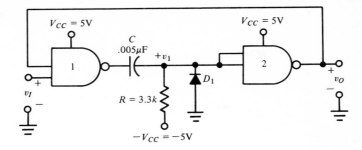

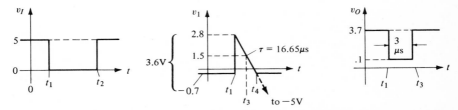

FIG. 8-23   A monostable IC arrangement with two-input TTL-NAND gates of Figs. 8-11 and 8-12.

multivibrator. This circuit has an advantage that the output pulse length does not depend upon the input pulse length. We remember that only if both inputs to the TTL NAND gate are high ("1"), then the output is low ("0"). Prior to time t_1, v_I is high and diode D_1 is conducting through R and $-V_{CC}$. Thus the input to gate 2 is low so that its output is high. Since the output of gate 2 is connected to the second input of gate 1, both inputs are high and the output of gate 1 is low. The conditions are consistent.

At time $t_1{}^+$, v_1 becomes high at $+2.8$ V from Fig. 8-12. Since the voltage across C, which is approximately $V_{D1} - V_{CEsat} = -0.7 - 0.1 = -0.8$ V, does not change from $t_1{}^-$ to $t_1{}^+$, the total change of about $+3.6$ V occurs at v_1 as shown in Fig. 8-23. Diode D_1 is reverse-biased and turns off so that gate 2 output becomes low. Capacitor C charges in the circuit shown by Fig. 8-24a. The equivalent circuit of the gate 1 output has been drawn in Fig. 8-24b. From Fig. 8-24b

$$i_1(t_1{}^+) = \frac{10 - 2.1}{3.33(10^3)} = 2.37 \text{ mA} \tag{8-9}$$

$$v_1(t_1{}^+) = -5 + i_1(t_1{}^+)R$$

$$= -5 + 7.8 \cong 2.8 \text{ V} \tag{8-10}$$

as shown in Fig. 8-23. Between times t_1 and t_4,

$$i_1(t) = 2.37(10^{-3})e^{-(t-t_1)/\tau} \tag{8-11}$$

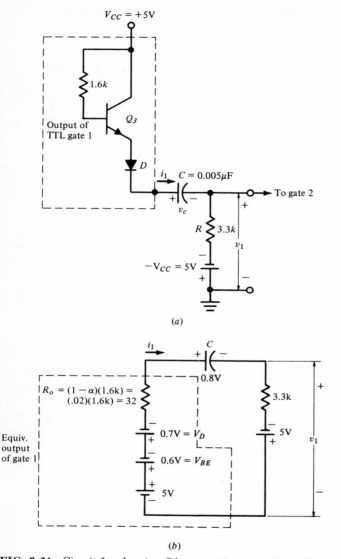

(a)

(b)

FIG. 8-24 Circuit for charging C between times t_1 and t_4 in Fig. 8-23.

where $\tau = (R + R_o)C = 3.33(10^{+3})(5)(10^{-9}) = 16.65\ \mu s$. Thus,

$$v_1(t) = -5 + i_1(t)R$$

$$= -5 + 2.37(10^{-3})(3.3)(10^3)\,e^{-(t-t_1)/16.65\ \mu s}$$

$$= -5 + 7.8e^{-(t-t_1)/16.65\ \mu s} \tag{8-12}$$

According to Fig. 8-12, when v_1 drops to the knee of the transfer curve at 1.5 V, switching will occur. Defining this time point as t_3, from Eq. (8-12),

$$1.5 = -5 + 7.8e^{-(t_3-t_1)/16.65 \ \mu s}$$

Solving for $t_3 - t_1$, we find $t_3 - t_1 \simeq 3.0 \ \mu s$. This pulse width can be controlled by changing $\tau \approx RC$ and $(-V_{CC})$.

We can see that the output of gate 1 remains high by either of the two inputs being held low until switching has occurred; i.e., until after t_3 regardless of the v_I level. However, if $t_2 < t_3$, capacitor C will discharge faster than shown in Fig. 8-23 between times t_3 and t_4 because the output of gate 1 is in saturation.

This application of ICs to a monostable multivibrator again emphasizes that we do need to know the transfer, input, and output equivalent circuits of a gate for detailed calculations of the operation.

Monostable multivibrators are also constructed in a single package. The 74121 is one such monostable multivibrator.

8-7 IC ASTABLE MULTIVIBRATOR

By modifying the circuit of Fig. 8-23, an astable multivibrator can be obtained. Square wave or a ramp type of output can be obtained for use in timing or driving applications. It can be seen from Fig. 8-25 that the astable multivibrator has been formed by cross coupling circuits similar to that of Fig. 8-23. The analysis for each timing period follows the procedure applied to the monostable

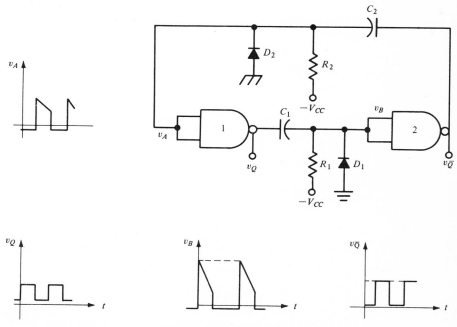

FIG. 8-25 Astable multivibrator with TTL NAND gates.

circuit (see Fig. 8-24). If $R_1 = R_2$ and $C_1 = C_2$, the v_Q and $v_{\bar{Q}}$ outputs will be positive going square waves. The period may be controlled by varying $R_1 = R_2$, $C_1 = C_2$, or $(-V_{cc})$.

8-8 *MSI* WAVEFORM GENERATOR

A monolithic functional block has been developed which can be applied to generate nearly any desired waveform [2]. This medium-scale-integrated circuit (MSI) is fabricated on a 2-mm square chip of silicon, and it is enclosed in a standard 16-pin IC package. A functional block diagram of the generator with pin connection numbers is shown in Fig. 8-26. Typical applications and performance characteristics are presented in Table 8-1. Some of the applications in Table 8-1 will not be discussed in this text, but the applications have been included in the table for completeness.

We shall briefly analyze the voltage-controlled-oscillator (VCO) section of Fig. 8-26 only. A basic type of multivibrator is modified to form this oscillator as in Fig. 8-27a. Transistors Q_3 and Q_4 are operated as constant-current sources whose current is controlled by an external voltage V_c. Transistors Q_1 and Q_2 are cross-coupled in the usual multivibrator manner with the exception that coupling is direct. At a given time, Q_1 and D_1 are on with Q_1 saturated, and Q_2 and D_2 are off. The capacitor C_o charges and discharges through the constant-current sources and causes switching to the second state with D_1 and Q_1 off and D_2 and Q_2 on. This progression repeats so that the operation is astable.

To follow the switching operation, we refer to Fig. 8-27b. This circuit is drawn for D_1 and Q_1 on at the instant switching to this state has occurred, say t_1^+.

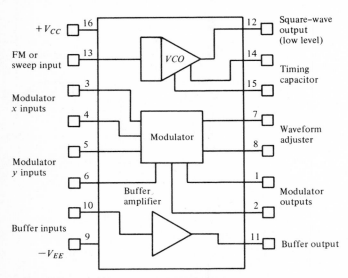

FIG. 8-26 A block diagram of the monolithic waveform generator. VCO is a voltage-controlled oscillator. (*By permission IEEE Spectrum, Institute of Electrical and Electronics Engineers, Inc.*, April, 1972.)

TABLE 8-1

Typical applications for a monolithic waveform generator	XR205 *Performance characteristics*
Waveform generation:	Supply voltage:
Sinusoidal Sawtooth	Single supply, 8 to 26 V
Triangular Ramp	Double supplies, ± 5 to ± 13 V
Square Pulse	Supply current: 10 mA
AM generation:	Operating temperature range: $-55°$C to
	$+125°$C
Double sideband	Frequency stability:
Suppressed carrier	Temperature, 300 ppm/°C
Sweep generation	Power supply, 0.2%/V
Tone-burst generation	Frequency range:
Simultaneous AM/FM	Sine wave and square wave, 0.1 Hz to
FSK and PSK signal generation	5 MHz
On/off–keyed oscillation	Triangle and ramp, 0.1 Hz to 500 kHz
	Frequency sweep range: 10 to 1
	Sweep (FM) nonlinearity: $<0.5\%$ for 10%
	deviation
	Output swing:
	Single-ended, 3 V peak to peak
	Differential, 6 V peak to peak
	Output impedance:
	With buffer stage, 50 Ω
	Without buffer stage, 4,000 Ω
	Output amplitude control range: 60 dB
	Sinusoidal output distortion: $<2.5\%$ (THD)
	Triangular, ramp, and sawtooth outputs:
	Nonlinearity, 1% for f_o <200 kHz
	Square-wave output:
	Amplitude, 3 V
	Duty cycle, 50% ($\pm 3\%$), adjustable from
	20 to 80%
	Modulation capability (all waveforms):
	AM, FM, FSK, PSK, tone burst

Capacitor C_o has the voltage V_{BE} across it with polarity as indicated. The reason for this initial voltage on C_o will be seen later. The voltage at v_1 is

$$v_1(t_1^+) = V_{10} = V_{CC} - V_{BE} - V_{CE\text{sat}} \tag{8-13}$$

Since switching has just occurred, by the symmetry of the circuit we see that

$$v_2(t_1^-) = v_1(t_1^+) = V_{10} = V_{20} \tag{8-14}$$

Now

$$v_2(t_1^+) = v_1(t_1^+) + v_{C_o} = v_1(t_1^+) + V_{BE} \tag{8-15}$$

and

$$v_{B2}(t_1^+) = V_{CE\text{sat}} - v_{C_o} = V_{CE\text{sat}} - V_{BE} \tag{8-16}$$

so that Q_2 is definitely turned off. The capacitor is discharging with constant current I_1 as

$$v_{C_o} = \frac{I_1}{C}(t - t_1) + V_{BE} \qquad \text{for } t_1 \leq t \leq t_2 \tag{8-17}$$

This state continues until $v_{B2} = V_{BEcutin}$ at which time t_2, Q_2 turns on driving Q_1 off. We now see that the voltage $v_{C_o}(t_2) = -V_{BE}$, and the initial condition for the next half period is just the reverse of that shown in Fig. 8-27b.

The waveforms are as in Fig. 8-27a. A square-wave or voltage ramp function is available. By taking the difference of v_2 and v_1 in a differential amplifier, a triangular waveform can be obtained.

The modulator section of Fig. 8-26 is designed to take the product of input signals to generate a desired wave or, in this case, change the ramp function to either a sine or triangular waveform. It also has inputs for different types of modulation. Figure 8-28 indicates the external connections for obtaining desired waveforms. Note that C_o is external and R_E is adjusted to obtain a triangular or sine wave. The buffer amplifier is not shown in Fig. 8-28. However, it may be connected between any output and a load to minimize loading effects.

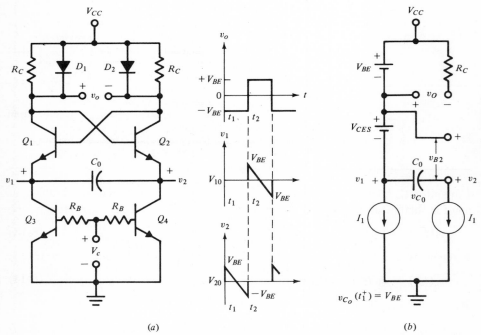

FIG. 8-27 (*a*) Voltage-controlled oscillator circuit and waveforms; (*b*) equivalent circuit of (*a*) when D_1 and Q_1 are conducting and D_2 and Q_2 are off.

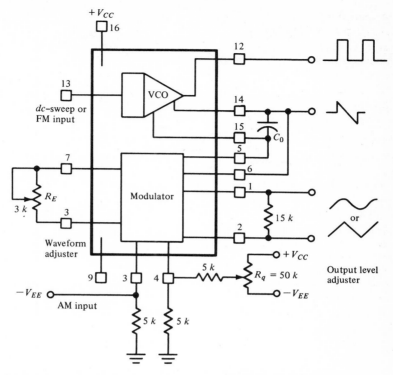

FIG. 8-28   Circuit connection of waveform generator for generation of four types of wave-forms. (*By permission IEEE, N.Y., N.Y.*)

8-9   CLOCKED FLIP-FLOPS

Bistable multivibraters which are switched with a timing wave are designated clocked flip-flops or binaries. Since we are not interested in designing IC bistable or counting circuits, these binaries will be discussed from the application standpoint.

8-9.1   GATED *SR* BISTABLE CIRCUITS

We saw that the *SR* bistable circuit acted as a short-term memory, remembering whether a set or reset pulse was last applied. Sometimes we wish to control the operation of the set–reset operations with a clock or timing pulse. The operation of a set (or reset) pulse is delayed until a clock pulse appears. This type of operation is defined as synchronous switching, allowing combinations of digital elements to all operate in the same time interval. Our previous discussion of the *SR* bistable circuit involved asynchronous operation.

Figure 8-29 shows a gated or clocked *SR* bistable circuit using AND and NOR gates. The internal connections shown in Fig. 8-29 aids in understanding the effect of the *R* and *S* inputs. Since the clock pulse and *R* or *S* inputs are intro-

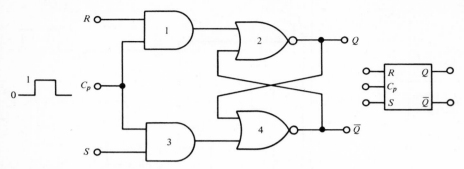

FIG. 8-29 A gated SR bistable circuit and commonly used symbol.

duced to the bistable circuit through AND gates, both must be present during the clock pulse to be effective. Thus, we may write

$$Q = \overline{\overline{Q} + RC_p} \qquad\qquad (8\text{-}18)$$

or

$$\overline{Q} = \overline{Q + SC_p} \qquad\qquad (8\text{-}19)$$

Equation 8-18 states that Q is "0" if $\overline{Q}$ or R and C_p are "1". Similarly, $\overline{Q} = $ "0" if Q, or S and C_p are "1" from Eq. (8-19).

The AND gates inhibit change. When C_p is at logical "0", R and S are inhibited from changing the state of the bistable. If a set pulse is applied, no change in Q will appear until $C_p = $ "1". When the clock returns to $C_p = 0$, the AND gate inhibits the set input and the multivibrator stays in the switched state, storing the set state. The truth table for the gated SR bistable is:

Inputs at t_n		Output at t_{n+1}	
R	S	Q_{n+1}	
0	0	Q_n	(same as before C_p)
0	1	1	
1	0	0	
1	1	ND	

Here, Q_{n+1} is the Q output in the $n+1$ time period (time during and after $C_p = 1$) and Q_n is the output in the preceding nth time period (time prior to $C_p = 1$). The meaning of the truth table may be seen more easily from Fig. 8-30. The nondefined state (ND) is avoided. The clock pulses may be of much shorter time length than shown.

8-9.2 MASTER-SLAVE FLIP-FLOP

Another type of SR flip-flop is made in IC form. The master-slave SR flip-flop is designed to store the input signal which will cause a change in output state during the presence of a clock pulse. Only when the clock pulse returns to the "0" state does the output change to the new state.

To indicate the sequential nature of toggle mode switching along with the effect of the S and R inputs, a truth table is expressed as follows:

Inputs at t_n		Output at t_{n+1}
S	R	Q_{n+1}
0	0	Q_n
0	1	0
1	0	1
1	1	ND

Again Q_{n+1} is the Q output in the $n + 1$ time period (after C_p returns to zero), and Q_n is the output in the preceding nth time period (time prior to $C_p = 1$). Figure 8-31 should clarify the table. Since the output does not change state until after the clock pulse, circuits which have the output connected to one of the S or R inputs will not affect operation during the clock pulse.

A second type of master-slave flip-flop is the JK bistable IC. In essence, the J and K inputs correspond to the S and R inputs respectively, of the SR master-slave bistable. The JK circuit differs from the SR in that the undefined state

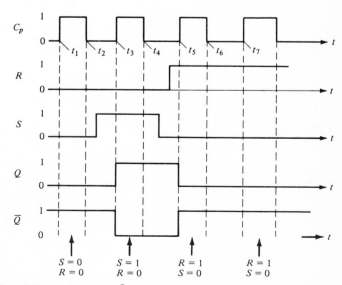

FIG. 8-30   Effect of R, S, and C_p on the Q and $\bar{Q}$ outputs for the gated SR bistable circuit of Fig. 8-29.

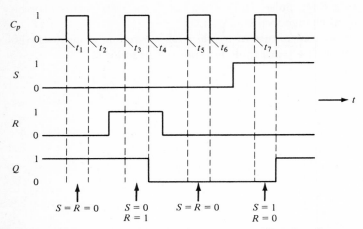

FIG. 8-31 *SR* master-slave bistable waveforms. Note that the output does not change state until C_p returns to 0.

occurring with $S = R = 1$ is eliminated. This may be seen from the following:

J	K	Q_{n+1}
0	0	Q_n
0	1	0
1	0	1
1	1	$\overline{Q_n}$

Comparing with the *SR* table above, we see that Q_{n+1} is now defined as $\bar{Q}_n$ when $J = K = $ "1".

Other types of bistable units are available but these will not be introduced since our applications do not require them.

8-10 BINARY COUNTER

To indicate one use of the clocked circuit, a binary-counter circuit is shown in Fig. 8-32. A binary counter counts events and indicates the number of events in a base 2 mathematical system. First, we follow the chart of events as shown in Fig. 8-32*b*. In essence, the frequency of the output of each flip-flop is one-half that of the previous bistable. Thus, each stage divides the number of input pulses (count) by 2, and the output at Q_4 is $\frac{1}{2^n} = \frac{1}{2^4} = \frac{1}{16}$ of the input frequency. If the input was derived from a sine or square wave, the output at Q_4 would be a positive-going square wave of $\frac{1}{16}$ the input frequency.

To demonstrate that the counting is to the base 2, Table 8-2 shows the decimal count of the input pulses and the state of each flip-flop after each pulse. Remembering our base 2 math, we see that the 13 count is 1101 or $1 \times 2^3 + $

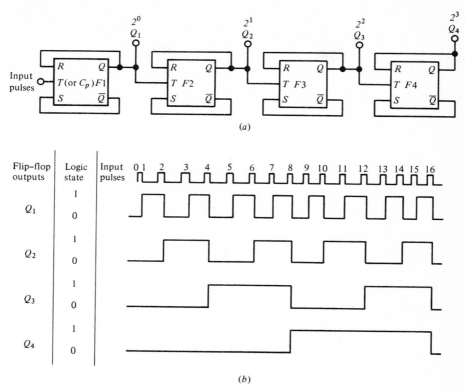

FIG. 8-32 (a) SR master-slave bistables connected as a JK binary counter (JK binaries may also be used without feedback); (b) sequence Q outputs for 16 input clock pulses.

$1 \times 2^{2} + 0 \times 2^{1} + 1 \times 2^{0} = 8 + 4 + 0 + 1 = 13$ in the base 10 system. Indicator lamps on each output would thus read the count in binary or base 2 up to a count of 16 and then repeat. By adding more blocks to Fig. 8-32a the count number could be increased.

8-10.1 DECADE COUNTER

The display system of the digital voltmeter of Fig. 7-1 must read in the decimal system. Thus, we need a counter that will indicate a decimal rather than a binary output. The scale-of-16 counter of Fig. 8-32a can be modified into a scale-of-10 counter as shown in Fig. 8-33a. All the S_D inputs are tied to ground (logic "0") as in Fig. 8-33c so that they are inoperative. To follow the operation we must refer to Fig. 8-33a and d. The flip-flops are of the JK type so that the K input (labeled C) controls the Q output when it is in the logic "1" state; e.g., if $K = $ "1", the output Q is "0" after the clock pulse returns to "0". The OR gate is connected so that the output Q_4 is kept at "0" during the first seven pulses of the clock input. During this same period, the first three flip-flops operate as in Fig. 8-32. Prior to clock pulse 8, the reset input C to FF-4 is 0 so that the return of the clock pulse to "0" changes the state of Q_4 to "1". The inverter connected

TABLE 8-2

	Base			
	2^3	2^2	2^1	2^0
Number of		Outputs		
Input Pulses	Q_4	Q_3	Q_2	Q_1
0	0	0	0	0
1	0	0	0	1
2	0	0	1	0
3	0	0	1	1
4	0	1	0	0
5	0	1	0	1
6	0	1	1	0
7	0	1	1	1
8	1	0	0	0
9	1	0	0	1
10	1	0	1	0
11	1	0	1	1
12	1	1	0	0
13	1	1	0	1
14	1	1	1	0
15	1	1	1	1
16	0	0	0	0

between Q_4 of FF-4 and C of FF-2 places a "1" on C off FF-1 so that on the next negative clock pulse, Q_2 is held to "0". However, at this tenth pulse, the C input to FF-4 is "1" (since $\bar{Q}_2$ or $\bar{Q}_3$ is "1") and Q_4 switches to "0" completing the count.

We note from Fig. 8-33d that the Q outputs are all correct for the binary equivalent of the decimal count; e.g., at the end of the ninth pulse, $Q_1 =$ "1", $Q_2 =$ "0", $Q_2 =$ "0" and $Q_4 =$ "1" yielding 1001 binary or 9 decimal. However, we still have the problem of decoding the binary outputs to a decimal reading. Several other systems for decade counting are available, but the basic idea is illustrated by the above example.

8-11 DECODING BINARY-TO-DECIMAL NUMBERS

A matrix of diodes may be used to decode binary to decimal numbers. The circuit of Fig. 8-34 illustrates a method for such decoding from the decade counter of Fig. 8-33. Since both Q and $\bar{Q}$ outputs are necessary, inverters have been added to furnish the $\bar{Q}$ outputs which are not available from the counter of Fig. 8-33. By remembering that when $Q = 1$, $\bar{Q} = 0$ and vice-versa, the diode connections are easily understood. Only one of the horizontal decimal output lines should be at a potential of approximately V_{CC} (logic "1") at one time. All the other lines should be at ground or logic "0". For a decimal line to be activated, all diodes

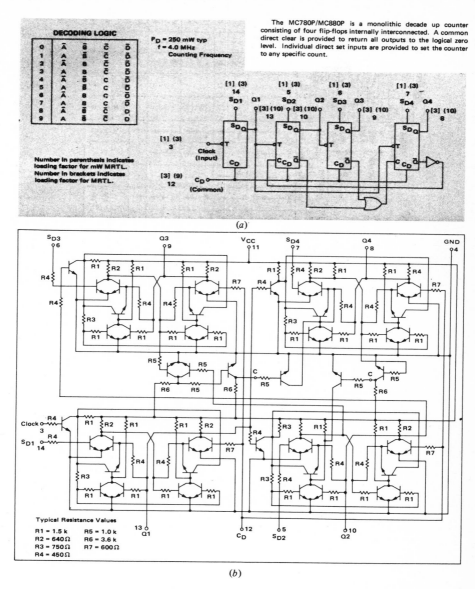

FIG. 8-33 *(By permission Motorola Semiconductor Products, Inc., Phoenix, Ariz.)*

connected to this line must be reverse-biased. As an example, consider decimal line 7. The decade binary output is 0111 so that Q_1, Q_2, Q_3, and $\bar{Q}_4$ are at a potential of V_{CC}. Each diode connected to the decimal 7 line is reverse-biased, and V_{CC} is connected (through R) to the output. Each of the other decimal lines has at least one diode connected to a logic "0" point so that all other decimal lines are at logic "0" or essentially ground. By considering each decimal line in turn, the reader can verify that only one decimal line is at logic "1" at any one time.

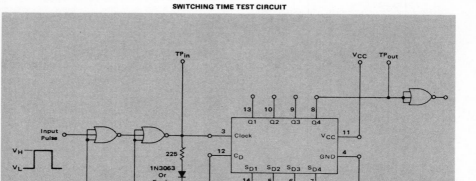

(c)

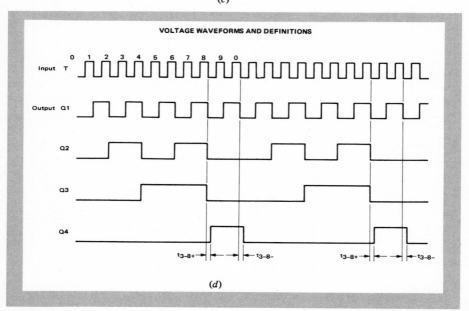

(d)

FIG. 8-33 Continued.

By connecting each of the decimal lines of Fig. 8-34 to a driver and light indicator, the number of the count can be indicated. Since the count shown by the light indicators may be too fast to follow by eye, a method must be included to hold the count during specific periods as shown schematically in Fig. 7-1.

An IC BCD-to-decimal decoder is given in Fig. 8-35. Remembering that all inputs to a NOR gate must be zero for a logic "1" output, the truth table can be verified. Note that the outputs are all "0" for counts greater than 10 or binary inputs greater than 1010. Other types of decoders may also include drivers which will operate the display device. A driver is necessary between the output of Fig. 8-35 and the display device.

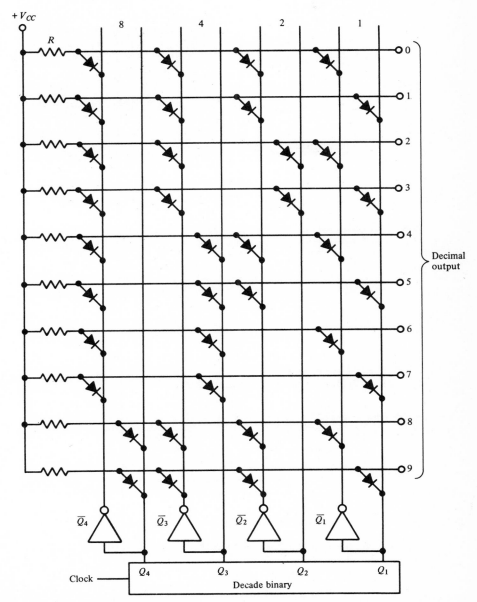

FIG. 8-34 Diode decoding matrix for use with decade counter of Fig. 8-33.

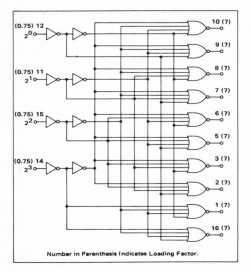

The MC770P/870P is a monolithic BCD-to-decimal decoder consisting of eight inverters and ten 4-input NOR gates which are utilized to convert binary coded decimal (8-4-2-1) input to an output, via the appropriate one of ten output lines.

Number in Parenthesis Indicates Loading Factor.

TRUTH TABLE

	INPUT (BCD)				OUTPUT (DECIMAL)									
Value	2^3	2^2	2^1	2^0	0	1	2	3	4	5	6	7	8	9
Pin No.	14	15	11	12	10	9	8	7	6	5	3	2	1	16
Logic Level	0	0	0	0	1	0	0	0	0	0	0	0	0	0
	0	0	0	1	0	1	0	0	0	0	0	0	0	0
	0	0	1	0	0	0	1	0	0	0	0	0	0	0
	0	0	1	1	0	0	0	1	0	0	0	0	0	0
	0	1	0	0	0	0	0	0	1	0	0	0	0	0
	0	1	0	1	0	0	0	0	0	1	0	0	0	0
	0	1	1	0	0	0	0	0	0	0	1	0	0	0
	0	1	1	1	0	0	0	0	0	0	0	1	0	0
	1	0	0	0	0	0	0	0	0	0	0	0	1	0
	1	0	0	1	0	0	0	0	0	0	0	0	0	1
	1	0	1	0	0	0	0	0	0	0	0	0	0	0
	1	0	1	1	0	0	0	0	0	0	0	0	0	0
	1	1	0	0	0	0	0	0	0	0	0	0	0	0
	1	1	0	1	0	0	0	0	0	0	0	0	0	0
	1	1	1	0	0	0	0	0	0	0	0	0	0	0
	1	1	1	1	0	0	0	0	0	0	0	0	0	0

$t_{pd} = 36$ ns
$P_D = 100$ mW typ (All inputs high)

4-INPUT "NOR" GATE (1-OF-10)

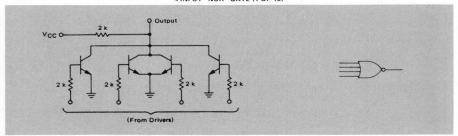

DUAL SERIES INVERTING DRIVER (1-OF-4)

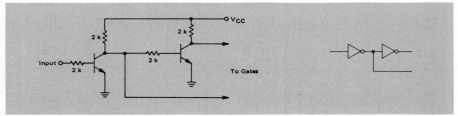

*P suffix = 16 pin dual-in-line plastic package, Case 612.

FIG. 8-35   (*By permission Motorola Semiconductor Products, Inc., Phoenix, Ariz.*)

8-12   COMPLETED DIGITAL VOLTMETER

We have now developed all the blocks necessary to complete the digital voltmeter of Fig. 7-1. A sketch of the remaining sections is given in Fig. 8-36 except for dc power supplies. The MC724P IC is applied to generate the square timing wave and the reset pulses. Note that the reset pulse sets all readouts to zero. One transistor develops the unblanking signal from the timing wave. When the

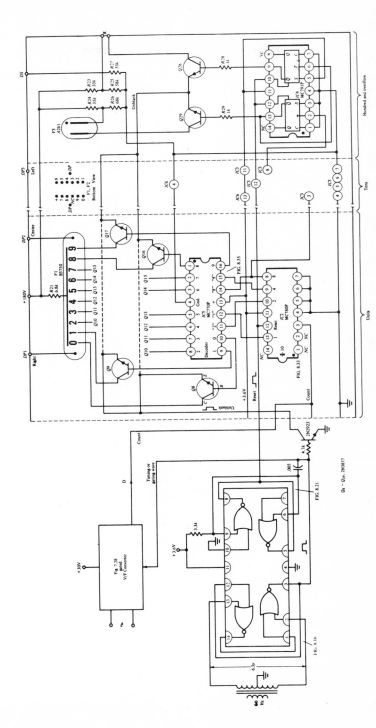

FIG. 8-36 Completed digital voltmeter. (*By permission of Popular Electronics, September, 1970. Ziff-Davis Publishing Co.*)

2N2923 is saturated, the unblanking line is connected to ground. When the 2N2923 is off, the unblank line is open so that all connected circuits are disabled.

Nixie tubes (such as Burroughs B-5750) and neon lamps constitute the readout. These are gas tubes that cause each number or element to glow when connected from $+200$ V to ground through a corresponding driver transistor. Each neon device must have a current limiting resistor. Only four of the ten driver transistors are shown for the units indicator. The hundreds indicator appears as a 1 so that count is only to 199. When count goes beyond 199, the MC791P dual JK flip-flop enables the over-range indicator x.

Note that the input to the tens counting section does not change until the units section has made 10 counts. The situation is similar for the hundreds indicator.

Commercial digital meters apply a dual-slope principle for voltage determination. These instruments can be made more accurate and can contain more digits than the circuit of Fig. 8-36. (A reasonable discussion of a dual-slope instrument is given in the Heathkit catalog obtainable from Heath Co., Benton Harbor, Mich.)

Our purpose here has been to indicate some basic digital gate characteristics and applications. We hope the reader will not assume that the digital voltmeter was the important topic, but rather a vehicle for discussion.

8-13 FET GATES

A linear or switching IC employing n-channel FETs was analyzed from Fig. 5-36. A series of FET gates in IC form designated as CMOS were also discussed in Chap. 5. Both n- and p-channel IC FETs are used in various combinations to form these gate circuits. When the CMOS gate is not switching, the power drawn from the supply voltage is relatively low. However, the upper limit on how fast these gates can switch is lower than that possible for bipolar gates at present.

8-13.1 INVERTER

Reviewing some of the discussion in Chap. 5, a representative connection of a CMOS gate as an inverter is sketched in Fig. 8-37. The p- and n-channel FET

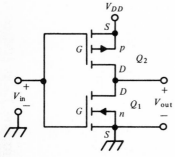

FIG. 8-37   A CMOS inverter (*By permission RCA Corporation, Somerville, N.J.*)

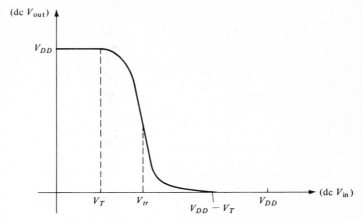

FIG. 8-38   The transfer characteristic for the inverter of Fig. 8-37. The tolerances between manufactured units is not shown [see Fig. 5-40(b)].

drains and gates are connected together, the output is obtained from the connected drains, and the substrate is connected to the source. When $V_{IN} = V_{DD}$(high) then $V_{GSp} = 0$ which is less than the threshold value V_T and Q_2 is off. At the same time, $V_{GSn} = V_{DD} \gg V_T$ and Q_1 is on and in its ohmic region. The output is thus connected to ground (low) through the low resistance of Q_1. The transfer characteristic for V_{IN} between zero and V_{DD} is plotted in Fig. 8-38.

When $V_{IN} = 0$, $V_{GSp} = -V_{DD}$ and Q_2 is on; $V_{GSn} = 0$, and Q_1 is off. The output is connected to V_{DD} through the low resistance of Q_2. Thus, for switching circuits, since either Q_1 or Q_2 is off, the power drawn from the supply is low. As an

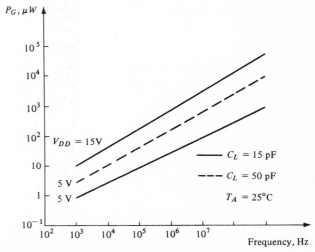

FIG. 8-39   The power dissipated per gate (NAND or NOR) vs. switching frequency and load capacitance. (*By permission RCA Corporation, Somerville, N.J.*)

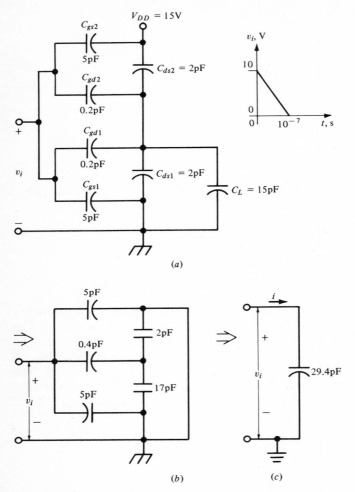

FIG. 8-40 (*a*) Representation of the FETs of Fig. 8-37 by equivalent internal capacitors applicable during switching; (*b*) and (*c*) combination of capacitors into one unit for determining input power during switching.

example, NAND and NOR gates dissipate from 0.25- to 10-μW power under quiescent conditions (not switching) depending upon applied voltage.

The power dissipated during switching increases because of the charging and discharging of internal and load capacitors. The power dissipated per gate for typical NAND and NOR gates with loading and frequency of switching is shown in Fig. 8-39.

To estimate the power dissipated during switching, we approximate the transistors by capacitive elements as indicated in Fig. 8-40*a*. Since V_{DD} does not change when the input signal is switched, the capacitors may be combined as shown in Fig. 8-40*b* and *c*. Assuming the input to switch linearly from 10 to 0 V in 10^{-7} s, the input current is constant at

$$|i| = C\frac{dv}{dt} = 29.4(10^{-12})\left(\frac{10}{10^{-7}}\right)$$

$$= 29.4(10^{-4})\ \text{A}$$

Representing the input voltage as

$$v = 10\left(1 - \frac{t}{10^{-7}}\right)\qquad 0 \le t \le 10^{-7}\ \text{s}$$

the power is

$$p = \frac{1}{T}\int_0^T vi\,dt = \frac{1}{10^{-7}}\int_0^{10^{-7}} 10\left(1 - \frac{1}{10^{-7}}\right)[29.4(10^{-4})]\,dt$$

$$= 29.4\cdot10^3\left(10t - \frac{10^8 t^2}{2}\right)_0^{10^{-7}}$$

$$= 14.2(10^{-3})\ \text{W}$$

We see that this result yields reasonable agreement with Fig. 8-39 when $f_i = 10^7$ Hz.

8-13.2   NOR AND NAND GATES

Typical NOR and NAND gate configurations are sketched in Fig. 8-41a and b, respectively. Considering the NOR gate of Fig. 8-41a, each input is connected to the gate of one n- and one p-type FET. When either the A or B input is V_{DD},

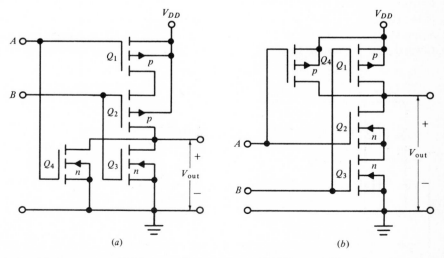

(a)   (b)

FIG. 8-41   (a) A two-input NOR gate; (b) a two-input NAND gate. The substrate is connected to the source of each FET. (*By permission RCA Corp., Somerville, N.J.*)

the output is near ground because the p-type transistors are turned off and the n-type units are turned on, connecting the output to ground through one (or both) n-type FETs. Similarly, when one input is at ground potential (or both at ground) both p-type units are on, and both n-type units are off. Then, the output is coupled to the V_{DD} terminal through the p-type FETs, and the output is high. We recognize this type of operation as that of a NOR gate.

The output of the NAND gate of Fig. 8-41b becomes low or near ground potential when both inputs are high (V_{DD}) so that the n-type FETs are on. If either input is low, the associated n-type FET is off, and the p-type unit is on, coupling the output to V_{DD} through this p-type unit.

Gates having several inputs (more than two) are possible. The parallel and series connections of FETs for each input are extended in the same manner as in Fig. 8-41. All functions of switching ICs which can be performed by bipolar transistor units can also be duplicated by CMOS ICs. In fact, many memory cells for use in computers are of the MOS type of construction.

8-14 CMOS MULTIVIBRATORS

The three types of multivibrators—astable, monostable, and bistable—can be formed with CMOS gates. To indicate the analysis techniques involved, we consider an example circuit of an astable unit.

Figure 8-42 shows an astable multivibrator circuit using two CMOS inverters, a resistor R and a timing capacitance C. Two diodes are connected to each input of the inverters. These are usually present in CMOS inverters to protect the input from excessive voltage levels. These diodes are included in the sketch because they affect the operation of the astable multivibrator.

The waveforms of the astable multivibrator are shown in Fig. 8-43. At $t = 0^+$ the output of inverter B is high or V_{DD}. The voltage across the capacitance is nearly zero having discharged through diode D_A. The input of inverter A is equal to the output of B or V_{DD}. The output of A is low. The capacitance charges toward $v_{cap} = -V_{DD}$ through the pMOS transistor No. 4, R, and the nMOS

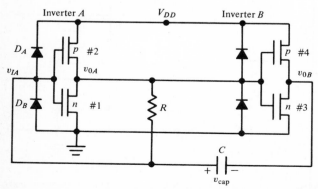

FIG. 8-42 A CMOS astable multivibrator. (*By permission RCA Corp., Somerville, N.J.*)

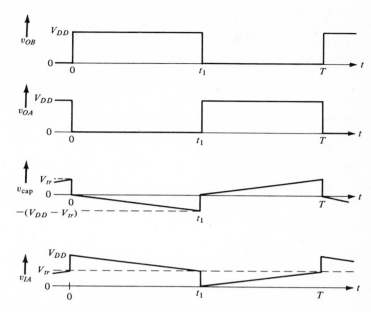

FIG. 8-43 Waveforms of the CMOS astable multivibrator of Fig. 8-42.

transistor No. 1. The input of inverter A decreases by the same amount. Eventually the input to A reaches the transfer voltage point (V_{tr}), and inverter A changes state. When the output of A goes high, the input of B goes high, and thus the output of B goes low. At that instant the voltage change at the output of B is $-(V_{DD} - V_{tr})$. Diode D_B charges the capacitance to zero volts quickly through the low output of inverter B.

For $t_1 < t < T$ the capacitance charges toward V_{DD} through pMOS tran-

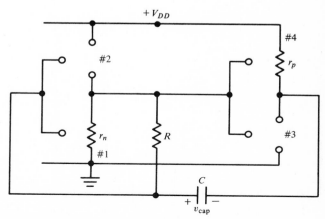

FIG. 8-44 Equivalent circuit of the CMOS astable multivibrator for $0 < t < t_1$.

sistor No. 2, R, and nMOS transistor No. 3. The input to inverter A is essentially the voltage across the capacitance since the output of B is low. When the capacitance charges to V_{tr}, inverter A will change state. The capacitance will discharge through diode D_A. The process is cyclic and thus astable.

Figure 8-44 shows the circuit which exists for $0^+ < t < t_1$. The capacitance begins with an initial charge of zero and charges toward $-V_{DD}$ with a time constant of $(R + r_n + r_p)C$ where r_n and r_p are the resistance of the conducting nMOS and pMOS transistors respectively. Thus,

$$v_{\text{cap}}(t) = -V_{DD} + V_{DD}e^{-t/\tau_1}$$

where $\tau_1 = (R + r_p + r_n)C \simeq RC$ if $R \gg (r_p + r_n)$.

The circuit changes state at $t = t_1$ when $v_{\text{cap}}(t_1) = -V_{DD} + V_{tr}$. Thus,

$$v_{\text{cap}}(t_1) = -V_{DD} + V_{tr} = -V_{DD} + V_{DD}e^{-t_1/\tau_1}$$

or

$$t_1 = \tau_1 \ln \frac{V_{DD}}{V_{tr}}$$

The circuit for $t_1^+ < t < T$ is shown in Fig. 8-45. The solution is

$$v_{\text{cap}}(t) = V_{DD} - V_{DD}e^{-(t-t_1)/\tau_1}$$

where $\tau_1 = (r_p + r_n + R)C \simeq RC$. The circuit changes state when $v_{\text{cap}} = V_{tr}$. Thus,

$$v_{\text{cap}}(T) = V_{tr} = V_{DD} - V_{DD}e^{-(T-t_1)/\tau_1}$$

or

$$T - t_1 = \tau_1 \ln \frac{V_{DD}}{V_{DD} - V_{tr}}$$

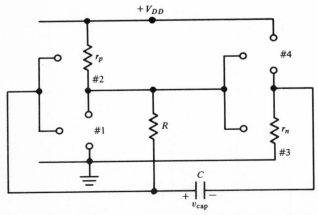

FIG. 8-45 Equivalent circuit of the CMOS astable multivibrator for $t_1 < t < T$.

The total period is T or

$$T = \tau_1 \left(\ln \frac{V_{DD}}{V_{DD} - V_{tr}} + \ln \frac{V_{DD}}{V_{tr}} \right)$$

$$T \cong RC \ln \frac{V_{DD}^2}{V_{tr}(V_{DD} - V_{tr})} \tag{8-20}$$

The transfer voltage V_{tr} is typically $(\frac{1}{2}) V_{DD}$. Therefore,

$$T \cong RC \ln \frac{V_{DD}^2}{\frac{V_{DD}}{2} \left(\frac{V_{DD}}{2} \right)} = RC \ln 4 = 1.39 RC \tag{8-21}$$

Equation (8-20) indicates the quality and stability of the circuit. The period is linearly related to R and C, which is a desirable feature. If $R \gg r_p + r_n$, the period is independent of the on resistance of the transistors. In practice r_n and r_p are not constant. However, they are usually specified by manufacturers as below some maximum value; e.g., less than 600 Ω. Thus, the variable nature of r_n and r_p can be made negligible by making R greater than $20(r_n + r_p)$ or greater than 24 k.

The period is also dependent on V_{DD} and V_{tr}. The power supply can be regulated thus eliminating V_{DD} as a variable. The transfer voltage point V_{tr} will produce variations in the period which are difficult to control. All other parameters R, C, and V_{DD} can be held to any desirable tolerance by precision components and regulated supplies, but V_{tr} varies considerably from one device to another. It can be shown that a 10-percent change in V_{tr} results in a 0.7-percent change in T.

One manufacturer specifies approximately ±20-percent variation for V_{tr} between the maximum and minimum device. This will result in a ±3-percent change in T.

REFERENCES

1. RCA Corporation, COS/MOS Integrated Circuits Manual, Princeton, N.J., 1971.
2. Grebene, A. B., Monolithic Waveform Generation, *IEEE Spectrum*, April, 1972.
3. Lancaster, D., Assembling the Popular Electronics Mini-DVM, *Popular Electronics*, September, 1970.
4. Strauss, L., "Wave Generation and Shaping," McGraw-Hill Book Company, New York, 1970.
5. Motorola Semiconductor Products, Inc., Motorola Digital Integrated Circuits, Technical Data MECL 2, MC1000/MC1200 series, Phoenix, Ariz., 1969.
6. Motorola Semiconductor Products, Inc., Motorola Digital Integrated Circuits, Technical Data, Plastic MRTL Low-Power and Medium-Power MC700P/MC800P series, Phoenix, Ariz., 1970.

PROBLEMS

8-1. In Fig. 8-10, diode D_1 is replaced by two similar diodes in series between emitter of Q_1 and base of Q_2. Determine and sketch the input-output voltage-transfer characteristics.

8-2. Using the assumptions of the text, determine the base current of transistor Q_2 in Fig. 8-10 assuming $V_A = V_B = V_C = 5\,V$. Assume $V_{CC} = 5\,V$, β_F's $= 30$; β_R's $= 2$, and a load between output y and ground of 20 kΩ.

8-3. In Fig. 8-11 determine the emitter, collector, and base currents of each transistor if (a) $V_A = V_B = 0$; (b) $V_A = V_B = 5\,V$. Assume $V_{CC} = 5\,V$, β_F's $= 30$; β_R's $= 2$, and a load between output y and ground of 20 kΩ. Assume $V_{BE} = 0.75\,V$.

8-4. In Fig. 8-13 assume all input voltages to be zero except one, say A. Calculate and plot the output curves shown in (b) of that figure; i.e., verify the OR and NOR voltage-transfer characteristics.

8-5. For the circuit of Fig. 8-20, plot v_C versus t.

8-6. In Fig. 8-21, change V_{CC} to $+5\,V$, and determine a new value for C so that $T_p = 3\,\mu s$. Sketch v_O versus t and the voltage across the capacitor versus t.

8-7. Refer to Fig. 8-23. Obtain an output pulse width of 5 μs by (a) changing C; (b) by changing $-V_{CC}$.

8-8. In Fig. 8-25, $R_1 = R_2 = 3.3$ kΩ, $C_1 = C_2 = 0.005$ μF. Label the magnitudes and time periods on the waveforms shown.

8-9. Explain how the MC791P dual JK flip-flop in Fig. 8-36 enables the over-range indicator x.

8-10. Buffers are sometimes employed to interface between logic gates and loads other than a load of a similar gate type. Figure P8-10 shows an RTL gate and a possible buffer stage. (a) For a load resistor R_L connected between the output and ground, show that $I_L = (\beta_1 + 1)(V_{CC} - V_{BE1})/(R_C + (\beta_1 + 1)R_L)$ when $v_1 = $ "0" and show that $I_L = 0$ when $v_1 = $ "1". (b) With R_L connected between the output and $+V_{CC}$, show that: $I_L = 0$ when $v_1 = $ "0"; $I_L = -(V_{CC} - V_{BE2})/R_L$, when $v_1 = $ "1". (*Note*: a negative load current I_L is now possible.)

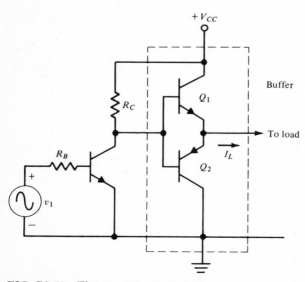

FIG. P8-10   The circuit for Prob. 8-10.

8-11. An emitter-coupled astable multivibrator needing only one capacitor is shown in Fig. P8-11a, (a) With the capacitor removed, calculate the bias conditions on Q_1 and Q_2 to show that they are in the linear active region. (b) With C in the circuit, one transistor conducts current and is in its active region while the second transistor is cut off. The cutoff transistor turns on when its emitter voltage is 0.7 V below its base voltage, thus turning off the previously conducting transistor. Verify the waveforms measured with respect to ground shown in Fig. P8-11b. Assume $V_{BE} = 0.7$ V for cutin and active region.

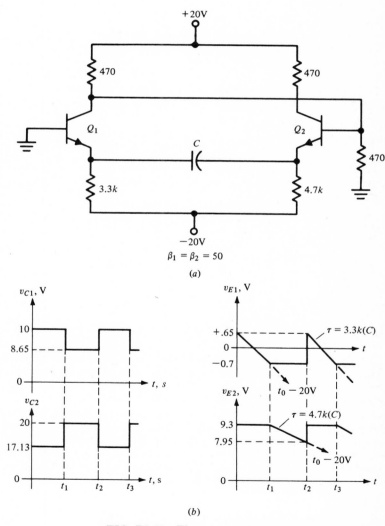

(a)

(b)

FIG. P8-11 The circuit and waveforms for Prob. 8-11.

8-12. In the bistable circuit of Fig. P8-12, silicon transistors with $h_{FE} = 100$ are used. Show that the circuit is bistable by finding V_O, when $V_I = 0$, and when $V_I = 6$ V. To simplify calculations, assume Q_1 or Q_2 is off when the other transistor is on. Verify this latter assumption.

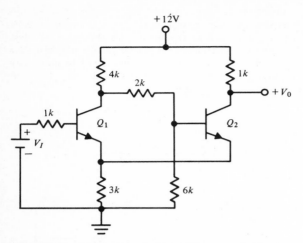

FIG. P8-12   Emitter-coupled bistable circuit for Prob. 8-12.

8-13. The value of R and C in Fig. 8-23 are changed to 2.2 k and 0.01 μF, respectively. Calculate, sketch, and label numerically v_1 and v_O if v_I is as shown in Fig. P8-13. Include any time constants in your sketch.

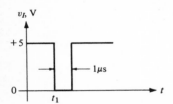

FIG. P8-13   Waveform for Prob. 8-13.

8-14. The monostable circuit of Fig. 8-23 can be modified to use RTL NOR gates. Draw such a monostable multivibrator using two two-input RTL NOR gates, (as in Fig. 8-7). Note that the input pulse must be inverted, and it is not necessary to analyze this circuit. Diodes may be added or deleted. Sketch the shape of the output pulse.

8-15. Figure P8-15 shows a high-threshold logic gate employing silicon transistors. The Zener diode has a breakdown value of 6.9 V which may be considered constant. The minimum $h_{FE} = 50$ for Q_1 and Q_2; $V_{CEsat} = 0.2$ V. (a) Sketch and label the transfer characteristic v_O versus v_A if $V_B = 15$ V. (b) Determine the maximum fanout if Q_2 is not to pull out of saturation. Assume the gate driven and loaded by similar gates. For Q_1 and Q_2, $V_{BEcutin} = 0.5$ V, $V_{BEsat} = 0.8$ V, $V_{BEactive} = 0.7$ V; for D_1 and D_2, $V_{cutin} = V_{active} = 0.7$ V.

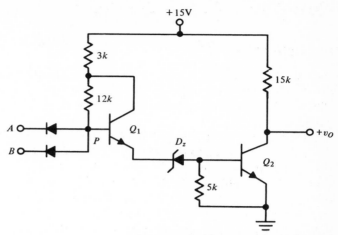

FIG. P8-15 The circuit for Prob. 8-15.

8-16. A variation of the circuit in Fig. 8-42 is shown in Fig. P8-16. (*a*) Find the period if $R_s \gg R$; (*b*) How much does T change with a 20-percent variation in V_{tr}?

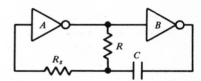

FIG. P8-16 The circuit for Prob. 8-16 with power supplies omitted.

8-17. (*a*) Sketch the equivalent circuit which exists when diode D_A in Fig. 8-42 is conducting. Assume transistor on resistance of r_n and r_p. (*b*) Find the transient response of the circuit in (*a*).

9

Computer-Aided Design

INTRODUCTION

In Chap. 6 we analyzed electronic circuits using general-purpose digital computer programs. These and other general-purpose computer programs are available which allow us to analyze a wide variety of electronic circuits, both linear and nonlinear.

On-line computer-aided design programs permit us to construct, test, and modify a circuit much as is done in breadboarding. In Chap. 6 we examined how the computer program can do some of the common laboratory measurements such as gain, input resistance, output resistance, rise-time, etc. If we think of the CAD programs in terms of simulated breadboarding, then other laboratory techniques may be applied. For example, the measurement of the h parameters, y parameters, or other parameters can be simulated very easily. These and other real "hands-on" techniques will be simulated in this chapter.

A most promising advantage of the design of circuits by computer involves the use of powerful numerical-analysis routines which can optimize a desired response. One such optimization routine is available in LINCAD and is described in Sec. 9-2.

We point out here that the discussion of computer-aided design programs has been restricted to general-purpose programs. No general techniques have been developed which can synthesize a wide variety of circuits. Synthesis techniques (and programs) are available which can generate a particular type of circuit. For example, the procedure for synthesizing filter circuits both passive and active has been known for years. The optimization of IF and RF amplifiers in terms of two-port (h parameters) was presented in 1956 by Linvill [1, 2]. These and similar techniques have been computerized and are valuable tools in the design of electronic circuits. However, these special-purpose programs will never replace the need for good, powerful, general-purpose programs.

9-1 VALIDITY OF THE HYBRID-π EQUIVALENT CIRCUIT OF A BIPOLAR TRANSISTOR

In Chap. 4 the hybrid-π equivalent circuit of the bipolar transistor was presented. In this section we shall examine the model in more detail. Manufacturers often give the y parameters of a transistor as a function of frequency for use in designing RF and IF amplifiers. The y parameters of a typical transistor in the RCA CA3018 integrated circuit is shown in Fig. 9-1.

In this section we shall compute the y parameters by use of LINCAD. A circuit for the calculation of the y parameters is shown in Fig. 9-2. The y parameters are given by

$$I_1 = y_{11}V_1 + y_{12}V_2 \qquad\qquad (9\text{-}1a)$$

$$I_2 = y_{21}V_1 + y_{22}V_2 \qquad\qquad (9\text{-}1b)$$

In Fig. 9-2 when $V_a \neq 0$ and $V_b = 0$, the output is essentially short-circuited by the 1-Ω resistance, or $V_2 \simeq 0$. Also, the 1-Ω resistor in the input makes $V_1 \simeq V_a$. Equations (9-1a) and (9-1b) become

$$I_1 = \frac{V_1 - V_2}{1} = y_{11}V_1 \simeq y_{11}V_a$$

$$I_2 = \frac{V_5 - V_4}{1} = y_{21}V_1 \simeq y_{21}V_a$$

or

$$y_{11} = \frac{I_1}{V_1} \simeq \frac{I_1}{V_a} = \frac{V_1 - V_2}{V_a}$$

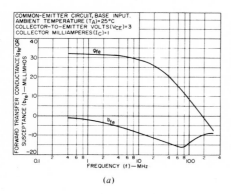

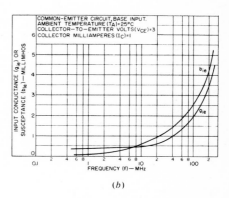

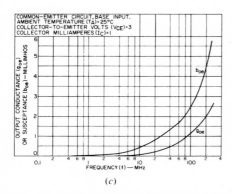

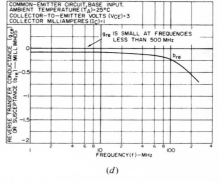

FIG. 9-1 The CA3018 typical admittance characteristics as a function of frequency. (a) Forward transfer admittance; (b) input admittance; (c) output admittance; (d) reverse transfer admittances. (*By permission RCA Inc., Somerville, N.J.*)

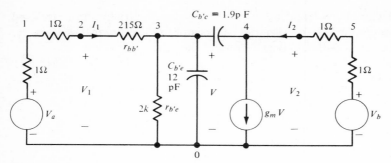

FIG. 9-2 Circuit for the measurement of the y parameters of the CA3018 transistor.

and

$$y_{21} = \frac{I_2}{V_1} \simeq \frac{I_2}{V_a} = \frac{V_5 - V_4}{V_a}$$

If $V_a = 1,000$ V then

$$V_1 - V_2 \cong y_{11} \qquad \text{Millimhos} \tag{9-2a}$$

$$V_5 - V_4 \cong y_{21} \qquad \text{Millimhos} \tag{9-2b}$$

Similarly when $V_b = 1,000$ V and $V_a = 0$

$$V_1 - V_2 \cong y_{21} \qquad \text{Millimhos} \tag{9-2c}$$

$$V_5 - V_4 \cong y_{22} \qquad \text{Millimhos} \tag{9-2d}$$

The output options of REAL and IMAGINARY in LINCAD permit the direct plotting of the real and imaginary part of the y parameters. The hybrid-π model for a transistor in the CA3018 is

$$g_m = 35I_C = 35 \times 1 \text{ ma} = .035 \text{ mho}$$

$$r_{b'e} = \frac{\beta}{g_m} = \frac{70}{0.035} = 2 \text{ k}$$

$$C_{b'c} = C_{cbo} = 1.9 \text{ pF}$$

$$C_{b'e} = \frac{g_m}{2\pi f_T} - C_{b'c} = \frac{0.035}{2\pi \times 400 \text{ MHz}} - 1.9 = 12 \text{ pF}$$

$$C_{SC} = 3.5 \text{ pF}$$

$$r_{bb'} = 215 \text{ }\Omega$$

The value of $r_{bb'}$ was found by making the real part of y_{21} at 60 MHz equal to that given in Fig. 9-1. The capacitance C_{CS} is the collector-to-substrate capacitance. Each transistor in the CA3018 is isolated from the substrate by a reverse-biased diode, as shown in Fig. 9-7. The capacitance C_{CS} is the capacitance of this reverse-biased diode, as specified by the manufacturer.

The results of the LINCAD simulation are shown in Fig. 9-3. The real and imaginary parts of each of the parameters were plotted on the same graph by turning the paper feed back and superimposing the printout for each.

A comparison of Figs. 9-1 and 9-3 shows a remarkable agreement between the measured and simulated y parameters. Better agreement no doubt could be made by adjusting the hybrid-π parameters.

9-2   OPTIMIZATION

Up to five variables (G, L, C, or the frequency) can be varied simultaneously in LINCAD to obtain a minimum, maximum, or desired value of a node voltage difference. The commands to select a variable are

TUNE F	To adjust the frequency
TUNE G	To adjust a resistance
TUNE L	To adjust an inductor
TUNE C	To adjust a capacitance

If G, L, or C is selected, the program requests the node connections of the element. The TUNE X command is repeated by the user as a reply to '$> =$' until all components to be adjusted simultaneously are inserted. The user then types one of the following:

MINIMIZE	To find the minimum of a node voltage
MAXIMIZE	To find the maximum of a node voltage
SET	To set a node voltage to a numerical value

The program requests the node numbers of the voltage difference. Two node numbers are required. If a node voltage is to be with respect to ground, zero (∅) is entered for the second node. The procedure will be applied in the next section.

9-3   CASCODE VIDEO AMPLIFIER

In this section we shall design a linear amplifier with the following characteristics:

1. Voltage gain, $A_v = 35$ dB
2. Frequency response 20 Hz to 10 MHz
3. Power supply, 5 V
4. Integrated circuit if possible
5. Minimum component cost

Requirements (1) and (2) are dictated by the application, and the power supply requirement is a result of the amplifier being included in a system which contains T^2L logic circuits. The final design is to have minimum component cost, a result of the desire to economize because of the large production anticipated.

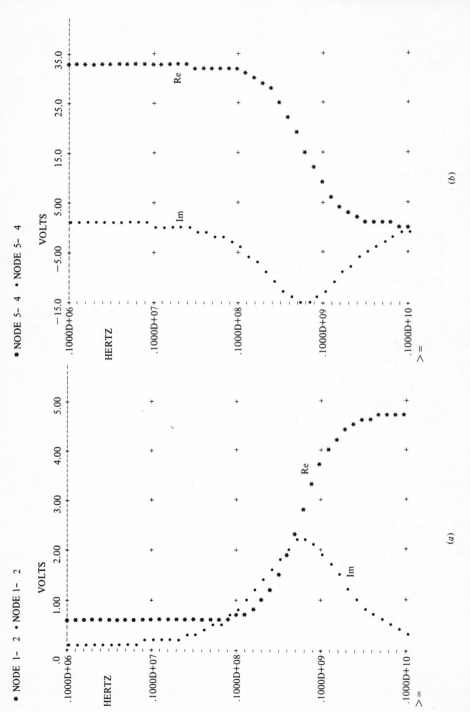

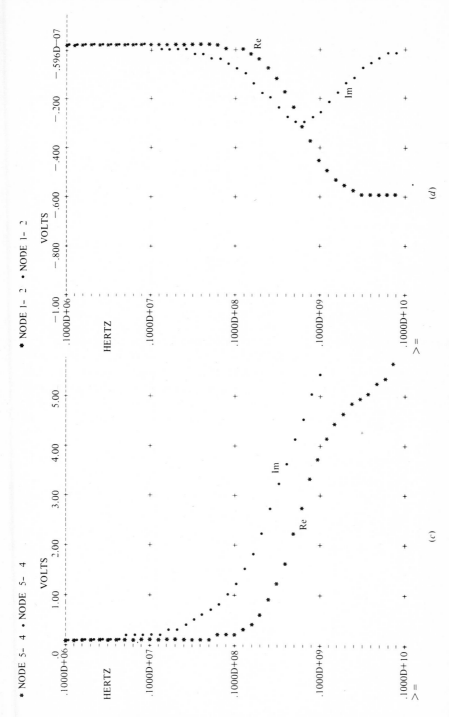

FIG. 9-3 The y parameters of the CA3018 as generated by LINCAD. (a) Forward transfer admittance y_{21} of a CA3018 transistor; (b) input admittance y_{11} of a CA3018 transistor; (c) output admittance y_{22} of a CA3018 transistor; (d) reverse transfer admittance y_{12} of a CA3018 transistor.

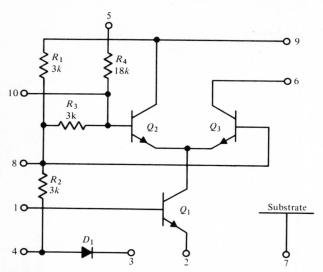

FIG. 9-4 The MC1550 *RF-IF* amplifier. (*By permission Motorola Semiconductor Products, Inc., Phoenix, Az.*)

We have defined the problem here in a simplified manner. The minimization is to be on final component costs for the design selected, whereas, in reality, the minimization of cost is complex. We should include all aspects of the problem from cost of design to the cost of final testing and packaging. A minimum cost might well be one in which the amplifier is designed quickly with minimum engineering time. A quickly designed circuit is likely to be overdesigned. Also, the

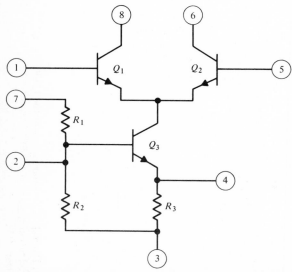

FIG. 9-5 The CA3028A differential-cascode amplifier. (*By permission RCA, Inc., Somerville, N.J.*)

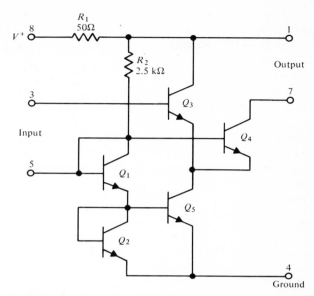

FIG. 9-6 The μA703 *RF-IF* amplifier. (*By permission Fairchild Semiconductor Corp., Mountain View, Cal.*)

cost of the printed circuit board, component insertion into the board and testing influence the decision. To avoid having this chapter larger than the preceding eight chapters, we have confined the minimization to that of component cost after the circuit has been selected.

We cannot use one of the common integrated-circuit operational amplifiers

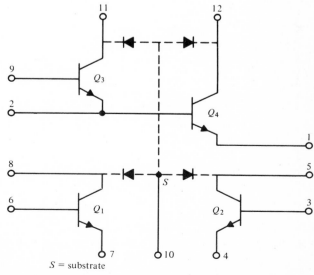

S = substrate

FIG. 9-7 The CA3018 transistor array. (*By permission RCA, Inc., Somerville, N.J.*)

because of the single-ended power supply and gain-bandwidth product required in the design problem. The gain-bandwidth product of IC operational amplifiers is less than the 560 MHz required here.

Integrated circuits designed for *IF–RF* amplifier applications appear to present a good possibility. We have shown a few of these in Figs. 9-4, 9-5, and 9-6. Another possible integrated circuit and the one we shall apply is the transistor array integrated circuit shown in Fig. 9-7.

Figure 9-8 shows the schematic diagram of a CA3018 cascode video amplifier. The basic cascode amplifier consists of a common-emitter amplifier Q_2 followed by a common-base amplifier Q_1. These are followed by cascaded emitter followers Q_3 and Q_4. The term cascode implies a common-emitter stage followed by a common-base stage.

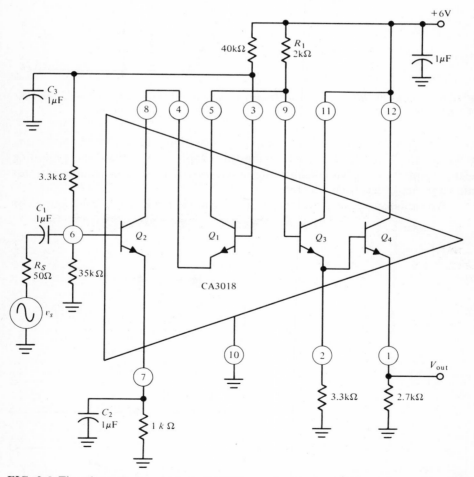

FIG. 9-8 The schematic diagram for a CA3018 cascode video amplifier. (*By permission RCA, Inc., Somerville, N.J.*)

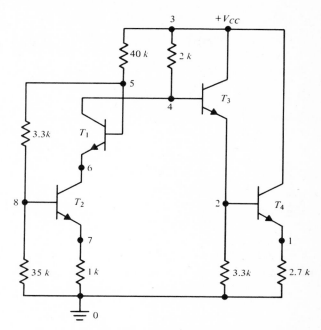

FIG. 9-9 The dc circuit for OSUCAD analysis.

The frequency breakpoints of the amplifier, as reported by the manufacturer in an application note [3], are stated to be 6 kHz and 11 MHz, with a mid-frequency gain of 37 dB ± 1 dB. The gain and upper 3-dB point meet the requirements, but the lower 3-dB point of 6 kHz does not.

The high-frequency response of the amplifier is determined by the transistor high-frequency characteristics. These characteristics are modeled by the hybrid-π parameters $C_{b'e}$ and $C_{b'c}$ as described in Chap. 4.

The low-frequency response of the amplifier is determined primarily by the capacitances C_1, C_2, and C_3. These must be selected to give a lower 3-dB point of 20 Hz. Herein lies the simplified cost-minimization request in the assignment. We could have a quick solution to the problem by selecting $C_1 = C_2 = C_3 \simeq$ 1 μF $\times$ 6,000/20 = 300 μF, where the factor of 6,000 Hz/20 Hz = 300 should reduce the lower 3-dB point to 20 Hz, assuming an inverse linear relationship between capacitance and lower 3-dB point. Such an engineering guess is quite adequate in many cases but not when large production is anticipated (nor is it in the spirit of our problem).

The dc circuit and OSUCAD analysis of the amplifier in Fig. 9-8 is shown in Figs. 9-9 and 9-10. We computed the dc operating points for $V_{CC} = 6$ V and $V_{CC} = 5$ V. With the operating point we can calculate g_m, while the hybrid-π parameters $r_{b'e}$ and $C_{b'e}$ can be calculated using the manufacturer's data of $h_{fe} = 70$ and $f_T = 400$ MHz. From Chap. 4, $g_m = 38 I_C$; $r_{b'e} = h_{fe}/g_m$; $C_{b'e} =$

```
DC SOLUTION OF A CA3018 CASCODE AMPLIFIER

DEVICE      VCE,VD OR VDS     IC OR ID      VBE OR VGS      IB OR IC
               (VOLTS)          (MA)         (VOLTS)     (MICROAMPERE)

    E CC  =   5.0000       VOLTS
    T 4          3.2948        .62257          .64648         8.9038
    T 3          2.6483        .71108          .64987        10.170
    T 1          1.7718        .98911          .65832        14.146
    T 2           .21069      1.0034           .65880        15.645
         V 1 - V 0 =   1.7052        V 2 - V 0 =   2.3516
         V 3 - V 0 =   5.0000        V 4 - V 0 =   3.0015
         V 5 - V 0 =   1.8881        V 6 - V 0 =   1.2298
         V 7 - V 0 =   1.0191        V 8 - V 0 =   1.6779

    E CC  =   6.0000       VOLTS
    T 4          3.8901        .77036          .65191        11.017
    T 3          3.2382        .83565          .65399        11.951
    T 1          1.8471       1.2802           .66498        18.309
    T 2           .25089      1.2987           .66539        19.124
         V 1 - V 0 =   2.1099        V 2 - V 0 =   2.7618
         V 3 - V 0 =   6.0000        V 4 - V 0 =   3.4158
         V 5 - V 0 =   2.2337        V 6 - V 0 =   1.5687
         V 7 - V 0 =   1.3178        V 8 - V 0 =   1.9832
```

FIG. 9-10 The OSUCAD dc solution of the cascode amplifier of Fig. 9-8.

$g_m/2\pi f_T - C_{b'c}$. The manufacturer specifies $C_{b'c} = 1.9$ pF and a collector-to-substrate capacitance of 3.5 pF. From Sec. 9-1, $r_{bb'} = 215\ \Omega$. The hybrid-π parameters are listed in Table 9-1 for $V_{CC} = 5$ V and $V_{CC} = 6$ V.

The ac equivalent circuit of the video amplifier circuit of Fig. 9-8 is shown in Fig. 9-11. The circuit contains 12 nodes and 34 components. The capacitances, especially the collector-to-substrate capacitances, make the problem rather difficult to simplify and solve by hand.

A plot of the frequency response need not be requested to find the two 3-dB points. We shall demonstrate the optimizing routines of LINCAD with one variable, the frequency. The results are shown below. We typed the underlined lowercase quantities.

TABLE 9-1

	$V_{CC} = 5$ V				$V_{CC} = 6$ V			
	I_C (mA)	gm (mho)	$r_{b'e}$ (kΩ)	$C_{b'e}$(pF)	I_C (mA)	g (mho)	$r_{b'e}$ (kΩ)	$C_{b'e}$ (pF)
$T1$	0.989	0.0376	1.86	13.05	1.28	0.0486	1.44	17.45
$T2$	1.0034	0.0381	1.84	13.3	1.2987	0.0494	1.42	17.7
$T3$	0.7111	0.027	2.59	8.85	0.8356	0.0318	2.2	10.7
$T4$	0.6226	0.0237	2.96	7.5	0.7704	0.0293	2.39	9.75

$C_{b'c} = 1.9$ pF; $C_{CS} = 3.9$ pF; $r_{bb'} = 215\ \Omega$

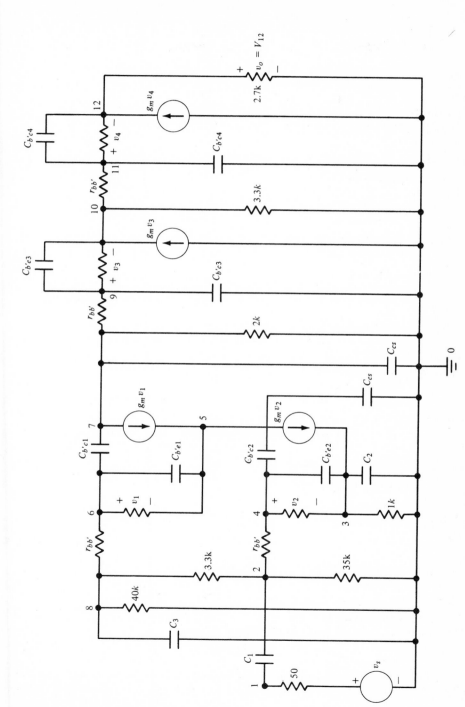

FIG. 9-11 The CA3018 cascode amplifier for ac analysis.

> =
tune *f*
ESTIMATED FREQ IN HZ
?
1000
> =
maximize
NODE DIFF TO MAXIMIZE
?
12   0
F =   .28090D+06 HZ
V   12-V   0   =   37.860   DECIBELS
> =
tune *f*
ESTIMATED FREQ IN HZ
?
1
> =
set
VALUE IN DECIBELS
?
34.86
NODE DIFF TO SET
?
12   0
F =   6964.1   HZ
V   12-V   0   =   34.850   DECIBELS
> =
tune *f*
ESTIMATED FREQ IN HZ
?
1*e*8
> =
set
VALUE IN DECIBELS
?
34.86
F =   .10391D+08 HZ
V   12-V   0   =   34.858   DECIBELS
> =

The procedure is self-explanatory. The maximum gain was found to be
37.86 dB: the lower 3-dB point (34.85 dB) 6,964.1 Hz and the upper 3-dB point

0.10391 × 10⁸ Hz or 10.391 MHz. The node difference referred to is the output or $V_{12} - V_0$ of Fig. 9-11. We can see that the hybrid-π model we applied is quite adequate since the measured values are 37. ± 1 dB, 6 kHz and 11 MHz, respectively. Without the two collector-to-substrate capacitances, the upper 3-dB point was found to be 17 MHz. Thus, we cannot neglect the collector-to-substrate capacitances and expect the results to compare with measurement.

Having confidence in the model we can now move on to the optimization which will compute the three capacitances C_1, C_2, and C_3 in Fig. 9-8 to determine the low-frequency response. First, the g_m, $C_{b'e}$, and $r_{b'e}$ for $V_{CC} = E_{CC} = 5$ V from Table 9-1 are inserted into the computer. For comparison we evaluate the new gain and bandwidth for the circuit of Fig. 9-8. Using the procedure described above, the voltage gain is

$$A_v = 35.88 \text{ dB}$$

and the 3-dB points are 5.63 kHz and 10.55 MHz.

When the results are compared to those calculated at $E_{CC} = 6$ V, we find that the low-frequency response decreases by 19 percent. The upper 3-dB point increases by 1.5 percent and the midband gain decreases by 2 dB. These computed values are very close to the actual manufacturer's measured values of 37 ± 1 dB, 6 kHz and 11 MHz.

Observing Figs. 9-8 and 9-9, the dc voltages across capacitance C_1, C_2, and C_3 are V_8, V_9, and V_5, respectively. Thus, C_1 should have a maximum voltage rating greater than $V_8 = 1.68$ V, C_2 should have a rating greater than 1.02 V, and C_3 a rating greater than 1.89 V. Electrolytic capacitances rated at 6 WVDC (working volts direct current) are chosen for the application.

Assume the cost of 6 WVDC electrolytic capacitances is based on a normalized value of 1 for a 10-μF capacitance as given below:

10, 20, 30	1 unit
50	1.14
100	1.27
150	1.5
250, 300	1.55
500	1.68
800	1.77

The tolerances of the above capacitances are −10 percent and +75 percent.

We shall start by performing a "brute force" optimization by tuning all three capacitances C_1, C_2, and C_3 simultaneously to produce 32.88 dB at 20 Hz. The procedure employed is shown below.

```
> =
tune c
```

```
NODES OF C
?
1   2
> =
tune c
NODES OF C
?
0   8
> =
tune c
NODES OF C
?
0   3
> =
set
VALUE IN DECIBELS
?
32.88
NODE DIFF TO SET
?
12   0
FREQ IN HZ
?
20
C     1   2     1241.5   MICROFARAD
C     0   8     18.490   MICROFARAD
C     0   3     280.13   MICROFARAD
V     12-V  0  =  32.857   DECIBELS
```

These capacitance values will produce a frequency response which is 3-dB down at 20 Hz. They are not the minimum cost, however. It is valuable to know just how small each of the capacitances can be to still produce 3-dB down at 20 Hz. This can be done easily by making C_2 and C_3 large (1 F) and tuning C_1 to produce 3-dB down gain at 20 Hz. We will find the minimum value of C_1. The procedure is shown below.

```
> =
c
NODES,MICROFARAD
?
0   8   1e6
```

```
> =
c
‾
NODES,MICROFARAD
?
0   3   1e6
> =
tune c
‾‾‾‾‾‾
NODES OF C
?
1   2
> =
set
‾‾‾
VALUE IN DECIBELS
?
32.88
‾‾‾‾‾
NODE DIFF TO SET
?
12   0
‾‾‾‾
FREQ IN HZ
?
20
‾‾
C      1   2    6.3043   MICROFARAD
V    12-V   0 = 32.852   DECIBELS
> =
```

Thus, C_1—the capacitance from node 1 to node 2—cannot be less than 6.3 μF to produce 3 dB down at 20 Hz.

Similarly, making $C_1 = C_3 = 10^6$ μF minimum capacitance for C_2 is found to be 279.47 μF. An attempt to perform the same procedure for $C_3(C_1 = C_2 = 10^6$ μF) produced the "no-convergence" message. Apparently, it is not possible to produce a 3 dB down gain with $C_3 = 0$ and C_1 and C_2 large.

Making $C_1 = C_3 = 30$ μF, the minimum cost capacitance, we canfind the value of C_2 which gives a 3 dB down gain at 20 Hz. The procedure is as follows.

```
> =
c
‾
NODES, MICROFARAD
?
1   2   30
c
‾
```

NODES, MICROFARAD
?
0   8   30
$>=$
tune c
NODES OF C
?
3   0
$>=$
set
VALUE IN DECIBELS
?
32.88
NODE DIFF TO SET
?
12   0
FREQ IN HZ
?
20
C   0   3   323.25   MICROFARAD
V   12-V   0 = 32.872   DECIBELS
$>=$

A 500-μF capacitance will be required for C_2. The two 30-μF and one 500-μF capacitances have a total cost of 3.68 units. Perhaps a lower cost could be achieved with $C_2 = 300\ \mu$F and increasing one of the other capacitances. Further application of the optimization routines produced the following capacitances for a 3 dB down gain at 20 Hz: (1) $C_1 = 50$ F, $C_2 = 300$ F, and $C_3 = 33\ \mu$F requiring the use of a 50-μF capacitance for C_3 resulting in a total cost of 3.83 units and (2) $C_1 = 30\ \mu$F, $C_2 = 300\ \mu$F, and $C_3 = 52\ \mu$F requiring the use of a 100-μF capacitance for C_3 resulting in a total cost of 3.82 units.

Thus the minimum cost appears to be 3.68 units with

$$C_1 = C_3 = 30\ \mu\text{F} \qquad \text{and} \qquad C_2 = 500\ \mu\text{F}$$

The response of the amplifier using these component values and $V_{CC} = E_{CC} = 5$ V is shown in Fig. 9-12. The final design has a midband gain of 35.88 dB and a bandwidth from 14.1 Hz to 10.55 MHz.

9-4   TUNED AMPLIFIERS

The y parameters, as presented in the previous section, are an excellent starting point in the design of tuned amplifiers. In this section we shall design a 30-MHz

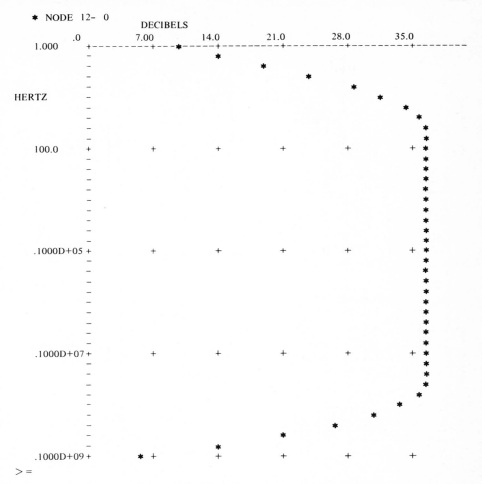

FIG. 9-12   LINCAD frequency response of the CA3018 cascode amplifier.

RF amplifier using the μA703 integrated circuit. The bandwidth is to be 1 MHz. The y parameters of the μA703 as a function of frequency are shown in Fig. 9-13.

Consider the equivalent circuit of the amplifier as expressed in terms of y parameters as shown in Fig. 9-14a where the load is an admittance Y_L. The Norton equivalent circuit of the source is a current source I_s and a source admittance Y_s.

The node equations for the circuit of Fig. 9-14a are simply Eq. (9-1a and b) with the addition of expressions for I_1 and I_2 or

$$I_1 = y_{11}V_1 + y_{12}V_2 = I_s - Y_sV_1$$

$$I_2 = y_{21}V_1 + y_{22}V_2 = -Y_LV_2$$

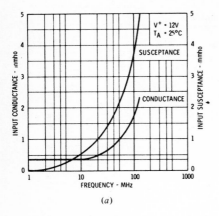

(a)

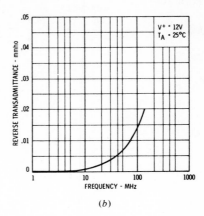

(b)

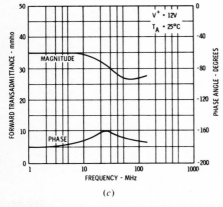

(c)

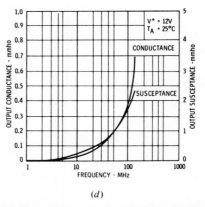

(d)

FIG. 9-13   The y parameters of the μA703. (a) Input admittance as a function of frequency; (b) maximum reverse transadmittance as a function of frequency; (c) forward transadmittance as a function of frequency; (d) output admittance as a function of frequency. (*By permission Fairchild Semiconductor Corp., Mountain View, Cal.*)

Rearranging these equations,

$$I_s = (y_{11} + Y_s)V_1 + y_{12}V_2 \tag{9-3a}$$

$$0 = y_{21}V_1 + (y_{22} + Y_L)V_2 \tag{9-3b}$$

Equation (9-3) can be solved for the voltage gain, current gain, power gain, input admittance, and output admittance. All these quantities will be complex since the y parameters source and load admittance are complex at high frequencies.

The various gains, the input admittance, and output admittance can be

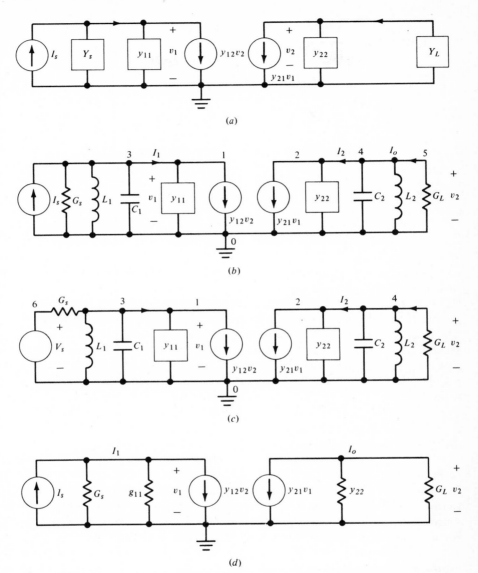

FIG. 9-14 (a) The equivalent circuit of the *RF* amplifier; (b) the tuned amplifier with a Norton source; (c) the tuned amplifier with a Thévenin source; (d) the equivalent circuit at f_0.

shown to be

$$Y_i = y_{11} - \frac{y_{12}y_{21}}{y_{22} + Y_L} \tag{9-4a}$$

$$Y_o = y_{22} - \frac{y_{12}y_{21}}{y_{11} + Y_s} \tag{9-4b}$$

$$A_v \equiv \frac{V_2}{V_1} = -\frac{y_{21}}{y_{22} + Y_L} \tag{9-5a}$$

$$A_i \equiv \frac{I_o}{I_1} = \frac{G_L y_{21}}{y_{11}(y_{22} + Y_L) - y_{12} y_{21}} = -\frac{G_L}{Y_i} A_v \tag{9-5b}$$

$$A_{is} \equiv \frac{I_o}{I_s} = \frac{G_L y_{21}}{(y_{11} + Y_s)(y_{22} + Y_L) - y_{12} y_{21}} = -\frac{G_L A_v}{Y_i + Y_s} \tag{9-5c}$$

$$A_{vs} \equiv \frac{V_2}{V_s} = \frac{-G_L^{-1} I_o}{G_s^{-1} I_s} = \frac{G_s}{G_L} A_{is}$$

$$G_p = \frac{P_o}{P_i} = \frac{G_L V_2^2}{G_i V_1^2} = \frac{G_L}{G_i} A_v^2 \tag{9-5d}$$

The voltage V_s is the voltage of the source in a Thévenin equivalent circuit replacing the Norton equivalent circuit I_s and G_s. The amplifier is unstable if $G_p \rightarrow \infty$. Stated another way, the amplifier is stable if the real part of Y_i is greater than zero, or $G_i > 0$. The amplifier is conditionally stable if $A_{is} \rightarrow \infty$. The amplifier is considered conditionally stable if the real part of $(Y_i + Y_s)$ is greater than zero or $G_i + G_s > 0$. In both cases, the value of y_{12} must be minimized for stable operation. In transistors, y_{12} is primarily a result of the base-to-collector capacitance $C_{b'c}$. Neutralization is a common way to reduce y_{12}. A simple form of neutralization is an inductor connected between output and input. The inductance resonates with $C_{b'c}$ producing a smaller effective value of y_{12}. The μA703 was designed to minimize y_{12}, and an amplifier can be designed using the μA703 with no external neutralization.

The gain of a tuned amplifier is defined as the gain at the design frequency f_o. In our case $f_o = 30$ MHz. The input and output are conjugately matched at f_o, accomplished by tuned circuits as shown in Fig. 9-14b and c. In Fig. 9-14b, the source is represented by a Norton equivalent circuit, whereas in Fig. 9-14c, the source is represented by a Thévenin equivalent circuit. A conjugate match at f_o is equivalent to resonance of the input and output circuits at f_o.

The parameters y_{11} and y_{22} and the source and load admittances can be expressed in real and imaginary parts as,

$$y_{11} = g_{11} + jb_{11}$$

$$y_{22} = g_{22} + jb_{22}$$

$$Y_L = G_L + jB_L = G_L + j\left(\omega C_2 - \frac{1}{\omega L_2}\right)$$

$$Y_s = G_s + jB_s = G_s + j\left(\omega C_1 - \frac{1}{\omega L_1}\right)$$

where $\omega = 2\pi f$. The input and output admittance are

$$Y_i = G_i + jB_i$$
$$Y_o = G_o + jB_o$$

At f_o when the input and output are conjugately matched, $G_s = G_i$, $G_L = G_o$, $B_s = -B_i$, and $B_L = -B_o$.

These matching conditions are rather complicated to solve exactly by hand since Y_s must be known to find Y_L, and vice-versa. A reasonable simplification is to consider y_{12} sufficiently small and conjugately match Y_s to y_{11} and Y_L to y_{22}, respectively, as shown in Fig. 9-14d.

The various gains given in Eqs. (9-4a) through (9-5d) for conjugate matching of Y_s to y_{11} and Y_L to y_{22} become

$$A_{vm} = \frac{y_{21}}{2g_{22}}$$

$$A_i = -\frac{G_L}{Y_i} A_v$$

$$A_{vsm} = -\frac{y_{21}}{4g_{22}} = \frac{1}{2} A_{vm}$$

$$A_{is} = \frac{g_{22}y_{21}}{4g_{11}g_{22} - y_{12}y_{21}} \tag{9-6a}$$

$$G_{pm} = \frac{g_{22}}{g_{11}} \frac{|y_{21}|^2}{4g_{22}^2} = \frac{|y_{21}|^2}{4g_{11}g_{22}} \tag{9-6b}$$

The amplifier is unconditionally stable if $G_i > 0$. Since $G_i = g_{11} + Re(y_{12}y_{21})/2g_{22}$ where "Re" is an abbreviation for "real part of", then

$$2g_{11}g_{22} > Re(y_{12}y_{21}) \tag{9-7a}$$

It is conditionally stable if $G_i + G_s > 0$. Since $G_s = g_{11}$, it is conditionally stable if

$$4g_{11}g_{22} > Re(y_{12}y_{21}) \tag{9-7b}$$

We shall now calculate the gain and required component values for a 30-MHz tuned amplifier using the μA703. From Fig. 9-13 at $f = 30$ MHz and $V_{CC} = 12$ V,

$$y_{11} = 0.7 + j1.4 \text{ Millimhos}$$

$$y_{12} = j.0035 \text{ mmho}$$

$$y_{21} = 30\underline{/-162°} \text{ mmhos}$$

$$y_{22} = 0.1 + j.6 \text{ mmhos} \tag{9-8}$$

The input admittance at 30 MHz with conjugate matching of Y_s to y_{11} and Y_L to y_{22} is,

$$Y_i = g_{11} - \frac{y_{12}y_{21}}{2g_{22}} = 0.86 + j.5 \text{ Millimhos}$$

We see that the real part of Y_i is greater than zero. Thus, the amplifier is unconditionally stable at 30 MHz without neutralization. The voltage, current, and power gain from Eq. (9-6) are

$$A_{vm} = \frac{y_{21}}{2g_{22}} = -150$$

$$A_{vsm} = -75$$

$$A_{ism} = 10.7$$

$$A_{im} = 15.08 \tag{9-9}$$

$$G_{pm} = \frac{|y_{21}|^2}{4g_{11}g_{22}} = 3214 \rightarrow 35 \text{ dB}$$

The conjugate match of y_{11} to Y_s and y_{22} to Y_L requires

$$G_s = 0.7 \text{ mmho}$$

and

$$G_L = 0.1 \text{ mmho}$$

or $R_s = 1/G_s = 1.43$ k and $R_L = 1/G_L = 10$ k.

The bandwidth requirement of 1 MHz requires that the power gain be 3 dB down at 29.5 and 30.5 MHz, that is,

$$G_p(30.5 \text{ MHz}) = \tfrac{1}{2}G_{po}$$

or

$$A_v(30.5 \text{ MHz}) = 0.707A_{vo}$$

The zero subscript indicates the center frequency solution at $f = f_o$.

In many designs the 3-dB loss at the upper or lower 3-dB points is associated with the output resonant circuit. The input circuit is made quite broad in bandwidth. Assuming a bandwidth of 1 MHz in the output circuit, the following analysis will produce the required component values.

The output circuit Q is given by

$$Q_2 = \frac{f_o}{BW} = \frac{1}{(G_L + g_{22})(\omega_o L_2)} = \frac{\omega_o C_2 + b_{22}}{G_L + g_{22}}$$

$$= \frac{\omega_o C_2 + b_{22}}{2g_{22}}$$

where BW is the bandwidth and f_o is the center frequency. The required external capacitance is

$$\omega_o C_2 = \frac{2f_o g_{22}}{BW} - b_{22} = 5.4 \text{ mmhos}$$

$$C_2 = \frac{5.4 \times 10^{-3}}{2\pi \times 30 \times 10^6} = 28.6 \text{ pF}$$

Then,

$$L_2 = \frac{1}{\omega_o(\omega_o C_2 + b_{22})} = 0.884 \ \mu\text{H}$$

The input bandwidth will be made ten times greater than the output circuit or 10 MHz. Following a similar approach

$$\omega_o C_1 = \frac{2f_o g_{11}}{BW} - b_{11}$$

$$\omega_o C_1 = 2 \times 3 \times 0.7 - 1.4 = 2.8 \text{ mmhos}$$

$$C_1 = \frac{2.8 \times 10^{-3}}{\omega_o} = 14.9 \text{ pF}$$

$$L_1 = \frac{1}{\omega_o(\omega_o C_1 + b_{11})} = 1.26 \ \mu\text{H}$$

The design at this point is summarized as follows:

$$R_s = 1.43 \text{ k}$$

$$R_L = 10 \text{ k}$$

$$C_1 = 14.9 \text{ pF}$$

$$L_1 = 1.26 \ \mu\text{H} \tag{9-10}$$

$$C_2 = 28.6 \text{ pF}$$

$$L_2 = 0.884 \ \mu\text{H}$$

In practice it is usually desired to have a source impedance of 50 Ω and a load impedance of 50 Ω for operation into coaxial lines. The circuits of Fig. 9-15 show how such a match can be accomplished and yet have the integrated circuit see a source resistance of 1.43 k.

The reader is referred to the literature [2] for a more complete discussion of impedance matching. It is worthwhile, however, to calculate the approximate turns ratio if an output transformer were employed. The load presented to the transistor by an ideal transformer is given by,

$$R_L = n_2^2 R'_L$$

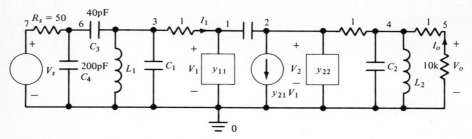

FIG. 9-15 The LINCAD circuit which includes the Input Matching Circuit.

Thus,

$$n_2{}^2 = \frac{R_L}{R_L'} = \frac{10\text{ k}}{50} = 200$$

or

$$n_2 = 14.1$$

Thus a transformer with a turns ratio of 14:1 will match the output to a 50-ohm load. The input matching circuit shown in Fig. 9-15 is completely accomplished by the two capacitances C_3 and C_4. The Norton equivalent admittance of the source resistance R_s and the two capacitances C_3 and C_4 is given (see Fig. 9-15) by:

$$Y_s' = \cfrac{1}{\cfrac{1}{j\omega C_3} + R_s \,\middle|\middle|\, \cfrac{1}{j\omega C_4}}$$

which becomes

$$Y_s' = \frac{\omega^2 R_s C_3{}^2}{1 + \omega^2 R_s{}^2 (C_3 + C_4)^2} + j\omega C_3 \frac{1 + \omega^2 R_s{}^2 (C_3 + C_4) C_4}{1 + \omega^2 R_s{}^2 (C_3 + C_4)^2} \tag{9-11}$$

For $C_3 = 40$ pF, $C_4 = 200$ pF, $R_s = 50$ Ω, and $\omega = \omega_o = 2\pi f_o = 1.88(10^8)$

$$Y_s' = 0.465 + j6.49 \text{ Millimhos}$$

The real part is $1/0.465 = 2.15$ kΩ which reasonably matches the 1.43 kΩ of g_{11}. The reactive part of Y_s' is capacitive and must be included in the total C_1 shown in Fig. 9-15.

This completes the hand calculations of the *RF* amplifier design. Actually, we could have computed the matching network components with LINCAD. However, to understand the principles, some hand calculations were considered useful here.

The entire circuit of the μA703 could be simulated by using hybrid-π models for each of the five transistors. Since the bandwidth is only 1 MHz, the *y*-parameter representation of Fig. 9-13 will be adequate. The *y* parameters change little considering the limited variation in frequency. The parameters in Fig. 9-13 are for a 12-V supply. If other than a 12-V supply were used, say 6 V, then the y parameters of Fig. 9-13 would not be accurate. We would have to evaluate the new y parameters in some way. LINCAD could be used for this y-parameter evaluation by using the hybrid-π model for each of the five transistors. (See Prob. 9-13.)

We shall design the RF amplifier using the y parameters of Fig. 9-13. Since the parameter y_{fe} is in polar form, we shall use the lower-frequency values and the 3-dB down point for f_{co}. Thus, from Fig. 9-13c

$g_{mo} = 0.035$ mho

$f_{co} = 55$ MHz

Thus,

$$y_{fe} = \frac{0.035}{1 + jf/55 \text{ MHz}}$$

The input and output admittances y_{11} and y_{22} are in rectangular form. We shall use the Y command of LINCAD to enter y_{11} and y_{22}, and then convert them to equivalent RC parallel circuits with the CONVERT TO RLC command. The procedure is as follows:

```
> =
y
NODES,REAL AND IMAG MHOS
?
0   1   .7E-3   1.4E-03
> =
y
NODES,REAL AND IMAG MHOS
?
0   2   .1E-3   .6E-3
> =
Convert to RLC
FREQ IN HZ
?
30E6
R    1    0    1.4286        KILOHMS
R    2    0    10.000        KILOHMS
C    1    0    .74272D-05  MICROFARAD
C    2    0    .31831D-05  MICROFARAD
> =
```

The reverse transfer admittance is quite small and could be omitted. We shall model y_{12} with a capacitance between input and output. (Input and output are nodes 2 and 3, respectively.) A circuit similar to Fig. 9-2 for the measurement of y_{12} was entered into LINCAD. The tune option was used to evaluate the capacitance so that $y_{12} = 0.0035$ mmhos. The result is that a capacitance of 0.0186 pF will produce a reverse transfer admittance of 0.0035 mmhos at 30 MHz. Thus circuit operation is rather independent of y_{12}, and the gain calculated in Eq. (9-9) for conjugate matches of Y_s to Y_i and Y_L to Y_o is to be expected.

The circuits in Fig. 9-14b and c are to be simulated using LINCAD. The component values are given in Eq. (9-10). One-Ω resistances are placed between nodes 1 and 3, nodes 2 and 4, and nodes 4 and 5 to permit measurement of current and separation of the tuned circuit from the y-parameter model. With $I_s = 1,000$ mA, the various gains are

$$A_{is} \equiv \frac{I_o}{I_s} = \frac{V_5 - V_4}{1} = V_5 - V_4$$

$$A_v \equiv \frac{V_2}{V_1}$$

The FIND BW command was used to find the center frequency and bandwidth of the various circuits. The following was printed when the circuit of Fig. 9-14b with the Norton equivalent source was analyzed:

```
> =
find bw
NODE DIFF TO FIND BW
?
5   0
ESTIMATED FREQ IN HZ
?
1E6
CENTER FREQ=       .29823D+08 HZ        99.407      DB
F3DB=        .30429D+08 HZ
F3DB=        .29251D+08 HZ        .11691D+07 = BW IN HZ
> =
vtvm
FREQ IN HZ
?
29.823E6
V  5-V  4   =   9.3398   V,   −24.18 DEG
V  2-V  0   =   93417.   V,    155.78 DEG
V  1-V  0   =   652.66   V,   −16.93 DEG
V  5-V  0   =   93398.   V,    155.82 DEG
> =
```

Thus,

$$A_{is} = V_5 - V_4 = 9.34\underline{/-24.18°}$$

$$A_v = V_2/V_1 = 143\underline{/172.71°}$$

The analysis with the Thévenin equivalent source of Fig. 9-14c was found to be

$$A_{vs} = \frac{V_2}{V_s} = \frac{V_2}{1} = V_2 = 76.4\underline{/151.36°}$$

very near the design value of $75\underline{/180°}$.

The current gains A_{is} and A_v are somewhat lower than predicted by the hand calculation neglecting y_{12}. The problem was also solved with the capacitance between nodes 1 and 2 removed, that is, with $y_{12} = 0$. The results were

$$A_{is} = 10.94\underline{/-28.61°}$$

$$A_v = 153\underline{/151.56°}$$

$$A_{vs} = 76.5\underline{/151.36°}$$

$$f_o = 30.024 \text{ MHz}$$

$$BW = 0.98 \text{ MHz}$$

These results are nearly the same as those calculated by hand. Even so, the hand calculations are within 15 percent of the LINCAD simulation which includes y_{12}. Of course, if y_{12} were larger, the differences would be much greater, causing us to have less confidence in the hand calculations which omit the effect of y_{12}.

One advantage of LINCAD is that components can be added, removed, or changed without leaving the program. Most CAD programs require the user to exit from the program, edit a data set, and then reload (and sometimes recompile) the computer program. The interactive nature of LINCAD eliminates these time-consuming and costly steps. In Fig. 9-14, the change in going from the circuit of (b) to that in (c) required five commands; two to remove I_s and G_s, and three to insert V_s, a zero resistance in series with V_s, and the new connections for G_s.

The calculation of the circuit in Fig. 9-15 with the components listed in Eq. (9-10) was

```
>=
find bw
NODE DIFF TO FIND BW
?
5   0
ESTIMATED FREQ IN HZ
?
1E6
CENTER FREQ=   .30010D+08 HZ        36.350       DB
F3DB=          .29447D+08 HZ
F3DB=          .30538D+08 HZ        .19860D+07=BW IN HZ
>=
vtvm
```

FREQ IN HZ
?
30.01E6
V 5-V 4 = .65686D-02 V, −82.57 DEG
V 3-V 1 = .90224D-03 V,   8.40 DEG
V 5-V 0 = 65.686   V,   97.48 DEG
V 1-V 0 = .42844   V, −54.76 DEG

The center frequency from this data is 30.01 MHz and the bandwidth 1.086 MHz, very near the desired values. The voltage gain is, with an input of 1 V,

$$A_{vs} = V_5 \cong V_2 = 65.7\underline{/97°}$$

somewhat lower than the design value of $75\underline{/180°}$ and the analysis of Fig. 9-14c. The current gain A_i is

$$A_i = \frac{(V_5 - V_4)/1\ \Omega}{(V_3 - V_1)/1\ \Omega} = \frac{V_5 - V_4}{V_3 - V_1} = 7.38\underline{/-90.97°}$$

Removing the capacitance between nodes 1 and 2 (making $y_{12} = 0$), we obtained

$$f_o = 30.007\ \text{MHz}$$

$$\text{bandwidth} = 1.0006\ \text{MHz}$$

$$A_{vs} = 70.8\underline{/97.5°}$$

The voltage gain is closer to the predicted values of $75\underline{/180°}$. The reason that the difference occurs is that the capacitance C_3 and C_4 contribute a net capacitance which is detuning the input circuit. The center frequency and bandwidth are very near the design value because we originally designed the circuit to have the bandwidth primarily determined by the output circuit. The input circuit was purposely designed with a broad 10-MHz bandwidth. The fact that the input is not conjugately matched can also be observed from the voltage at node 1 in Fig. 9-15, 0.428 V. When the circuit is conjugately matched, node 1 voltage should be 0.5 V.

The TUNE C will now be applied to C_1 with the value of V_1 to be optimized at 0.5 V. The following was obtained:

> =
tune c
NODES OF C
?
3   0
> =
set

VALUE IN VOLTS
?
<u>.5</u>
<u>NODE DIFF TO SET</u>
?
<u>1 0</u>
<u>FREQ IN HZ</u>
?
<u>30E6</u>
C 0 3 .12227D-04 MICROFARAD
V 1-V 0 = .49996 VOLTS
> =
<u>find bw</u>
<u>NODE DIFF TO FIND BW</u>
?
<u>5 0</u>
<u>ESTIMATED FREQ IN HZ</u>
?
<u>1E5</u>
CENTER FREQ = .30007D+08 HZ 37.681 DB
F3DB = .29491D+08 HZ
F3DB = .39491D+08 HZ .10005D+07 = BW IN HZ
> =
<u>vtvm</u>
<u>FREQ IN HZ</u>
?
<u>30.007E6</u>
V 5-V 4 = .76565D-02 V, −81.66 DEG
V 3-V 1 = .78232D-03 V, 8.23 DEG
V 5-V 0 = 76.565 V, 98.34 DEG
V 1-V 0 = .49971 V, −55.21 DEG

The capacitance which produces a conjugate match at the input is

$$C_1 = 12.227 \text{ pF}$$

The voltage gain is

$$A_{vs} = V_5 = 76.565\underline{/98.34°}$$

nearly the same as that obtained from the analysis of Fig. 9-14c without y_{12}. Actually y_{12} was also detuning the circuit of Fig. 9-14b and 9-14c as applied in the hand-calculated values of Eq. (9-10).

The results show that the hand calculation was quite good under the condition that y_{12} is negligible. Proper tuning produced the predicted voltage gain. The center frequency and bandwidth changed little with y_{12} or input tuning since both were a function of the output circuit.

9-5 SINUSOIDAL OSCILLATORS

In Chap. 2 it was shown that feedback in operational amplifiers could cause instability. The instability shows up as oscillation, an undesirable effect in amplifiers. On the other hand, when we desire a source generator of some given frequency, we purposely introduce feedback to produce oscillation at that frequency.

There are many approaches to the presentation of sinusoidal oscillator design. One approach shown in Fig. 9-16 considers the oscillator as an amplifier with feedback. The circuit is divided into two parts, the amplifier and the feedback network. The amplifier has an output of

$$v_o = | A_v | (v_s - v_f) \tag{9-12}$$

and the feedback network is described by the transfer function

$$a(j\omega) = a \equiv \frac{v_f}{v_o} \tag{9-13}$$

Substituting $v_f = a v_o$ into Eq. (9-12),

$$v_o = | A_v | (v_s - av_o)$$

rearranging terms,

$$v_o = \frac{| A_v | v_s}{1 + a | A_v |} \tag{9-14}$$

or

$$A_{vf} \equiv \frac{v_o}{v_s} = \frac{| A_v |}{1 + a | A_v |} \tag{9-15}$$

where A_{vf} is the voltage gain with feedback. If $(1 + a | A_v |) > 1$, then $A_{vf} < | A_v |$, and we say that negative feedback was introduced. If $0 < (1 + a | A_v |) < 1$,

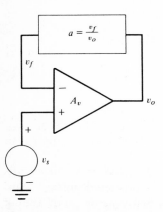

FIG. 9-16   The representation of a sinusoidal oscillator as an amplifier with feedback.

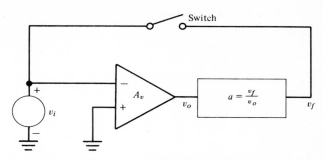

FIG. 9-17 Block diagram of an oscillator arranged for a linear analysis program.

then $A_{vf} > |A_v|$, and we say that the circuit has positive feedback. If $(1 + a|A_v|) = 0$, $A_{vf} \to \infty$, a result which can be interpreted as v_o being non-zero for $v_s = 0$. Stated another way, we can say that if $(1 + a|A_v|) = 0$ and $v_s = 0$ in Fig. 9-16, then from Eq. (9-14) $v_o = 0/0$ or v_o is indeterminate and can be nonzero. Thus, with $v_s = 0$, no signal source is required to produce an output. Of course, we know that the device is converting dc power to sinusoidal power.

The frequency of oscillation is dependent on the feedback network. At f_o, the frequency of oscillation, $(1 + a|A_v|) = 0$ or

$$a = a(j\omega_o) = a(j2\pi f_o) = -\frac{1}{|A_v|} \tag{9-16}$$

We cannot simulate the circuit of Fig. 9-16 exactly with a linear ac analysis program since making $v_s = 0$ will result in $v_o = 0$. The circuit is modified slightly by opening the feedback loop, as shown in Fig. 9-17. The gain of the amplifier is

$$\frac{v_o}{v_i} = -|A_v|$$

and the output of the feedback network is

$$v_f = av_o = -a|A_v|v_i \tag{9-17}$$

If $v_f = v_i$, then

$$v_f = v_i = -a|A_v|v_i$$

or

$$(1 + a|A_v|)v_i = 0$$

For use as an oscillator, $v_i \neq 0$. Thus,

$$(1 + a|A_v|) = 0 \tag{9-18}$$

the same as derived for the circuit of Fig. 9-16. Now if $v_f = v_i$ with the switch open, then no change would occur in the circuit if the switch were closed. (The switch will have zero current in it.) The source v_i could also be removed (not in the simulation), and no change would occur.

Thus a sinusoidal oscillator circuit can be analyzed by a linear-analysis program if the loop is broken and a source is placed at the input of the amplifier. The circuit will oscillate if $v_f = v_i$ or if Eq. (9-16) is satisfied. Conversely, the gain required for oscillation is

$$|A_v| = -\frac{1}{a(j2\pi f_o)} = -\frac{1}{a(j\omega_o)} \qquad (9\text{-}19)$$

We have assumed the amplifier to have a 180° phase difference between output and input. Equation (9-19) specifies that, at the frequency of oscillation, the feedback network must produce 180° phase shift. Thus the net phase shift between v_i and v_f is 360°. Actually the requirement for $v_i = v_f$ is that the net phase shift between v_i and v_f be $360n$ where n is an integer and that the magnitude of v_i be equal to the magnitude of v_f.

The feedback network can be a transmission line, an LC circuit, an RC circuit, or any circuit which can produce 180° phase shift.

Consider the RC ladder network in Fig. 9-18. Three capacitances are required since each can produce, at the most, slightly less than a 90° phase shift. (We need 180°.) The circuit was entered into the computer to find the frequency at which the output was 180° out of phase with the input and to find the gain required of the amplifier. The LINCAD commands and results are as follows:

> =
phase
> =
tune f
ESTIMATED FREQ IN HZ
?
10
> =
set
VALUE IN DEGREES
?
−180

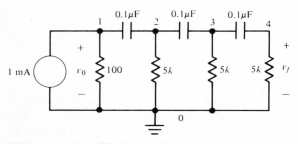

FIG. 9-18   An RC ladder feedback network.

NODE DIFF TO SET
?
4 0
$\overline{\text{F}=}$ 129.00   HZ
V 4-V 0   =   −180.04   DEGREES
$\overline{>=}$
vtvm
$\overline{\text{FREQ IN HZ}}$
?
129
$\overline{\text{V}}$ 1-V   0   =   .99793D−01   V,   −0.38   DEG
V   4-V   0   =   .33893D−02   V, . −179.96   DEG

At 129 Hz the transfer function of the feedback network is

$$a = v_4/v_1 = 0.03396\underline{/-179.58} \simeq -0.03396$$

The gain of the amplifier required for operation as an oscillator is

$$|A_v| = -\frac{1}{-0.03396} = 29.44$$

Figure 9-19 shows the entire circuit with the resistors $R_f = 560$ k and $R_1 = 19.02$ k to produce an amplifier with a $|A_v| = 29.44$. As in Chap. 6, the transconductance g_m was calculated from the operational amplifier gain of 100,000 and the output resistance of 100 Ω to give

$$g_m = \frac{|A_{vo}|}{R_o} = \frac{100,000}{100} = 1,000 \text{ mho}$$

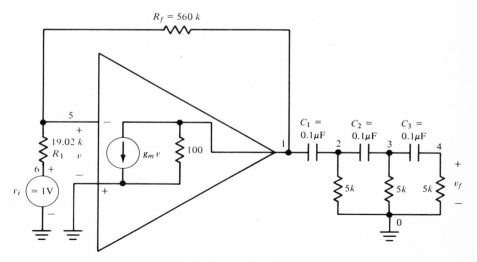

FIG. 9-19   The oscillator circuit.

The results of the TUNE F command to produce $v_f = v_i$, that is, $v_6 - v_4 = 0$ are as follows:

> =
magnitude
> =
tune f
ESTIMATED FREQ IN HZ
?
10
> =
minimize
NODE DIFF TO MINIMIZE
?
6   4
F =   129.20   HZ
V   6-V   4   =   .652970−02   VOLTS
> =
vtvm
FREQ IN HZ
?
129.9
V   6-V   0   =   1.0000   V,   0.00   DEG
V   4-V   0   =   1.0029   V,   0.34   DEG

The frequency of oscillation is 129.2 Hz. The slight difference between the analysis of the feedback network and the entire circuit is a result of the difference between the convergence criterion of SET and MINIMIZE.

The circuit in Fig. 9-19 can be adjusted to oscillate at other frequencies by employing the TUNE C command of LINCAD. If the frequency of oscillation desired is 500 Hz, the following is obtained:

> =
tune c
NODES OF C
?
1   2
> =
tune c
NODES OF C
?
2   3
> =
tune c

```
NODES OF C
?
3   4
> =
minimize
NODE DIFF TO MINIMIZE
?
6   4
FREQ IN HZ
?
500
NO CONVERGENCE,MORE ITERATIONS.YES OR NO
yes
C   1   2      .27037D−01   MICROFARAD
C   2   3      .22002D−01   MICROFARAD
C   3   4      .28798D−01   MICROFARAD
V  6-V  4  =   .63947D−08   VOLTS
> =
vtvm
FREQ IN HZ
?
500


V  1-V  0  =   29.434  V,   180.00  DEG
V  4-V  0  =   1.0000  V,     0.00  DEG
V  6-V  0  =   1.0000  V,     0.00  DEG
```

The actual circuit is constructed by omitting the 1-V source at node 6 and connecting node 4 to node 6.

The circuit was constructed using a 741 operational amplifier operating from ± 15-V power supplies. The circuit with $C_1 = C_2 = C_3 = 0.1\ \mu\text{F}$ and $R_1 = 5$ k oscillated at 137 Hz. The resistance R_1 was adjusted to 17.2 k for best operation. The output was sinusoidal with a peak-to-peak output of 26 V.

The capacitances were changed to those values computed for oscillation at 500 Hz. The actual oscillator operated at 508 Hz when R_1 was adjusted to 16.1 k. Peak-to-peak output was 26 V.

The results are within experimental error. The variation of R_1 from the computed value of 19 k is a result of the nonideal nature of the 741 operational amplifier. The gain of the amplifier was not quite high enough. Also the gain is frequency dependent requiring some trimming of R_1 as the frequency changed. In practice some type of automatic gain control would be included to compensate for component variation.

REFERENCES

1. Linvill, J. G. and L. G. Schimpf, The Design of Tetrode Transistor Amplifiers, *Bell System Tech. J.*, vol. 35, pp. 813–814, July, 1956.
2. Linvill, J. G. and J. F. Gibbons, "Transistors and Active Circuits," McGraw-Hill Book Company, New York, 1961.
3. Theriault, G. E., Leidich, A. J. and T. H. Campbell, Application of the RCA-CA3018 Integrated-Circuit Transistor Array, RCA Application Note ICAN-5296, October, 1967.

PROBLEMS

9-1. Obtain the y parameters of the CA3018 as a function of frequency for $I_C = 5$mA. The circuit shown in Fig. P9-2 could fulfill the requirements for the amplifier requested in Sect. 9-3. Measurement for $V_{CC} = 6$ V gave a midfrequency gain of 49 dB and breakpoints of 800 Hz and 21 MHz. Problems 9-2 through 9-6 deal with the design.

9-2. Find the dc solution for $V_{CC} = 6$ V and $V_{CC} = 5$ V for the circuit in Fig. P9-2.

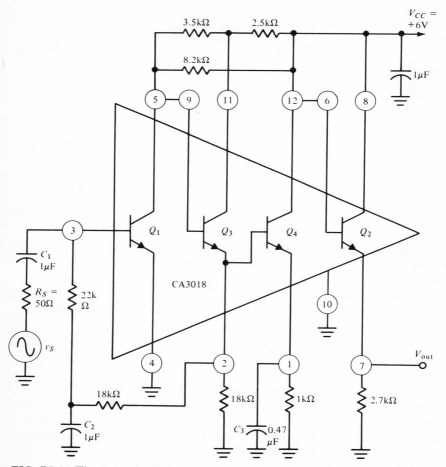

FIG. P9-2 The circuit for Prob. 9-2.

9-3. Sketch the equivalent circuit of the amplifier shown in Figure P9-2 using the hybrid-π model for each transistor. $V_{CC} = 6$ V.

9-4. Simulate the circuit of Fig. P9-2 for $V_{CC} = 6$ V using LINCAD or a similar CAD program. Compare the performance with actual measurements.

9-5. Repeat Prob. 9-4 with $V_{CC} = 5$ V.

9-6. Find the minimum capacitance unit cost for C_1, C_2, and C_3 to give 3-dB down at 20 Hz for the amplifier of Fig. P9-2 with $V_{CC} = 5$ V.

9-7. Will the CA3018 produce a stable RF amplifier at 30 MHz? Use the y parameters of Fig. 9-1.

9-8. Design an RF amplifier using the CA3018 with a center frequency of 30 MHz and a bandwidth of 1 MHz. Use the y parameters of Fig. 9-1. Calculate R_S, R_L, C_1, C_2, L_1, L_2, and the power gain.

9-9. Using the hybrid-π model for the CA3018 at $I_C = 1$ mA simulate the amplifier designed in Prob. 9-8. Have the computer calculate A_i, A_{is}, A_{vs}, and the bandwidth.

9-10. Assuming that the transistors in the μA703 are reasonably modeled by the 2N2369 compute the dc operating points of the transistors when $V^+ = 12$ V.

9-11. Assuming $r_{bb'} = 125$ Ω, $C_c = 1$ pF, $f_T = 500$ MHz, $g_m = 38I_C$, and $h_{fe} = 150$ calculate the hybrid-π models of the transistors in the μA703 at $V^+ = 12$ V.

9-12. Obtain the y parameters of the μA703 as a function of frequency when $V^+ = 12$ V by simulation of the small-signal equivalent circuit with the hybrid-π models of Prob. 9-11. Compare the results with the manufacturer's measured parameters in Fig. 9-13.

9-13. Repeat Probs. 9-10, 9-11, and 9-12 for $V^+ = 6$ V.

9-14. Obtain the y parameters of the μA703 at 30 MHz for $V^+ = 6$ V (see Prob. 9-13). Repeat the RF amplifier design of the text calculating R_S, R_L, C_1, C_2, L_1, and L_2.

Lincad-users Manual

LINCAD analyzes linear active networks given the elements by type, node connection, and value.

The nodes must be numbered with consecutive integers beginning with zero (∅) as the ground node. The elements are described to the computer by symbols (such as, R for resistance, C for capacitance) and numerical data. Any analysis may be obtained at any time. Of course, to obtain meaningful results the circuit must be inserted first.

The user communicates with the program by selecting a command and then entering numerical data. The teletype types the symbol pair $> =$ to indicate that it's ready for a command. The teletype types the symbol '?' when numerical data are to be entered.

The commands to insert components, select the type of analysis, and list the circuit are given in the following summary:

TO INPUT COMPONENTS

E	Independent voltage source—positive voltage has plus on the higher-numbered node
EI	Imaginary part of an independent voltage source
ER	Real part of an independent voltage source
I	Independent current sources—positive current from the higher- to lower-numbered node
IR	Real part of an independent current source
II	Imaginary part of an independent current source
R	Resistance
G	Conductance
YR	Real part of the admittance
L	Inductance
C	Capacitance
YI	Imaginary part of the admittance
GM	Voltage-controlled current generator—frequency-dependent form
YM	Voltage-controlled current generator—admittance form

SERVICE COMMANDS

LIST	To list the components of an RLC circuit
LIST(Y)	To list the components of an admittance circuit
SAVE	To save a circuit in a previously allocated file. Consult your own computer facility regarding the storage of data
OLD	To retrieve a previously saved circuit
NEW	To begin a new circuit—not required on entering LINCAD
END	To exit gracefully from LINCAD

NODE SELECTION

NODES   To select specific node voltages or node voltage differences to be plotted or printed—the node voltages of nodes 1, 2, 3, 4, and 5 are automatic on entering LINCAD

COMMANDS FOR UNITS SELECTION

The following options determine the output units:

MAGNITUDE	To select the magnitude of voltages as the output; automatic on entering LINCAD
DB	To select voltage in DB
PHASE	To select the phase of node voltages
REAL	To select the real part of the node voltages
IMAGINARY	To select the imaginary part of the node voltages

LINEAR ANALYSIS

ALL	To print all node voltages of an RLC circuit at one frequency
ALL(Y)	To print all the node voltages of an admittance circuit
VTVM	To print the node voltages or voltage difference specified by NODES of an RLC circuit at one frequency
VTVM(Y)	To print the node voltages specified by NODES for an admittance circuit
PLOT F	To plot the frequency response on a linear scale
PLOT LOG	To plot the frequency response on a log scale
PLOT R	
PLOT G	
PLOT L	
PLOT C	To plot the node voltage or voltage differences as a function of the magnitude of an element
PLOT YR	To plot the node voltages as a function of the real part of the admittance
PLOT YI	To plot the node voltages as a function of the imaginary part of the admittance
CONVERT TO Y	To convert the RLC circuit to an admittance circuit
CONVERT TO RLC	To convert the admittance circuit to an RLC circuit

COMMANDS FOR CIRCUIT OPTIMIZATION

TUNE X	To select a variable for optimization
MINIMIZE	To minimize a node voltage difference
MAXIMIZE	To maximize a node voltage difference
SET	To set a node voltage difference to a value

The paragraphs to follow explain the use of the various commands in more detail.

INSTRUCTIONS FOR THE INPUT OF COMPONENTS

Only one resistor, one inductor, one capacitor, and one independent current source are permitted between a pair of nodes. Similar parallel components (R, L, C, and I) must be combined before use in the program unless additional nodes connected by negligibly small resistors are used.

The voltage source specified for a node pair is placed in series with the parallel combination of the R, L, and C specified for the same node pair. Figure 6-1 shows the connection which results when an R, L, C, and E and I are given node numbers of J and K. The voltage source is connected in series with the parallel combination of R, L, C, (or YI) specified for the same node pair. Current generators are connected between the nodes. Note that if a voltage source is specified between nodes J and K then at least one R, L, or C must also be specified between the same nodes. Otherwise the voltage source will not be connected to the rest of the circuit.

The last R (or G or YR), L, C, YI, E, (or ER and EI) I (or IR and II), GM, or YM specified for a node pair is used.

The independent voltage sources are addressed by either E to express them in magnitude or phase or by ER and EI to express them in real and imaginary parts. The same holds for independent current sources.

The current of positive current generators flows from the higher- to the lower-numbered node. The voltage of a positive voltage source has the plus on the higher node. The entry for a $2\emptyset,\emptyset\emptyset\emptyset$-$\Omega$ resistance connected between nodes 3 and 4 is

```
> = R
NODES, KILOHMS
?
3 4 2Ø
```

The node connections and value are typed after the request (NODES, KILOHMS?). The nodes must be integer but can be in any order.

The other elements are entered in the same manner except that the units are microfarads for capacitance, millihenries for inductance, volts and degrees for voltage sources, milliamperes and degrees for current sources, and mhos for voltage-controlled current generators.

The node connections of the controlling voltage of a dependent current generator are requested immediately after inserting the generator. In the case of GM, the cutoff frequency is also requested. In the case of YM, the imaginary part of YM is also requested at that time. If there is no cutoff frequency, entering $\emptyset$ causes transfer around the frequency calculations of the GM term with a saving in computation time.

Note that GM and YM are independent and can be set equal by the CONVERT command.

The sign of GM (or YM) may be found by the following convention. If the controlled generator is in a direction from the higher- to the lower-numbered node and the controlling voltage is positive on the higher-numbered node, then the sign is positive. If one of these directions is changed, the sign is negative. If both are changed the sign is positive.

The last GM (and FCO or YM) specified for the same set of node pairs is used. Thus, the user may correct or change the value by retyping the information.

Elements can be removed by retyping them with zero magnitude, for example, $E = 0$, $I = 0$, $C = 0$, and $L = 0$. The inductance equal to zero is a sign to LINCAD that no inductor is desired between that set of nodes. The resistors which are addressed by R or G are removed by making $G = 0$. A value of $R = 0$ is implemented by a resistance of $10^{-7}\ \Omega$. A controlled generator can be removed by retyping all the information with zero (Ø) mho.

The circuit can be modified, that is, elements changed in magnitude, new elements added, new nodes added, and elements deleted at any time.

USE OF THE COMMAND "NODES"

Specific node voltages or differences between node voltages can be selected by NODES for use in PLOT X, VTVM, and SENSITIVITIES. The node numbers are typed after the statement:

NODES TO BE PLOTTED (FILL TO 8 NO. WITH ZEROS)
?

Zeros are used to fill to eight numbers for two purposes:
1. To satisfy the input requirement of eight numbers
2. To allow the program to compute the number of curves

Node voltage differences can be plotted or printed. The following example will show the method:

The user desires the voltages at node 3, and the difference between the voltage at nodes 4 and 2. He replies with

3 4 −2 Ø Ø Ø Ø Ø

The 3 indicates that he wished the node voltage at node 3. The 4 −2 indicates that he wishes the voltage of node 4 minus the voltage of node 2. Eight numbers have been entered. Blanks separate each. The minus sign is a message to the program to subtract that voltage from the previous one.

USE OF THE COMMANDS "VTVM" AND "ALL"

All node voltages at one frequency are computed via the ALL option for an RLC circuit.

The node voltages and node voltage differences specified in NODES may be printed at one frequency via VTVM. The frequency, in hertz, is typed after the following statement:

> FREQ IN HZ
> ?

If NODES had not been selected prior to this option, the node voltages of nodes 1, 2, 3, 4, and 5 will be printed.

USE OF THE PLOT COMMANDS

The PLOT command is used to plot the node voltages or node voltage differences as a function of the magnitude of a component (R, G, L, C) or the frequency on a log or linear scale.

The node voltages (or node voltage differences) as selected via NODES are plotted as a function of an element using this option. If NODES was not selected prior to this option, then the node voltages of nodes 1, 2, 3, 4, and 5 will be plotted.

The element type to be varied is typed after the word plot—R for resistance, C for capacitance, and L for inductance; for example, PLOT R.

The node numbers of an element to be varied are typed on request. The scale dimensions and number of intervals are typed after the request:

> INITIAL AND FINAL (UNITS), MAX AND MIN VOLTAGE,
> NO. OF INTERVALS
> ?
> where

INITIAL AND FINAL (UNITS) refers to the scale values to be plotted down the typewriter page; MAX AND MIN VOLTAGE refers to the scale for the node voltages to be plotted across the typewriter page, and NO. OF INTERVALS refers to the number of intervals to be plotted (no decimal point on the number of intervals).

To obtain a plot vs. the frequency on a linear scale

> PLOT F

To obtain a plot of the frequency on a log scale

> PLOT LOG

Since the plot is semilog, it is convenient to specify the initial and final hertz in multiples of 1Ø. For a neat plot the number of points should be an integral multiple of the number of decades.

Decibels or phase can be plotted by typing the desired output selection prior to this option.

INSTRUCTIONS FOR OPTIMIZATION

Up to five variables can be varied simultaneously to obtain a minimum, maximum, or desired value of node voltage difference.
The commands to select a variable are:

TUNE F	To adjust the frequency
TUNE G	To adjust a resistance
TUNE L	To adjust an inductor
TUNE C	To adjust a capacitance

If a G, L, or C is selected, the program requests the node connections of the element. The TUNE X command is repeated by the user as a reply to '$> =$' until all components to be adjusted simultaneously are inserted. The user then types one of the following:

MINIMIZE	To find the minimum of a node voltage
MAXIMIZE	To find the maximum of a node voltage
SET	To set a node voltage to a numerical value

The program requests the node number of the voltage difference. Two node numbers are required. Thus, if a node voltage is to be with respect to ground, type zero ($\emptyset$) for the second node.
Two selections are available in this routine:

ACCURACY	To select the error test used by MAXIMUM and SET. The accuracy is 0.1 percent unless changed.
ERROR	To select the error used by MINIMUM. The error is 0.$\emptyset\emptyset\emptyset$1 unless changed by this command.

USE OF THE CONVERT TO XXX COMMANDS

The resistor between a node pair can be addressed by R, G, or YR. In listing the circuit, the only difference between LIST and LIST(Y), as far as resistors are concerned, is whether the units are ohms, mhos, or real mhos.

The LC combination is stored separately from the imaginary part of Y (YI). The GM (FCO) and YM for a node pair set are also stored separately. The L, C, GM (FCO) are listed by LIST, whereas the YI and YM are listed by LIST(Y).

The routines described here convert the LC combination to a YI (imaginary part) and the GM (FCO) to a YM (real and imaginary part of a voltage controlled current generator and vice-versa).

CONVERT TO Y	The LC combination is converted to an imaginary admittance YI; the GM (FCO) controlled generators are converted to the transadmittance YM

The conversion is as follows:

R is unchanged but is expressed in mhos with LIST (Y)

$YM = GM/(1 + j f/f_{co})$ where f is the frequency and f_{co} is the cutoff frequency

$YI = \omega C - 1/(\omega L)$ where $\omega = 2\pi f$

CONVERT TO RLC L and C are computed from YI and GM and f_{co} from YG

The conversion is as follows: YR is unchanged but is expressed in ohms by LIST. If **YI** is positive,

$C = (YI + 1/(\omega L))/\omega$ where L is that given under $> = L$

If YI is negative,

$L = 1/((abs(YI) + \omega C)\omega)$ where C is that given under $> = C$

$GM/(1 + j f/f_{co}) = YM$

COMMAND FOR SENSITIVITY ANALYSIS

To print the sensitivities of voltage, phase, or decibels, the user types SENSITIVITY in response to $> =$.

The sensitivity of the node voltage or voltage difference is defined as the percentage change in the voltage divided by the percentage change in the element value. For a resistance, the sensitivity is

$$\text{Sensitivity} = \frac{dv/v}{dR/R}$$

The sensitivities of the node voltages or voltage differences as specified by NODES are printed for each element. The sensitivities of phase or decibels can be printed by typing PHASE or DB prior to SENSITIVITIES.

These sensitivities are computed by changing the element value by 0.1 percent.

Osucad-users Manual

The utility program OSUCAD analyzes nonlinear electronic networks given the elements in type, node connection, and value.

The nodes must be numbered with consecutive integers beginning with zero (Ø) as the ground node. The user communicates with the program by selecting a command and then entering numerical data. Commands are requested by the symbols " =," " +," and " ..". Numerical data is to be typed after the symbol "?". The various commands are listed below.

REPLIES TO =

The symbol " =" indicates that the user is in the main controlling program.

TO ENTER LINEAR COMPONENT

E *name*
C *name*
R *name*
L *name*
I *name*

where *name* is an alphanumeric name of five characters or less. Upon typing the above, the computer requests the node connections and value. The nodes are typed without a decimal point. The value and (or) node connections can be changed by retyping the information beginning with X *name*. The element can be removed by REMOVE X *name*. One blank separates the type from its *name*.

TO ENTER TRANSISTORS AND DIODES

T *name*
MF *name*
D *name*

where *name* is as described above. Do not use the name MODEL as that is the request for inserting model parameters.

The model number (for example, 2N2368) is typed after the request MODEL NO.?. If the device is in the library, the connections are requested. These are:

NODES, (B, C, E)?   Base, collector, emitter for a bipolar transistor
NODES, (G, D, S)?   Gate, drain, source for an FET
NODES, (C, A)?   Cathode, anode for a diode

To change the transistor model number and/or the connections, the user retypes the information beginning with *T name*. The transistor or diode can be removed by REMOVE X *name*.

OFF   To initialize a transistor in the cutoff state, the reply must follow immediately after the base, collector, and emitter nodes are typed

OTHER REPLIES TO "="

SAVE	To save a circuit on the disk
OLD	To retrieve a previously saved circuit
	To use the SAVE and OLD features, consult your own computer facility as to the storage of data
NEW	To begin a new circuit—automatic on entering OSUCAD
LIST	To list the circuit
DEVICES	To list the device library contents
PARAMETERS	To print the device parameters; the units are in volts, amperes, ohms, farads, and farads/ampere.
DC	To obtain the dc solution of active networks; the Ebers-Moll model is used for transistors; diodes are represented by the exponential diode equation
DYNAMIC	To obtain the dynamic response, Ebers-Moll charge control model is used for transistors
EQUATIONS	To print the state variable equations
REMOVE X *name*	To remove an element where X is E, I, R, L, C, T, D, or MF and *name* is the name given under "=".
NODES	To select specific nodes or node voltage differences to be plotted later
D MODEL	To insert the parameters of a new device or to change the parameters of a library device
T MODEL	

REPLIES TO "+"

The "+" indicates the user is in the dc analysis portion of the program.

DC   Repeats the solution obtained on entering the dc analysis
A carriage return for an entry switches control back to the main controlling program "="

VARIATION OF INDEPENDENT SOURCES

E *name*   To vary an independent source voltage
I *name*   To vary an independent current source
The name must be given exactly as entered under "=".

VARIATION OF DEVICE PARAMETERS

ALPHA
ALPHAI
IES
ICS

The computer requests the device to be varied. The reply to DEVICE?
is T *name* as typed under " = ".

TEMPERATURE VARIATION

TEMP

The computer requests the devices to be temperature controlled by DEVICE?.
The reply is T *name* or D *name* as typed under " = ". The computer continues to
type DEVICE? to obtain more names. To end the entry of temperature-
sensitive devices, the user types a carriage return.

If all devices are to be temperature sensitive simultaneously, the user may
type ALL as an answer to the first DEVICE?.

Following a successful description of the parameter to be varied, the
computer will request the minimum, maximum, and incremental values of the
quantity to be varied. If zero (∅, ∅) is typed for the maximum and incremental
value, the program assumes that the user desires to change the parameter
permanently to the value given as the minimum (on return to " = " the original
value typed in reply to " = " is reinserted). Temperature cannot be changed
permanently in this manner.

SELECTION OF DEVICE VOLTAGES AND CURRENTS FOR PLOTTING

To select device voltages or currents to be plotted, the user types the name of the
device as given under " = "; for example, *T*1. The computer will then request the
parameter to be varied. The user replies with one of the following:

V CE	To plot collector-to-emitter voltage
V BE	To plot base-to-emitter voltage
I C	To plot collector current
I E	To plot emitter current
I B	To plot base current
V D	To plot diode voltage
I A	To plot diode current

The selections are erased when the user leaves the dc portion of OSUCAD.
The selections must be made every time the user goes from " = " to " + ." The
selections can be erased in the dc mode by typing CLEAR in response to " + ."

REPLIES TO OBTAIN A PLOT OF THE DC CHARACTERISTICS

PLOT   To obtain a plot, this command must be typed prior to one of the
variation commands listed above. If PLOT is not typed between

the last list of operating points or the last plot, it is assumed that a list of operating points is desired

CLEAR   To clear the selection of device voltages and currents

ITERATIONS   To select the number of iterations in the Newton Raphson solution before the message NO CONVERGENCE is printed. The number of iterations is set at 7Ø if this command is not used

REPLIES TO ".."

The ".." indicates that the user is in the dynamic analysis portion of the program.

SOURCE WAVEFORM SPECIFICATIONS FOR THE DYNAMIC RESPONSE

E *name*   To insert waveform data of the voltage source with the name given under "="

I *name*   To insert current waveform data of the current source with the name given under "=".

The waveform is as shown below

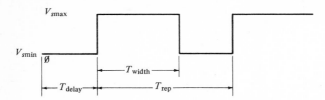

SELECTION OF DEVICE VOLTAGES AND CURRENTS FOR PLOTTING

To select device voltages or currents to be plotted, the user types the name of the device as given under "="; for example, $T1$. The computer will then request the parameter to be varied. The user replies with one of the following:

V CE   To plot collector-to-emitter voltage
V BE   To plot base-to-emitter voltage
I C   To plot collector current
I E   To plot emitter current
I B   To plot base current
V D   To plot diode voltage
I A   To plot diode current

REPLIES TO OBTAIN A PLOT OF THE DYNAMIC RESPONSE

PLOT   To obtain the dynamic response plot

CONTINUE   To continue the calculation begun with PLOT, sources may be changed between PLOT and CONTINUE

TO CHANGE OR ENTER A DEVICE MODEL

Device parameters of stored models can be changed or new devices entered by the D and T MODEL commands.

Following the D MODEL and T MODEL commands, the program requests the device name. If it is a new device name, the program will print

TYPE?

The replies are NPN for *npn* bipolar transistors and *n*-channel depletion-mode FETs, PNP for *pnp* bipolar transistors and *p*-channel depletion mode FETs, and EMP for enhancement mode *p*-channel FETs.

If the device name is a previously stored device, the TYPE request is omitted since the type is already known. Following the entry of the type or device name, the program prints:

PARAMETER NO., PARAMETER
?

The user replies with one of the numbers given below with the appropriate parameter in the units listed. To return to "=", the user replies to PARAMETER NO., PARAMETER with 99,0.

Parameter number	Diode	Bipolar transistor	MOSFET	Enhancement-mode MOSFET
1	R_B (kΩ)	R_B (kΩ)	R_G (kΩ)	R_G (kΩ)
2	C_D (pF)	C_{OE} (pF)	$C_{G'S}$ (pF)	$C_{G'S}$ (pF)
3	ϕ (V)	ϕ_E (V)		
4	n	n_E		
5	I_S (nA)	I_{ES} (mA)	K (mA/V²)	K (mA/V²)
6	θ (volts⁻¹)	θ_E (V⁻¹)	V_p (V)	$V_{BB'}$ (V)
7	R_S (kΩ)	α_N	1.	1.
8	K_D (pF/mA)	K_E (pF/mA)		
11		R_C (kΩ)	R_D (kΩ)	
12		C_{OC} (pF)	$C_{G'D}$ (pF)	$C_{G'D}$ (pF)
13		ϕC (V)		
14		n_C		
15		I_{CS} (mA)	K (mA/V²)	K (mA/V²)
16		θ_C (V⁻¹)	V_p (V)	V_p (V)
17		α_I	.9999	.9999
18		K_C (pF/mA)		

New device parameters will be saved with the SAVE commands. The parameters can be printed via the PARAMETERS command. The units of the parameters with this command are volts, amperes, farads, ohms, and farads/ampere.

appendix C

```
C   LINCAD---LINEAR COMPUTER AIDED-DESIGN   F E BATTOCLETTI 2/4/74       00000010
C       MAIN PROGRAM CONTAINING THE INPUT AND SERVICE ROUTINES           00000020
        IMPLICIT REAL*8(A-H,O-Z)                                         00000030
        DIMENSION SY2(2), CN(4), XTYP(4), SY(2)                          00000040
        COMMON YI(60),ER(60),XIR(60),YR(60),CAP(60),HENRI(60),EI(60)     00000050
        COMMON XII(60),YGM(20),FGM(20),GGM(20),BGM(20),TITLE(14)         00000060
        COMMON VNODER(30),VNODEI(30),FREQ,PRTURB(5),X(5),EPS,EPS2        00000070
        COMMON VALUE,F,XMIN,XMAX,YMIN,YMAX,XX1,WIDTH,NJ(20)              00000080
        COMMON NJJ(20),NK(20),NKK(20),JN(60),KN(60),NP(13),LPAR(5)       00000090
        COMMON NNODES,NBR,NGM,IPLT,MM,MODE,NDE,NDE2,KSIGN,NN             00000100
        COMMON IBW,IWIDTH,M,LL,NPS,IFREQ,ICOORD,IPX                      00000110
        COMMON /BDATA/ TYPE(52),VAL(17),VAL1(17),CONS(8),HD(7)           00000120
        DATA XTYP(1),XTYP(2),XTYP(3),XTYP(4)/2HBW,3HIZE,3HIZE,1H /       00000130
        DATA CN(1),CN(2),CN(3),CN(4)/5HCONTR,5HOLLIN,5HG NOD,2HES/       00000140
        DATA XLOG,YM,YRSYM,RLC/3HLOG,2HYM,2HYR,3HRLC/                    00000150
        DATA PLUSJ,XMINJ/2H+J,2H-J/                                      00000160
        CALL ERRSET (207,300,-1,1,0,225)                                 00000170
        CALL ERRSET (241,300,-1,1,0,301)                                 00000180
        WRITE (6,143) HD                                                 00000190
  1     DO 2 I=1,236                                                     00000200
  2     NJ(I)=0                                                          00000210
        DO 3 I=1,654                                                     00000220
  3     YI(I)=0.                                                         00000230
        DO 4 I=1,14                                                      00000240
        TITLE(I)=VAL1(9)                                                 00000250
        NP(I)=I                                                          00000260
  4     CONTINUE                                                         00000270
        EPS=.001                                                        00000280
        EPS2=1.E-8                                                       00000290
        NP(6)=0                                                          00000300
        IPLT=-1                                                          00000310
  5     IWIDTH=115                                                       00000320
        ICOORD=20                                                       00000330
        WIDTH=100.                                                      00000340
  6     WRITE (6,135)                                                    00000350
        READ (5,136) OP,OP2,OP3                                          00000360
        IPX=IPLT                                                         00000370
        MODE=8                                                          00000380
        NEL1=61                                                          00000390
        NEL2=360                                                        00000400
        IF(OP-YRSYM) 8,7,8                                               00000410
  7     MM=4                                                             00000420
        GO TO 93                                                         00000430
  8     IF(OP-YM) 10,9,10                                                00000440
  9     MODE=-8                                                          00000450
        MM=9                                                             00000460
        GO TO 93                                                         00000470
 10     MM=1                                                             00000480
 11     IF(OP-TYPE(MM)) 12,14,12                                         00000490
 12     MM=MM+1                                                          00000500
        IF(MM-52) 11,11,13                                               00000510
 13     WRITE (6,138)                                                    00000520
        GO TO 6                                                          00000530
 14     IF(MM-10) 93,93,15                                               00000540
```

```
15      IF(MM-12) 6,76,16                                          00000550
16      IF(MM-14) 35,5,17                                          00000560
17      IF(MM-16) 13,1,18                                          00000570
18      IF(MM-18) 111,54,19                                        00000580
19      IF(MM-20) 67,36,20                                         00000590
20      IF(MM-22) 65,42,21                                         00000600
21      IF(MM-24) 63,50,22                                         00000610
22      IF(MM-26) 50,49,23                                         00000620
23      IF(MM-27) 49,6,24                                          00000630
24      IF(MM-29) 132,132,25                                       00000640
25      IF(MM-31) 132,105,26                                       00000650
26      IF(MM-34) 105,42,27                                        00000660
27      IF(MM-36) 114,122,28                                       00000670
28      IF(MM-38) 123,124,29                                       00000680
29      IF(MM-40) 113,112,30                                       00000690
30      IF(MM-46) 13,64,33                                         00000700
33      IF(MM-51) 64,37,34                                         00000710
34      IF(MM-53) 37,13,13                                         00000720
35      IWIDTH=70                                                  00000730
        ICOORD=10                                                 00000740
        WIDTH=50.                                                 00000750
        GO TO 6                                                   00000760
36      STOP                                                      00000770
37      MM=MM-27                                                  00000780
        M=-1                                                      00000790
        MODE=-8                                                   00000800
        GO TO 66                                                  00000810
42      XX1=OP2                                                   00000820
        M=1                                                       00000830
        IF(XLOG.EQ.XX1) GO TO 52                                  00000840
        IF(YRSYM.NE.XX1) GO TO 44                                 00000850
        M=4                                                       00000860
43      MODE=-8                                                   00000870
        GO TO 47                                                  00000880
44      IF(TYPE(1).EQ.XX1) GO TO 43                               00000890
45      IF(TYPE(M).EQ.XX1) GO TO 46                               00000900
        M=M+1                                                     00000910
        IF(M-12) 45,13,13                                         00000920
46      IF(M.NE.11) GO TO 47                                      00000930
        IF(MM-22) 120,53,115                                      00000940
47      WRITE (6,151) TYPE(M)                                     00000950
        READ (5,*) JJ,KK                                          00000960
        CALL LFIND (JJ,KK,LL,&6)                                  00000970
        IF(MM-22) 115,48,115                                      00000980
48      IF(MODE) 53,53,51                                         00000990
49      M=-2                                                      00001000
        GO TO 51                                                  00001010
50      M=-1                                                      00001020
51      WRITE (6,156)                                             00001030
        READ (5,*) FREQ                                           00001040
        IF(MM-23) 53,53,66                                        00001050
52      MM=23                                                     00001060
        M=11                                                      00001070
53      WRITE (6,141) VAL(M),VAL1(M),VAL(IPLT+13),VAL1(IPLT+13)   00001080
        READ (5,*) XMIN,XMAX,YMAX,YMIN,NPS                        00001090
        IF(MM.NE.23) GO TO 66                                     00001100
        XMIN=DLOG10(XMIN)                                         00001110
        XMAX=DLOG10(XMAX)                                         00001120
        M=0                                                       00001130
        GO TO 66                                                  00001140
54      IF(NBR) 6,6,55                                            00001150
55      XX1=OP3                                                   00001160
```

```
        WRITE (6,156)                                      00001170
        READ (5,*) FREQ                                    00001180
        W=6.28318*FREQ                                     00001190
        IF(XX1-RLC) 56,67,56                               00001200
56      DO 58 I=1,NBR                                      00001210
        YI(I)=W*CAP(I)                                     00001220
        WHENR=W*HENRI(I)                                   00001230
        IF(DABS(WHENR)-1.0E-50) 58,58,57                   00001240
57      YI(I)=YI(I)-1.0/WHENR                              00001250
58      CONTINUE                                           00001260
        IF(NGM) 63,63,59                                   00001270
59      DO 62 I=1,NGM                                      00001280
        FF=0.                                              00001290
        FF2=YGM(I)                                         00001300
        IF(FGM(I)-1.0E-10) 61,61,60                        00001310
60      FF=FREQ/FGM(I)                                     00001320
        FF2=FF2/(1.0+FF*FF)                                00001330
61      BGM(I)=-FF*FF2                                     00001340
        GGM(I)=FF2                                         00001350
62      CONTINUE                                           00001360
63      NEL1=1                                             00001370
        NEL2=180                                           00001380
        MODE=-8                                            00001390
        GO TO 75                                           00001400
64      IPLT=MM-47                                         00001410
        GO TO 6                                            00001420
65      WRITE (6,142)                                      00001430
        READ (5,*) (NP(N),N=1,8)                           00001440
        GO TO 6                                            00001450
66      CALL LINCD2                                        00001460
        GO TO 6                                            00001470
67      DO 72 I=1,NBR                                      00001480
        YII=YI(I)                                          00001490
        IF(YII) 71,72,68                                   00001500
68      WHENR=W*HENRI(I)                                   00001510
        IF(DABS(WHENR)-1.0E-50) 70,70,69                   00001520
69      YII=YII-1.0/WHENR                                  00001530
70      CAP(I)=YII/W                                       00001540
        GO TO 72                                           00001550
71      HENRI(I)=1./(W*(-YII+W*CAP(I)))                    00001560
72      CONTINUE                                           00001570
        IF(NGM) 75,75,73                                   00001580
73      DO 74 I=1,NGM                                      00001590
        FF=-BGM(I)/GGM(I)                                  00001600
        FGM(I)=FREQ/FF                                     00001610
74      YGM(I)=GGM(I)*(1.+FF*FF)                           00001620
75      MM=12                                              00001630
C       LIST CIRCUIT                                       00001640
76      WRITE (6,143) TITLE                                00001650
        DO 88 III=NEL1,NEL2                                00001660
        R=YI(III)                                          00001670
        IF(DABS(R)-1.0E-30) 77,80,80                       00001680
77      IF(MODE) 78,78,88                                  00001690
78      IF(III-60) 79,79,88                                00001700
79      IF(DABS(YR(III))-1.E-30) 88,80,80                  00001710
80      II=(III-1)/60                                      00001720
        I=III-II*60                                        00001730
        JJ=JN(I)                                           00001740
        KK=KN(I)                                           00001750
        II=II+1                                            00001760
        IF(MODE) 83,83,81                                  00001770
81      IF(II-4) 83,82,83                                  00001780
```

```
  82    R=.001/R-1.0E-10                                                    00001790
         II=10                                                              00001800
         GO TO 87                                                           00001810
  83    R=R/CONS(II)                                                        00001820
         IF(II-1) 6,84,85                                                   00001830
  84    AA=YR(III)                                                          00001840
         SIGN=PLUSJ                                                         00001850
         IF(R)160,162,161                                                   00001860
 160    SIGN=XMINJ                                                          00001870
         R=-R                                                               00001880
 161    WRITE(6,158)JJ,KK,AA,SIGN,R                                         00001890
         GO TO 88                                                           00001900
 162    WRITE(6,158)JJ,KK,AA                                                00001910
         GOTO 88                                                            00001920
  85    IF(II-3) 86,86,87                                                   00001930
  86    AA=EI(III-60)/CONS(II)                                              00001940
         F=57.2958*DATAN2(AA,R)                                             00001950
         R=R/DCOS(F/57.2958)                                               00001960
         WRITE (6,144) TYPE(II+29),JJ,KK,R,VAL1(II),VAL1(9),F,VAL(13),VAL1(00001970
        113)                                                                00001980
         GO TO 88                                                           00001990
  87    WRITE (6,144) TYPE(II),JJ,KK,R,VAL(II),VAL1(II)                     00002000
  88    CONTINUE                                                            00002010
         DO 92 I=1,NGM                                                      00002020
         FGMI=FGM(I)/1.0E6                                                  00002030
         YGMI=YGM(I)                                                        00002040
         IF(MODE) 89,89,90                                                  00002050
  89    FGMI=BGM(I)                                                         00002060
         YGMI=GGM(I)                                                        00002070
  90    IF(DABS(FGMI)+DABS(YGMI)-1.E-50)92,92,91                           00002080
  91    WRITE (6,144) TYPE(9),NJ(I),NK(I),YGMI,VAL(9),VAL1(9),FGMI,VAL(MOD00002090
        1E+9),VAL1(MODE+9)                                                  00002100
         WRITE(6,157)NJJ(I),NKK(I),CN                                       00002110
  92    CONTINUE                                                            00002120
         GOTO 6                                                             00002130
C       INPUT OF ER, EI, IR, II, R, G, L, C, GM                            00002140
  93    WRITE (6,145) VAL(MM),VAL1(MM)                                      00002150
         READ (5,*) J,K,AA                                                  00002160
         CALL LFIND (J,K,L,&6)                                              00002170
         IF(MM-10) 95,94,95                                                 00002180
  94    YR(L)=.001/(AA+1.0E-10)                                            00002190
         GO TO 6                                                            00002200
  95    IF(MM-9) 96,97,6                                                    00002210
  96    L=60*(MM-1)+L                                                       00002220
         YI(L)=AA*CONS(MM)                                                  00002230
         GO TO 6                                                            00002240
  97    WRITE (6,146) VAL(MODE+9),VAL1(MODE+9)                             00002250
         READ (5,*) JJ,KK,F                                                 00002260
         CALL LFIND (JJ,KK,L,&6)                                            00002270
         DO 101 I=1,NGM                                                     00002280
         IF(NJJ(I)-JJ) 101,98,101                                          00002290
  98    IF(NKK(I)-KK) 101,99,101                                          00002300
  99    IF(NJ(I)-J) 101,100,101                                           00002310
 100    IF(NK(I)-K) 101,102,101                                           00002320
 101    CONTINUE                                                            00002330
         NGM=NGM+1                                                          00002340
         I=NGM                                                              00002350
 102    NJJ(I)=JJ                                                           00002360
         NKK(I)=KK                                                          00002370
         NJ(I)=J                                                            00002380
         NK(I)=K                                                            00002390
         IF(MODE) 103,103,104                                              00002400
```

```
    103   BGM(I)=F                                                    00002410
          GGM(I)=AA                                                   00002420
          GO TO 6                                                     00002430
    104   FGM(I)=F*1.0E6                                              00002440
          YGM(I)=AA                                                   00002450
          GO TO 6                                                     00002460
C         INPUT OF E, I, AND Y                                        00002470
    105   IF(MM.GT.32) GO TO 106                                      00002480
          WRITE (6,148) VAL1(MM-29)                                   00002490
          GO TO 107                                                   00002500
    106   WRITE (6,147)                                               00002510
    107   CONTINUE                                                    00002520
          READ (5,*) J,K,AA,F                                         00002530
          CALL LFIND (J,K,L,&6)                                       00002540
          IF(MM-32) 109,108,110                                       00002550
    108   L=L+60                                                      00002560
          AA=.001*AA                                                  00002570
    109   F=F/57.2958                                                 00002580
          ER(L)=AA*DCOS(F)                                            00002590
          EI(L)=AA*DSIN(F)                                            00002600
          GO TO 6                                                     00002610
    110   YI(L)=F                                                     00002620
          YR(L)=AA                                                    00002630
          GO TO 6                                                     00002640
    111   READ (5,137) TITLE                                          00002650
          GO TO 6                                                     00002660
C         OPTIMIZATION SETUP                                          00002670
    112   WRITE (6,149)                                               00002680
          READ (5,*) EPS                                              00002690
          EPS=.01*EPS                                                 00002700
          GO TO 6                                                     00002710
    113   WRITE (6,150)                                               00002720
          READ (5,*) EPS2                                             00002730
          GO TO 6                                                     00002740
    114   NN=1                                                        00002750
          IPX=1                                                       00002760
          IFREQ=1                                                     00002770
          IBW=0                                                       00002780
          GO TO 125                                                   00002790
    115   IF(NN-5) 117,116,116                                        00002800
    116   WRITE (6,152)                                               00002810
          GO TO 6                                                     00002820
    117   NN=NN+1                                                     00002830
          IF(M-10) 119,118,120                                        00002840
    118   M=4                                                         00002850
    119   LL=60*(M-4)+LL                                              00002860
          X(NN)=YR(LL)                                                00002870
          PRTURB(NN)=.1*YR(LL)                                        00002880
          LPAR(NN)=LL                                                 00002890
          GO TO 6                                                     00002900
    120   IFREQ=1                                                     00002910
    121   WRITE (6,153)                                               00002920
          READ (5,*) FREQ                                             00002930
          X(NN)=FREQ                                                  00002940
          PRTURB(NN)=.1*FREQ                                          00002950
          LPAR(NN)=-1                                                 00002960
          IF(MM-35) 6,130,6                                           00002970
    122   KSIGN=-1                                                    00002980
          GO TO 125                                                   00002990
    123   KSIGN=1                                                     00003000
          GO TO 125                                                   00003010
    124   KSIGN=0                                                     00003020
```

```
          WRITE (6,154) VAL(IPLT+13),VAL1(IPLT+13)                          00003030
          READ (5,*) VALUE                                                  00003040
    125   WRITE (6,155) TYPE(MM),XTYP(MM-34)                                 00003050
          READ (5,*) NDE,NDE2                                               00003060
          IF(NDE) 126,126,127                                               00003070
    126   JJ=NDE                                                            00003080
          NDE=NDE2                                                          00003090
          NDE2=JJ                                                           00003100
    127   IF(IFREQ) 128,128,129                                             00003110
    128   WRITE (6,156)                                                     00003120
          READ (5,*) FREQ                                                   00003130
    129   IFREQ=0                                                           00003140
          IF(MM-35) 131,121,131                                            00003150
    130   KSIGN=-1                                                          00003160
    131   CALL LINCD3                                                       00003170
          IFREQ=0                                                           00003180
          NN=0                                                              00003190
          GO TO 6                                                           00003200
    132   REWIND 1                                                          00003210
          IF(MM-29) 6,133,134                                              00003220
    133   CONTINUE                                                          00003230
          WRITE (1) (YI(I),I=1,655)                                         00003240
          WRITE (1) (NJ(I),I=1,235)                                         00003250
          GO TO 6                                                           00003260
    134   CONTINUE                                                          00003270
          READ (1) (YI(I),I=1,655)                                          00003280
          READ (1) (NJ(I),I=1,235)                                          00003290
          GO TO 6                                                           00003300
    135   FORMAT ('+>=')                                                    00003310
    136   FORMAT (2A5,1X,A5)                                                00003320
    137   FORMAT (14A5)                                                     00003330
    138   FORMAT ('+ILLEGAL COMMAND')                                       00003340
    141   FORMAT (19H INITIAL AND FINAL ,2A5,13H,MAX AND MIN ,2A5,17H,NO. OF00003350
         1 INTERVALS)                                                       00003360
    142   FORMAT (47H NODES TO BE PLOTTED(FILL TO 8 NO.S WITH ZEROS))        00003370
    143   FORMAT (1X,14A5)                                                  00003380
    144   FORMAT (1X,A2,2I4,2X,G12.5,1X,2A5,2X,G12.5,1X,2A5)                00003390
    145   FORMAT (7H NODES,,2A5)                                            00003400
    146   FORMAT (19H CONTROLLING NODES,,2A5)                               00003410
    147   FORMAT (26H NODES,REAL AND IMAG MHOS )                            00003420
    148   FORMAT (7H NODES,,A5,4H,DEG)                                      00003430
    149   FORMAT (20H ACCURACY IN PERCENT)                                  00003440
    150   FORMAT ('+ENTER ERROR')                                          00003450
    151   FORMAT (10H NODES OF ,A2)                                         00003460
    152   FORMAT (17H ONLY 5 VARIABLES)                                     00003470
    153   FORMAT (21H ESTIMATED FREQ IN HZ)                                 00003480
    154   FORMAT (10H VALUE IN ,2A5)                                        00003490
    155   FORMAT (14H NODE DIFF TO ,2A5)                                    00003500
    156   FORMAT (11H FREQ IN HZ)                                           00003510
    157   FORMAT(3X,2I4,2X,4A5)                                             00003520
    158   FORMAT(' Y ',2I4,2X,G12.5,1X,A2,G12.5,' MHOS')                    00003530
          END                                                               00003540
          SUBROUTINE LINCD2                                                 00003550
          IMPLICIT REAL*8(A-H,O-Z)                                          00003560
          DIMENSION SYM(9), YO(8), N2(8), SY(2), N1(8)                      00003570
          COMMON YI(60),ER(60),XIR(60),YR(60),CAP(60),HENRI(60),EI(60)      00003580
          COMMON XII(60),YGM(20),FGM(20),GGM(20),HGM(20),TITLE(14)          00003590
          COMMON VNODER(30),VNODEI(30),FREQ,PRTURB(10),EPS,EPS2             00003600
          COMMON VALUE,F,X,XMAX,YMIN,YMAX,XX1,WIDTH,NJ(20)                  00003610
          COMMON NJJ(20),NK(20),NKK(20),JN(60),KN(60),NP(13),LPAR(5)        00003620
          COMMON NNODES,NBR,NGM,IPLT,MM,MODE,NDE,NDE2,KSIGN,NN              00003630
          COMMON IBW,IWIDTH,M,LL,NPS,IFREQ,ICOORD,IPX                       00003640
```

```
          COMMON /AC/ X110(110),YLABEL(6),Y(8)                          00003650
          COMMON /BDATA/ TYPE(52),VAL(17),VAL1(17),CONS(8),HD(7)        00003660
          DATA SYM(1),SYM(2),SYM(3),SYM(4)/1H*,1H ,1HX,1HO/             00003670
          DATA SYM(5),SYM(9),PLUS,OFF/1H-,1H ,1H+,1HO/                  00003680
          DATA SYM(6),SYM(7),SYM(8)/1HI,1HH,1HZ/                        00003690
          DATA FCO,PLUSJ,XMJ/3HFCO,2H J,3H -J/                          00003700
          XMAG(VR,VI)=DSQRT(VR*VR+VI*VI)                                00003710
          IF(MM-24) 2,9,2                                               00003720
    1     RETURN                                                        00003730
    2     IF(MM-23) 3,8,4                                               00003740
    3     IF(M-10) 7,6,8                                                00003750
    4     IF(MM-25) 9,5,9                                               00003760
    5     WRITE (6,90)                                                  00003770
          GO TO 24                                                      00003780
    6     XSAV=YR(LL)                                                   00003790
          GO TO 8                                                       00003800
    7     LL=60*(M-1)+LL                                                00003810
          XSAV=YI(LL)                                                   00003820
    8     XI=(XMAX-X)/FLOAT(NPS)                                        00003830
          XMAX=XMAX+.5*XI                                               00003840
    9     NVD=0                                                         00003850
          I=0                                                           00003860
   10     I=I+1                                                         00003870
          IF(NP(I)) 11,13,12                                            00003880
   11     NZ(NVD)=-NP(I)                                                00003890
          GO TO 10                                                      00003900
   12     NVD=NVD+1                                                     00003910
          N1(NVD)=NP(I)                                                 00003920
          N2(NVD)=0                                                     00003930
          GO TO 10                                                      00003940
   13     IF(M) 24,14,14                                                00003950
   14     NNT=5                                                         00003960
          WRITE (6,91) TITLE,(SYM(I),N1(I),N2(I),I=1,NVD)               00003970
          NCOUNT=9                                                      00003980
          SCALE=WIDTH/(YMAX-YMIN)                                       00003990
          NLAB=0                                                        00004000
          DO 15 I=1,6                                                   00004010
   15     YLABEL(I)=FLOAT(I-1)*.2*WIDTH/SCALE+YMIN                      00004020
          IF(IWIDTH-100) 16,17,17                                       00004030
   16     WRITE (6,93) VAL(IPLT+13),VAL1(IPLT+13),YLABEL                00004040
          GO TO 18                                                      00004050
   17     WRITE (6,92) VAL(IPLT+13),VAL1(IPLT+13),YLABEL                00004060
   18     XX=X                                                          00004070
          IF(M) 19,19,20                                                00004080
   19     FREQ=10.**X                                                   00004090
          SY(1)=VAL(11)                                                 00004100
          SY(2)=VAL1(11)                                                00004110
          XX=FREQ                                                       00004120
          GO TO 24                                                      00004130
   20     SY(1)=VAL(M)                                                  00004140
          SY(2)=VAL1(M)                                                 00004150
          IF(M-10) 22,21,23                                            00004160
   21     YR(LL)=.001/X                                                 00004170
          GO TO 24                                                      00004180
   22     YI(LL)=X*CONS(M)                                              00004190
          GO TO 24                                                      00004200
   23     FREQ=X                                                        00004210
   24     CALL ACSLN                                                    00004220
          II=0                                                          00004230
   25     II=II+1                                                       00004240
          IF(MM-25) 26,39,26                                            00004250
   26     J=N1(II)                                                      00004260
```

```
           VR=VNODER(J)                                          00004270
           VI=VNODEI(J)                                          00004280
           K=N2(II)                                              00004290
           IF(K) 28,28,27                                        00004300
     27    VR=VR-VNODER(K)                                       00004310
           VI=VI-VNODEI(K)                                       00004320
     28    IF(M+1) 29,40,29                                      00004330
     29    IF(IPLT) 30,31,33                                     00004340
     30    Y(II)=DSQRT(VR*VR+VI*VI)                              00004350
           GO TO 54                                              00004360
     31    Y(II)=0.0                                             00004370
           IF(VI*VR) 32,54,32                                    00004380
     32    Y(II)=57.2958*DATAN2(VI,VR)                           00004390
           GO TO 54                                              00004400
     33    IF(IPLT-2) 36,34,35                                   00004410
     34    Y(II)=VR                                              00004420
           GO TO 54                                              00004430
     35    Y(II)=VI                                              00004440
           GO TO 54                                              00004450
     36    DB=DSQRT(VR*VR+VI*VI)                                 00004460
           IF(DB-1.0E-30) 37,37,38                               00004470
     37    Y(II)=-600.                                           00004480
           GO TO 54                                              00004490
     38    Y(II)=20.0*DLOG10(DB)                                 00004500
           GO TO 54                                              00004510
     39    VR=VNODER(II)                                         00004520
           VI=VNODEI(II)                                         00004530
     40    F=DSQRT(VR*VR+VI*VI)                                  00004540
           ANGLE=0.0                                             00004550
           IF(VR*VI) 41,42,41                                    00004560
     41    ANGLE=57.2958*DATAN2(VI,VR)                           00004570
     42    IF(F-1.0E-30) 43,44,44                                00004580
     43    DB=-600.                                              00004590
           GO TO 45                                              00004600
     44    DB=20.0*DLOG10(F)                                     00004610
     45    IF(MM-25) 47,46,47                                    00004620
     46    WRITE (6,94) II,VR,VI,F,ANGLE,DB                      00004630
           IF(II-NNODES) 25,1,1                                  00004640
     47    IF(IPLT-1) 53,48,49                                   00004650
     48    WRITE (6,96) J,K,DB,ANGLE                             00004660
           GO TO 54                                              00004670
     49    SIGN=PLUSJ                                            00004680
           IF(VI) 50,52,51                                       00004690
     50    SIGN=XMJ                                              00004700
           VI=DABS(VI)                                           00004710
     51    WRITE (6,97) J,K,VR,SIGN,VI                           00004720
           GO TO 54                                              00004730
     52    WRITE (6,97) J,K,VR                                   00004740
     53    WRITE (6,95) J,K,F,ANGLE                              00004750
     54    IF(II-NVD) 25,55,55                                   00004760
     55    IF(M) 56,57,57                                        00004770
     56    IF(MM-26) 1,75,1                                      00004780
     57    DO 58 I=1,IWIDTH                                      00004790
     58    X110(I)=SYM(NNT)                                      00004800
           X110(11)=SYM(5)                                       00004810
           I=1                                                   00004820
     59    L=(Y(I)-YMIN)*SCALE+11.5                              00004830
           IF(L+1) 60,60,61                                      00004840
     60    X110(1)=OFF                                           00004850
           GO TO 63                                              00004860
     61    IF(L-IWIDTH) 62,63,63                                 00004870
     62    X110(L+1)=SYM(I)                                      00004880
```

```
                GO TO 63                                                    00004890
    63      I=I+1                                                           00004900
            IF(I-II) 59,59,64                                               00004910
    64      NCOUNT=NCOUNT+1                                                 00004920
            NLAB=NLAB+1                                                     00004930
            IF(NLAB-5) 66,65,66                                            00004940
    65      WRITE (6,103) SY,(X110(I),I=11,IWIDTH)                         00004950
            GO TO 70                                                        00004960
    66      IF(NCOUNT-10) 67,68,67                                         00004970
    67      WRITE (6,98) (X110(I),I=1,IWIDTH)                              00004980
            GO TO 70                                                        00004990
    68      DO 69 I=11,IWIDTH,ICOORD                                       00005000
    69      X110(I)=PLUS                                                    00005010
            NCOUNT=0                                                        00005020
            WRITE (6,99) XX,(X110(I),I=11,IWIDTH)                          00005030
    70      NNT=9                                                           00005040
            X=X+XI                                                          00005050
            IF(XMAX-X) 71,18,18                                            00005060
    71      IF(M) 1,1,72                                                    00005070
    72      IF(M-10) 74,73,1                                               00005080
    73      YR(LL)=XSAV                                                     00005090
            GO TO 1                                                         00005100
    74      YI(LL)=XSAV                                                     00005110
            GO TO 1                                                         00005120
C           SENSITIVITIES MM=26                                            00005130
    75      IF(M+2) 85,76,76                                               00005140
    76      DO 77 I=1,NVD                                                   00005150
    77      YO(I)=Y(I)                                                      00005160
            M=-3                                                            00005170
            III=0                                                          00005180
            WRITE (6,100) TITLE,VAL(IPLT+13),VAL1(IPLT+13),VAL(IPLT+13),VAL1(I00005190
           1PLT+13),(N1(I),N2(I),I=1,NVD)                                  00005200
    78      III=III+1                                                       00005210
            IF(III-120) 79,79,83                                            00005220
    79      RR=YR(III)                                                      00005230
            IF(DABS(RR)-1.0E-30) 78,78,80                                   00005240
    80      IX=(III-1)/60                                                   00005250
            I=III-IX*60                                                     00005260
            JJ=JN(I)                                                        00005270
            KK=KN(I)                                                        00005280
            IX=IX+4                                                         00005290
            IF(IX-4) 82,81,82                                              00005300
    81      R=.001/RR                                                       00005310
            YR(III)=RR/1.001                                                00005320
            IX=10                                                          00005330
            GO TO 24                                                        00005340
    82      R=RR/CONS(IX)                                                   00005350
            YR(III)=1.001*RR                                                00005360
            GO TO 24                                                        00005370
    83      III=0                                                          00005380
            IX=9                                                           00005390
    84      IF(III.EQ.NGM) GO TO 89                                        00005400
            III=III+1                                                       00005410
            JJ=NJ(III)                                                      00005420
            KK=NK(III)                                                      00005430
            R=YGM(III)                                                      00005440
            YGM(III)=1.001*R                                                00005450
            GO TO 24                                                        00005460
    85      DO 86 I=1,NVD                                                   00005470
    86      Y(I)=1000.0*(Y(I)/YO(I)-1.0)                                   00005480
            WRITE (6,101) TYPE(IX),JJ,KK,R,(Y(I),I=1,NVD)                  00005490
            IF(IX-9) 88,87,88                                             00005500
```

```
 87     YGM(III)=R                                                    00005510
        WRITE (6,101) FCO,NJJ(III),NKK(III),FGM(III)                  00005520
        GO TO 84                                                      00005530
 88     YR(III)=RR                                                    00005540
        GO TO 78                                                      00005550
 89     WRITE (6,102) (YO(I),I=1,NVD)                                 00005560
        GO TO 1                                                       00005570
 90     FORMAT (6X,9HREAL PART,2X,9HIMAG PART,3X,9HMAGNITUDE,7X,5HPHASE,2X00005580
       1,11H2O*LOG(MAG)/)                                             00005590
 91     FORMAT (1X,14A5/(5(2X,A1,5H NODE,I3,1H-,I2)))                 00005600
 92     FORMAT (20X,2A5/1X,6G20.4)                                    00005610
 93     FORMAT (20X,2A5/7X,6G10.3)                                    00005620
 94     FORMAT (2H V,I3,1X,3(1X,G11.4),2F10.3)                        00005630
 95     FORMAT (2H V,I3,2H-V,I3,2H =,G13.5,3H V,,F8.2,4H DEG)         00005640
 96     FORMAT (2H V,I3,2H-V,I3,2H =,G13.5,4H DB,,F8.2,4H DEG)        00005650
 97     FORMAT (2H V,I3,2H-V,I3,2H =,G13.5,A4,G13.5,2H V)             00005660
 98     FORMAT (2X,115A1)                                             00005670
 99     FORMAT (1X,G11.4,115A1)                                       00005680
100     FORMAT(1X,14A5/17X,'D(',2A5,')/',2A5/'    SENSITIVITY = ',24(1H-)/ 00005690
       1 28X,'DX/X'/23X,4(6X,I2,1H-,I2))                             00005700
101     FORMAT (1X,A3,2I3,1X,G11.4,2X,4(1X,G10.3))                    00005710
102     FORMAT (16H CENTER SOLUTION,8X,4(1X,G10.3))                   00005720
103     FORMAT (1X,2A5,1X,115A1)                                      00005730
        END                                                           00005740
        SUBROUTINE LINCD3                                             00005750
        IMPLICIT REAL*8(A-H,O-Z)                                      00005760
        DIMENSION S(6,6), SC(5)                                       00005770
        COMMON YI(60),ER(60),XIR(60),YR(60),CAP(60),HENRI(60),EI(60)  00005780
        COMMON XII(60),YGM(20),FGM(20),GGM(20),BGM(20),TITLE(14)      00005790
        COMMON VNODER(30),VNODEI(30),FREQ,PRTURB(5),X(5),EPS,EPS2      00005800
        COMMON VALUE,F,XMIN,XMAX,YMIN,YMAX,XX1,WIDTH,NJ(20)           00005810
        COMMON NJJ(20),NK(20),NKK(20),JN(60),KN(60),NP(13),LPAR(5)    00005820
        COMMON NNODES,NBR,NGM,IPLT,MM,MODE,NDE,NDE2,KSIGN,NN          00005830
        COMMON IBW,IWIDTH,M,LL,NPS,IFREQ,ICOORD,IPX                   00005840
        COMMON /BDATA/ TYPE(52),VAL(17),VAL1(17),CONS(8),HO(7)        00005850
        DATA YES/3HYES/                                               00005860
        FNN=NN                                                        00005870
        IF(NDE2) 1,2,2                                                00005880
 1      NDE2=-NDE2                                                    00005890
 2      NNP1=NN+1                                                     00005900
        DO 4 I=1,NN                                                   00005910
        DO 3 J=1,NNP1                                                 00005920
 3      S(I,J)=X(I)                                                   00005930
 4      S(I,I)=X(I)+PRTURB(I)                                         00005940
        DO 6 J=1,NNP1                                                 00005950
        DO 5 I=1,NN                                                   00005960
 5      X(I)=S(I,J)                                                   00005970
        CALL FUNCT                                                    00005980
 6      S(NNP1,J)=F                                                   00005990
 7      DO 35 NUM=1,100                                               00006000
        FL=S(NNP1,1)                                                  00006010
        FH=FL                                                         00006020
        FH2=-1.E+30                                                   00006030
        JH=1                                                          00006040
        JL=1                                                          00006050
        DO 13 J=2,NNP1                                                00006060
        BURP=S(NNP1,J)                                                00006070
        IF(FH-BURP) 8,9,9                                             00006080
 8      FH2=FH                                                        00006090
        FH=BURP                                                       00006100
        JH=J                                                          00006110
        GO TO 13                                                      00006120
```

```
 9      IF(FH2-BURP) 10,11,11                                     00006130
10      FH2=BURP                                                  00006140
11      IF(FL-BURP) 13,13,12                                      00006150
12      FL=BURP                                                   00006160
        JL=J                                                      00006170
13      CONTINUE                                                  00006180
        IF(KSIGN) 14,15,16                                        00006190
14      IF(DABS(FH-FL)-DABS(EPS2*F)) 37,17,17                     00006200
15      IF(DABS(F)-DABS(EPS*VALUE)) 37,37,17                      00006210
16      IF(FH-FL-EPS2) 37,17,17                                   00006220
17      DO 19 I=1,NN                                              00006230
        SUM=0.                                                    00006240
        DO 18 J=1,NNP1                                            00006250
18      SUM=SUM+S(I,J)                                            00006260
19      SC(I)=(SUM-S(I,JH))/FNN                                   00006270
        DO 20 I=1,NN                                              00006280
20      X(I)=2.0*SC(I)-S(I,JH)                                    00006290
        CALL FUNCT                                                00006300
        IF(F-FL) 31,21,21                                         00006310
21      IF(F-FH2) 33,22,22                                        00006320
22      IF(F-FH) 23,25,25                                         00006330
23      DO 24 I=1,NN                                              00006340
24      S(I,JH)=X(I)                                              00006350
        S(NNP1,JH)=F                                              00006360
        FH=F                                                      00006370
25      DO 26 I=1,NN                                              00006380
26      X(I)=.5*(S(I,JH)+SC(I))                                   00006390
        CALL FUNCT                                                00006400
        IF(F-FH) 33,27,27                                         00006410
27      DO 30 J=1,NNP1                                            00006420
        IF(J-JL) 28,30,28                                         00006430
28      DO 29 I=1,NN                                              00006440
        BURP=.5*(S(I,J)+S(I,JL))                                  00006450
        S(I,J)=BURP                                               00006460
29      X(I)=BURP                                                 00006470
        CALL FUNCT                                                00006480
        S(NNP1,J)=F                                               00006490
30      CONTINUE                                                  00006500
        GO TO 35                                                  00006510
31      DO 32 I=1,NN                                              00006520
        BURP=X(I)                                                 00006530
        S(I,JH)=BURP                                              00006540
32      X(I)=BURP+BURP-SC(I)                                      00006550
        S(NNP1,JH)=F                                              00006560
        FL=F                                                      00006570
        CALL FUNCT                                                00006580
        IF(F-FL) 33,35,35                                         00006590
33      DO 34 I=1,NN                                              00006600
34      S(I,JH)=X(I)                                              00006610
        S(NNP1,JH)=F                                              00006620
35      CONTINUE                                                  00006630
        WRITE (6,54)                                              00006640
        READ (5,55) ANS                                          00006650
        IF(ANS-YES) 37,7,37                                       00006660
37      IF(MM-35) 38,48,38                                        00006670
38      DO 44 I=1,NN                                              00006680
        ANS=S(I,JL)                                               00006690
        LL=LPAR(I)                                                00006700
        IF(LL) 39,40,40                                           00006710
39      WRITE (6,56) ANS                                          00006720
        GO TO 44                                                  00006730
40      M=(LL-1)/60                                               00006740
```

```
        L=LL-M*60                                                        00006750
        YR(LL)=ANS                                                       00006760
        IF(M) 36,41,42                                                   00006770
41      ANS=.001/ANS                                                     00006780
        MM=10                                                            00006790
        GO TO 43                                                         00006800
42      MM=M+4                                                           00006810
        ANS=ANS/CONS(MM)                                                 00006820
43      WRITE (6,57) TYPE(MM),KN(L),JN(L),ANS,VAL(MM),VAL1(MM)           00006830
44      CONTINUE                                                         00006840
        F=S(NNP1,JL)                                                     00006850
        IF(KSIGN) 45,46,47                                               00006860
45      F=-F                                                             00006870
        GO TO 47                                                         00006880
46      F=VALUE-F                                                        00006890
47      WRITE (6,58) NDE,NDE2,F,VAL(IPLT+13),VAL1(IPLT+13)               00006900
        GO TO 36                                                         00006910
48      IF(IBW) 50,49,53                                                 00006920
49      FCENT=S(1,JL)                                                    00006930
        ANS=S(2,JL)                                                      00006940
        ANS=-ANS                                                         00006950
        VALUE=ANS-3.                                                     00006960
        PRTURB(1)=.02*FCENT                                              00006970
        X(1)=.8*FCENT                                                    00006980
        WRITE (6,59) FCENT,ANS                                           00006990
        IBW=-1                                                           00007000
        KSIGN=0                                                          00007010
        GO TO 2                                                          00007020
50      FLWR=S(1,JL)                                                     00007030
        WRITE (6,60) FLWR                                                00007040
        BW=FCENT-FLWR                                                    00007050
        X(1)=FCENT+BW                                                    00007060
        IF(X(1)) 51,52,52                                                00007070
51      X(1)=.01*FCENT                                                   00007080
52      PRTURB(1)=-.02*BW                                                00007090
        IBW=1                                                            00007100
        GO TO 2                                                          00007110
53      FUPP=S(1,JL)                                                     00007120
        BW=DABS(FUPP-FLWR)                                               00007130
        WRITE (6,60) FUPP,BW                                             00007140
36      RETURN                                                           00007150
54      FORMAT (41H NO CONVERGENCE,MORE ITERATIONS.YES OR NO)            00007160
55      FORMAT (14A5)                                                    00007170
56      FORMAT (4H F= ,G12.5,3H HZ)                                      00007180
57      FORMAT (1X,A3,2I3,G12.5,1X,2A5)                                  00007190
58      FORMAT (2H V,I3,2H-V,I3,2H =,G12.5,1X,2A5)                       00007200
59      FORMAT (14H CENTER FREQ= ,G12.5,3H HZ,5X,G12.5,3H DB)            00007210
60      FORMAT (7H F3DB= ,G12.5,3H HZ,5X,G12.5,9H=BW IN HZ)              00007220
        END                                                             00007230
        SUBROUTINE ACSLN                                                 00007240
        IMPLICIT REAL*8(A-H,O-Z)                                         00007250
        COMMON YI(60),ER(60),XIR(60),YR(60),CAP(60),HENRI(60),EI(60)     00007260
        COMMON XII(60),YGM(20),FGM(20),GGM(20),BGM(20),TITLE(14)         00007270
        COMMON VNODER(30),VNODEI(30),FREQ,PRTURB(5),X(5),EPS,EPS2        00007280
        COMMON VALUE,F,XMIN,XMAX,YMIN,YMAX,XX1,WIDTH,KJ(20)              00007290
        COMMON NJJ(20),NK(20),NKK(20),JN(60),KN(60),NP(13),LPAR(5)       00007300
        COMMON NNODES,NBR,NGM,IPLT,MM,MODE                               00007310
        COMMON /AC/ G(31,30),B(31,30)                                    00007320
        NN=NNODES+1                                                      00007330
        W=6.28318*FREQ                                                   00007340
        DO 1 I=1,NN                                                      00007350
        DO 1 J=1,NNODES                                                  00007360
```

```
          G(I,J)=0.                                                     00007370
  1       B(I,J)=0.                                                     00007380
          DO 7 I=1,NBR                                                  00007390
          IF(MODE) 4,4,2                                                00007400
  2       YII=W*CAP(I)                                                  00007410
          WHENR=W*HENRI(I)                                              00007420
          IF(DABS(WHENR)-1.0E-50) 5,5,3                                 00007430
  3       YII=YII-1.0/WHENR                                             00007440
          GO TO 5                                                       00007450
  4       YII=YI(I)                                                     00007460
  5       J=JN(I)                                                       00007470
          K=KN(I)                                                       00007480
          YRI=YR(I)                                                     00007490
          G(J,J)=G(J,J)+YRI                                             00007500
          B(J,J)=B(J,J)+YII                                             00007510
          VRR=ER(I)*YRI-EI(I)*YII-XIR(I)                                00007520
          VII=ER(I)*YII+EI(I)*YRI-XII(I)                                00007530
          G(NN,J)=G(NN,J)+VRR                                           00007540
          B(NN,J)=B(NN,J)+VII                                           00007550
          IF(K) 7,7,6                                                   00007560
  6       G(K,K)=G(K,K)+YRI                                             00007570
          B(K,K)=B(K,K)+YII                                             00007580
          G(J,K)=G(J,K)-YRI                                             00007590
          B(J,K)=B(J,K)-YII                                             00007600
          G(K,J)=G(K,J)-YRI                                             00007610
          B(K,J)=B(K,J)-YII                                             00007620
          G(NN,K)=G(NN,K)-VRR                                           00007630
          B(NN,K)=B(NN,K)-VII                                           00007640
  7       CONTINUE                                                      00007650
          IF(NGM) 20,20,8                                               00007660
  8       DO 19 I=1,NGM                                                 00007670
          IF(MODE) 9,9,10                                               00007680
  9       FF2=GGM(I)                                                    00007690
          BGMI=BGM(I)                                                   00007700
          GO TO 13                                                      00007710
 10       FF=0.0                                                        00007720
          FF2=YGM(I)                                                    00007730
          IF(FGM(I)-1.0E-10) 12,12,11                                   00007740
 11       FF=FREQ/FGM(I)                                                00007750
          FF2=FF2/(1.0+FF*FF)                                           00007760
 12       BGMI=-FF*FF2                                                  00007770
 13       JJ=NJ(I)                                                      00007780
          J=NJJ(I)                                                      00007790
          K=NKK(I)                                                      00007800
          KK=NK(I)                                                      00007810
          G(J,JJ)=G(J,JJ)+FF2                                           00007820
          B(J,JJ)=B(J,JJ)+BGMI                                          00007830
          IF(K) 15,15,14                                                00007840
 14       G(K,JJ)=G(K,JJ)-FF2                                           00007850
          B(K,JJ)=B(K,JJ)-BGMI                                          00007860
 15       IF(KK) 17,17,16                                               00007870
 16       G(J,KK)=G(J,KK)-FF2                                           00007880
          B(J,KK)=B(J,KK)-BGMI                                          00007890
 17       IF(K) 19,19,18                                                00007900
 18       G(K,KK)=G(K,KK)+FF2                                           00007910
          B(K,KK)=B(K,KK)+BGMI                                          00007920
 19       CONTINUE                                                      00007930
C         CROUT METHOD FOR SOLVING LINEAR EQUATIONS                     00007940
 20       DO 24 L=1,NNODES                                              00007950
          LLL=L-1                                                       00007960
          DO 24 I=L,NNODES                                              00007970
          II=I+1                                                        00007980
```

```
            AAR=G(II,L)                                              00007990
            AAI=B(II,L)                                              00008000
            IF(LLL) 23,23,21                                         00008010
    21      BBR=G(L,I)                                               00008020
            BBI=B(L,I)                                               00008030
            DO 22 K=1,LLL                                            00008040
            AR=G(K,I)                                                00008050
            AI=B(K,I)                                                00008060
            BR=G(L,K)                                                00008070
            BI=B(L,K)                                                00008080
            BBR=BBR-AR*BR+AI*BI                                      00008090
            BBI=BBI-AR*BI-AI*BR                                      00008100
            AR=G(K,L)                                                00008110
            AI=B(K,L)                                                00008120
            BR=G(II,K)                                               00008130
            BI=B(II,K)                                               00008140
            AAR=AAR-AR*BR+AI*BI                                      00008150
    22      AAI=AAI-AR*BI-AI*BR                                      00008160
            G(L,I)=BBR                                               00008170
            B(L,I)=BBI                                               00008180
    23      AR=G(L,L)                                                00008190
            AI=B(L,L)                                                00008200
            A2=AR*AR+AI*AI                                           00008210
            G(II,L)=(AAR*AR+AAI*AI)/A2                               00008220
    24      B(II,L)=(AAI*AR-AAR*AI)/A2                               00008230
            DO 26 L=2,NNODES                                         00008240
            I=NN-L                                                   00008250
            VR=G(NN,I)                                               00008260
            VI=B(NN,I)                                               00008270
            II=I+1                                                   00008280
            DO 25 K=II,NNODES                                        00008290
            AR=G(K,I)                                                00008300
            AI=B(K,I)                                                00008310
            BR=G(NN,K)                                               00008320
            BI=B(NN,K)                                               00008330
            VR=VR-AR*BR+AI*BI                                        00008340
    25      VI=VI-AR*BI-AI*BR                                        00008350
            G(NN,I)=VR                                               00008360
            B(NN,I)=VI                                               00008370
            VNODER(I)=VR                                             00008380
    26      VNODEI(I)=VI                                             00008390
            VNODER(NNODES)=G(NN,NNODES)                              00008400
            VNODEI(NNODES)=B(NN,NNODES)                              00008410
            RETURN                                                   00008420
            END                                                      00008430
            SUBROUTINE LFIND (J,K,L,*)                               00008440
            IMPLICIT REAL*8(A-H,O-Z)                                 00008450
            COMMON YI(60),ER(60),XIR(60),YR(60),CAP(60),HENRI(60),EI(60)  00008460
            COMMON XII(60),YGM(20),FGM(20),GGM(20),BGM(20),TITLE(14)  00008470
            COMMON VNODER(30),VNODEI(30),FREQ,PRTURB(5),X(5),EPS,EPS2  00008480
            COMMON VALUE,F,XMIN,XMAX,YMIN,YMAX,XX1,WIDTH,NJ(20)       00008490
            COMMON NJJ(20),NK(20),NKK(20),JN(60),KN(60),NP(13),LPAR(5)  00008500
            COMMON NNODES,NBR                                        00008510
            IF(K-J) 2,14,1                                           00008520
    1       I=J                                                      00008530
            J=K                                                      00008540
            K=I                                                      00008550
    2       L=1                                                      00008560
            IF(30-J) 11,3,3                                          00008570
    3       IF(JN(L)-J) 5,4,5                                        00008580
    4       IF(KN(L)-K) 5,12,5                                       00008590
    5       L=L+1                                                    00008600
```

```
         IF(L-NBR) 3,3,6                                         00008610
   6     NBR=NBR+1                                               00008620
         L=NBR                                                   00008630
         IF(60-NBR) 11,7,7                                       00008640
   7     IF(J-NNODES) 9,8,8                                      00008650
   8     NNODES=J                                                00008660
   9     JN(NBR)=J                                               00008670
         IF(K) 11,12,10                                          00008680
  10     KN(NBR)=K                                               00008690
  12     RETURN                                                  00008700
  11     WRITE (6,13)                                            00008710
  16     RETURN 1                                                00008720
  14     WRITE(6,15)J,K                                          00008730
         GOTO 16                                                 00008740
  13     FORMAT (37H SORRY,ONLY 30 NODES AND 60 BRANCHES.)       00008750
  15     FORMAT(32H ELEMENT CONNECTED BETWEEN NODES ,I3,4H AND,I3, 00008760
        1 16H IS MEANINGLESS.)                                   00008770
         END                                                     00008780
         SUBROUTINE FUNCT                                        00008790
         IMPLICIT REAL*8(A-H,O-Z)                                00008800
         COMMON YI(60),ER(60),XIR(60),YR(60),CAP(60),HENRI(60),EI(60) 00008810
         COMMON XII(60),YGM(20),FGM(20),GGM(20),BGM(20),TITLE(14) 00008820
         COMMON VNODER(30),VNODEI(30),FREQ,PRTURB(5),X(5),EPS,EPS2 00008830
         COMMON VALUE,F,XMIN,XMAX,YMIN,YMAX,XX1,WIDTH,NJ(20)     00008840
         COMMON NJJ(20),NK(20),NKK(20),JN(60),KN(60),NP(13),LPAR(5) 00008850
         COMMON NNODES,NBR,NGM,IPLT,MM,MODE,NDE,NDE2,KSIGN,NN    00008860
         COMMON IBW,IWIDTH,M,LL,NPS,IFREQ,ICOORD,IPX             00008870
         DO 3 I=1,NN                                             00008880
         LL=LPAR(I)                                              00008890
         IF(1-(LL-1)/60)17,16,17                                 00008900
  16     X(I)=DABS(X(I))                                         00008910
  17     IF(LL) 1,1,2                                            00008920
   1     FREQ=X(I)                                               00008930
         GO TO 3                                                 00008940
   2     YR(LL)=X(I)                                             00008950
   3     CONTINUE                                                00008960
         CALL ACSLN                                              00008970
         VR=VNODER(NDE)                                          00008980
         VI=VNODEI(NDE)                                          00008990
         IF(NDE2) 5,5,4                                          00009000
   4     VR=VR-VNODER(NDE2)                                      00009010
         VI=VI-VNODEI(NDE2)                                      00009020
   5     IF(IPX) 9,11,6                                          00009030
   6     IF(IPX-2) 9,7,8                                         00009040
   7     F=VR                                                    00009050
         GO TO 12                                                00009060
   8     F=VI                                                    00009070
         GO TO 12                                                00009080
   9     F=DSQRT(VR*VR+VI*VI)                                    00009090
         IF(IPX) 12,12,10                                        00009100
  10     F=20.*DLOG10(F)                                         00009110
         GO TO 12                                                00009120
  11     F=57.2958*DATAN2(VI,VR)                                 00009130
  12     IF(KSIGN) 13,14,15                                      00009140
  13     F=-F                                                    00009150
         GO TO 15                                                00009160
  14     F=DABS(VALUE-F)                                         00009170
  15     CONTINUE                                                00009180
         RETURN                                                  00009190
         END                                                     00009200
         BLOCK DATA                                              00009210
         IMPLICIT REAL*8(A-H,O-Z)                                00009220
```

```
      COMMON /BDATA/ TYPE(52),VAL(17),VAL1(17),CONS(8),HD(7)            00009230
      DATA HD/5HLINCA,5HD-LIN,5HEAR C,5HOMPUT,5HER AI,5HDED-D,5HESIGN/   00009240
      DATA TYPE/2HYI,2HER,2HIR,1HG,1HC,1HL,2HEI,2HII,2HGM,1HR,1HF,4HLIST00009250
     1,2HTT,3HIBM,1HX,3HNEW,5HTITLE,5HCONVE,3HRLC,3HEND,5HNODES,4HPLOT,500009260
     2HLIST(,4HVTVM,3HALL,5HSENSI,1HX,1HX,4HSAVE,3HOLD,1HE,1HI,1HY,4HTUN00009270
     3E,4HFIND,5HMAXIM,5HMINIM,3HSET,5HERROR,5HACCUR,5*3HXYZ,            00009280
     45HMAGNI,5HPHASE,2HDB,4HREAL,5HIMAGI,5HVTVM(,5HALL(Y/              00009290
      DATA VAL/5HIMAG ,5HREAL ,5HREAL ,5HREAL ,5HMICRO,5HMILLI,5HIMAG ,500009300
     1HIMAG ,4HMHOS,5HKILOH,5HHERTZ,5HVOLTS,5HDEGRE,5HDECIB,4HREAL ,4HIMA00009310
     2G,5HFCO I/                                                        00009320
      DATA VAL1/4HMHOS,5HVOLTS,2HMA,4HMHOS,5HFARAD,5HHENRY,5HVOLTS,2HMA,00009330
     11H ,2HMS,1H ,1H ,2HES,3HELS,2*5HVOLTS,5HN MHZ/                    00009340
      DATA CONS/1.,1.,.001,1.,1.E-6,.001,1.,.001/                       00009350
      END                                                              00009360
```

appendix D

FORTRAN LISTING OF OSUCAD

```
C        TSO-OSUCAD   F E BATTOCLETTI   2/4/74                          00000010
C        READS,STORES THE CIRCUIT DESCRIPTION 12/8/72                   00000020
         IMPLICIT REAL*8(A-H,O-Z)                                       00000030
         DIMENSION OP(28), C1(9), C2(9), C3(5), C4(8)                   00000040
         COMMON NWRD(2),IPLOT,LDS,NTOT,NEL,NT2(3)                       00000050
         COMMON LPR,NOP,NODES(11),NTBP,LEVB,LQGG(5),LZLL,LZLC(16)       00000060
         COMMON NDFLG,NSS(3),ND(8),N9,N10                               00000070
         COMMON JNV(500),JNT(60),NDS(240),VLIST(240),WORD(300),XLIST(3030)  00000080
         COMMON /DEVMOD/ TITLE(14),AXX(30,8),X11(30),X22(30),NLIB       00000090
         COMMON /SAV/ JSAVE(620)                                        00000100
         DATA OP/1HE,1HC,1HR,1HL,1HI,2HMF,1HT,1HD,1HD,2HRE,2HLI,2HPA,   00000110
        12HOF,2HDC,2HDE,2HNE,2HPO,2HDY,2HNO,2HEN,2HEQ,2HTI,2HOL,2HSA,   00000120
        2 2HTT,2HIB/                                                    00000130
         DATA C1(1),C2(1),C1(2),C2(2),C1(3),C2(3),C1(4),C2(4)/5HVOLTS,5H(T=  00000140
        10),5HMICRO,5HFARAD,5HKILOH,2HMS,5HMILLI,5HHENRY/               00000150
         DATA C1(5),C2(5)/5HMILLI,4HAMPS/                               00000160
         DATA C3(1),C3(2),C3(3),C3(4),C3(5)/1.0,1.0E-6,1000.,.001,.001/ 00000170
         DATA C4(1),C4(2),C4(3),C4(4),C4(5),C4(6),C4(7),C4(8)/1000.,1.0E-12  00000180
        1,1.,1.,,.001,1.,1.,1.0E-9/                                     00000190
         DATA C1(6),C2(6),C1(7),C2(7),C1(8),C2(8),C1(9),C2(9)/5H(G,D,,2HS),  00000200
        15H(B,C,,2HE),5H(C,A),1H ,5H(C,A),1H /                          00000210
         DATA XMODL,OFF,BLANK,PLOT/5HMODEL,3HOFF,1H ,4HPLOT/            00000220
         CALL ERRSET(207,300,-1,1,0,225)                               00000230
         CALL ERRSET(241,300,-1,1,0,301)                               00000240
         WRITE(6,1000)                                                  00000250
1000     FORMAT(' OSUCAD---OSU COMPUTER-AIDED DESIGN')                  00000260
C        CLEAR COMMON                                                   00000270
3        DO 4 I=1,620                                                   00000280
4        NWRD(I)=0                                                      00000290
         LDS=1                                                          00000300
         LZLL=2025                                                      00000310
         LEVB=800                                                       00000320
202      NWRD(1)=118                                                    00000330
         GOTO 5                                                         00000340
1        DO 2 I=1,620                                                   00000350
2        NWRD(I)=JSAVE(I)                                               00000360
5        III=0                                                          00000370
         NODE3=0                                                        00000380
         LPR=0                                                          00000390
         WRITE (6,96)                                                   00000400
         READ (5,97) TYPE,XX1,XXX,XX2,YY2                               00000410
6        DO 7 NOP=1,26                                                  00000420
         IF(OP(NOP)-TYPE) 7,8,7                                         00000430
7        CONTINUE                                                       00000440
         WRITE (6,98)                                                   00000450
         GO TO 5                                                        00000460
8        IF(III) 9,9,38                                                 00000470
9        IF(NOP-5) 41,41,10                                             00000480
10       IF(NOP-10) 21,37,11                                            00000490
11       IF(NOP-12) 81,90,13                                            00000500
12       STOP                                                           00000510
13       IF(NOP-14) 36,117,14                                           00000520
14       IF(NOP-16) 55,3,15                                             00000530
15       IF(NOP-19) 118,76,16                                           00000540
```

```
   16      IF(NOP-21) 12,80,17                                           00000550
   17      IF(NOP-23) 79,115,200                                         00000560
  200         IF(NOP-25)116,201,202                                      00000570
  201         NWRD(1)=70                                                 00000580
            GOTO 5                                                       00000590
C         OLD AND SAVE OPTIONS                                           00000600
  115     REWIND1                                                        00000610
          READ(1)(NWRD(I),I=1,860)                                       00000620
          READ(1)VLIST                                                   00000630
          READ(1)NLIB,(TITLE(I),I=1,314)                                 00000640
          GOTO 5                                                         00000650
  116     REWIND1                                                        00000660
          WRITE(1)(NWRD(I),I=1,860)                                      00000670
          WRITE(1)VLIST                                                  00000680
          WRITE(1)NLIB,(TITLE(I),I=1,314)                                00000690
          GOTO 5                                                         00000700
  117     IPLOT=0                                                        00000710
          IF(XX1-PLOT)93,118,93                                          00000720
  118     IPLOT=1                                                        00000730
          GOTO 93                                                        00000740
   21     IF(XX1-XMODL) 57,22,57                                         00000750
C         DEVICE PARAMETER INPUT                                         00000760
   22     WRITE (6,99)                                                   00000770
          READ (5,109) XXX,YYY                                          00000780
          ITMF=0                                                         00000790
          ITEMP=0                                                        00000800
          DO 24 I=1,NLIB                                                 00000810
          ILIB=I                                                         00000820
          IF(X11(I)-XXX) 24,23,24                                        00000830
   23     IF(X22(I)-YYY) 24,25,24                                        00000840
   24     CONTINUE                                                       00000850
          NLIB=NLIB+1                                                    00000860
          ILIB=NLIB                                                      00000870
          ITMF=1                                                         00000880
          GO TO 26                                                       00000890
   25     ILIB=ILIB+NOP/8-1                                              00000900
          GO TO 28                                                       00000910
   26     IF(NOP-8) 27,28,27                                             00000920
   27     WRITE (6,100)                                                  00000930
          READ (5,109) ZZZ                                              00000940
   28     WRITE (6,101)                                                  00000950
          READ (5,*) I,AA                                               00000960
          IF(I-19) 29,32,32                                              00000970
   29     JJ=(I-1)/10                                                    00000980
          IF(JJ) 31,31,30                                                00000990
   30     ITEMP=1                                                        00001000
   31     I=I-10*JJ                                                      00001010
          I1J=JJ+ILIB                                                    00001020
          AXX(I1J,I)=AA*C4(I)                                           00001030
          GO TO 28                                                       00001040
   32     IF(ITMF) 5,5,33                                                00001050
   33     IF(ITEMP) 35,35,34                                             00001060
   34     X11(ILIB)=BLANK                                                00001070
          X22(ILIB)=ZZZ                                                  00001080
          ILIB=ILIB+1                                                    00001090
          NLIB=NLIB+ITMF                                                 00001100
   35     X11(ILIB)=XXX                                                  00001110
          X22(ILIB)=YYY                                                  00001120
          GO TO 5                                                        00001130
   36     NDS(LSAV+3)=1                                                  00001140
          GO TO 5                                                        00001150
   37     III=1                                                         00001160
```

```
           TYPE=XXX                                                    00001170
           XX1=XX2                                                     00001180
           GO TO 6                                                     00001190
     C         REMOVE OPTION                                           00001200
      38   J1=100*(NOP-1)-90*(NOP/7)-90*(NOP/8)+1                      00001210
           J2=J1-90*(NOP/6)+10*(NOP/8)+99                              00001220
           DO 39 J=J1,J2                                               00001230
           LL=JNV(J)                                                   00001240
           IF(XX1-VLIST(LL)) 39,40,39                                  00001250
      39   CONTINUE                                                    00001260
           GO TO 7                                                     00001270
      40   JNV(J)=-LL                                                  00001280
           NEL=NEL+NOP/6-1                                             00001290
           GO TO 5                                                     00001300
      41   WRITE (6,102) C1(NOP),C2(NOP)                               00001310
           READ (5,*) NODE1,NODE2,VALUE                                00001320
           IF(NODE2-NODE1) 42,43,43                                    00001330
      42   L=NODE1                                                     00001340
           NODE1=NODE2                                                 00001350
           NODE2=L                                                     00001360
      43   IF(NTOT-NODE2) 44,45,45                                     00001370
      44   NTOT=NODE2                                                  00001380
      45   DO 48 J=1,100                                               00001390
           K=100*NOP-J+1                                               00001400
           L=JNV(K)                                                    00001410
           IF(L) 46,46,47                                              00001420
      46   KSAV=K                                                      00001430
           LSAV=L                                                      00001440
           GO TO 48                                                    00001450
      47   IF(VLIST(L)-XX1)48,54,48                                    00001460
      48   CONTINUE                                                    00001470
           NEL=NEL+1                                                   00001480
           IF(LSAV) 49,50,50                                           00001490
      49   L=-LSAV                                                     00001500
           GO TO 53                                                    00001510
      50   DO 51 K=1,500                                               00001520
           L=-JNV(K)                                                   00001530
           IF(L) 51,51,52                                              00001540
      51   CONTINUE                                                    00001550
           L=LDS                                                       00001560
           LDS=LDS+2                                                   00001570
           GO TO 53                                                    00001580
      52   JNV(K)=0                                                    00001590
      53   JNV(KSAV)=L                                                 00001600
           VLIST(L)=XX1                                                00001610
      54   NDS(L)=NODE2                                                00001620
           NDS(L+1)=NODE1                                              00001630
           VLIST(L+1)=VALUE*C3(NOP)                                    00001640
           GO TO 5                                                     00001650
      65   WRITE(6,104)XX2,YY2                                         00001660
           NOP=15                                                      00001670
      55   WRITE (6,103)                                               00001680
           GO TO 58                                                    00001690
     C        DEVICE STATEMENTS                                        00001700
      57   IF(XX2-BLANK)58,571,58                                      00001710
     571   WRITE (6,99)                                                00001720
           READ (5,109) XX2,YY2                                        00001730
      58   I=1                                                         00001740
     581   XXX=X11(I)                                                  00001750
           YYY=X22(I)                                                  00001760
           ZZZ=BLANK                                                   00001770
           IF(NOP-15) 62,59,62                                         00001780
```

```
62      IF(XXX-XX2)64,63,64                                      00001790
63      IF(YYY-YY2)64,67,64                                      00001800
59      IF(XXX-BLANK) 61,60,61                                   00001810
60      ZZZ=YYY                                                  00001820
        I=I+1                                                    00001830
61      WRITE (6,112) X11(I),X22(I),ZZZ                          00001840
64      I=I+1                                                    00001850
        IF(I-NLIB)581,581,641                                    00001860
641     IF(NOP-15) 65,5,65                                       00001870
67      L=4                                                      00001880
        ITEMP=10+10*(NOP/8)                                      00001890
        LSAV=LDS                                                 00001900
        DO 71 J=1,ITEMP                                          00001910
        K=J+10*(NOP-6)                                           00001920
        LL=JNT(K)                                                00001930
        IF(LL) 68,68,69                                          00001940
68      KSAV=K                                                   00001950
69      IF(VLIST(LL)-XX1) 71,70,71                               00001960
70      LSAV=LL                                                  00001970
        L=0                                                      00001980
        KSAV=K                                                   00001990
        GO TO 72                                                 00002000
71      CONTINUE                                                 00002010
        VLIST(LSAV)=XX1                                          00002020
72      VLIST(LSAV+1)=XX2                                        00002030
        VLIST(LSAV+2)=YY2                                        00002040
        JNT(KSAV)=LSAV                                           00002050
        LDS=LDS+L                                                00002060
        WRITE (6,102) C1(NOP),C2(NOP)                            00002070
        IF(NOP-7) 73,73,74                                       00002080
73      CONTINUE                                                 00002090
        READ (5,*) NODE1,NODE2,NODE3                             00002100
        NDS(LSAV+2)=NODE3                                        00002110
        NDS(LSAV+3)=0                                            00002120
        GO TO 75                                                 00002130
C       DIODES                                                   00002140
74      CONTINUE                                                 00002150
        READ (5,*) NODE1,NODE2                                   00002160
75      NDS(LSAV)=NODE1                                          00002170
        NDS(LSAV+1)=NODE2                                        00002180
        NTOT=MAXO(NTOT,NODE1,NODE2,NODE3)                        00002190
        GO TO 5                                                  00002200
76      NTBP=0                                                   00002210
        WRITE (6,105)                                            00002220
        READ(5,*)ND                                              00002230
77      IF(ND(NTBP+1)) 78,5,78                                   00002240
78      NTBP=NTBP+1                                              00002250
        GO TO 77                                                 00002260
79      READ (5,109) TITLE                                       00002270
        GO TO 5                                                  00002280
80      LPR=-1                                                   00002290
        GO TO 93                                                 00002300
C       LIST CIRCUIT                                             00002310
81      WRITE (6,112) TITLE                                      00002320
        DO 89 J=1,540                                            00002330
811     LL=JNV(J)                                                00002340
        IF(LL) 89,89,82                                          00002350
82      IF(J-500) 83,83,84                                       00002360
83      K=(J-1)/100+1                                            00002370
        VALUE=VLIST(LL+1)/C3(K)                                  00002380
        WRITE (6,113) OP(K),VLIST(LL),NDS(LL),NDS(LL+1),VALUE,C1(K),C  00002390
       12(K)                                                     00002400
```

```
            GO TO 89                                                   00002410
  84        K=(J-501)/10+6                                             00002420
            IF(K-8) 86,85,85                                           00002430
  85        WRITE(6,106) OP(K),VLIST(LL),VLIST(LL+1),VLIST(LL+2),C1(K),C2(K),  00002440
           1NDS(LL),NDS(LL+1)                                          00002450
            GOTO 89                                                    00002460
  86        XX1=BLANK                                                  00002470
            IF(NDS(LL+3)) 88,88,87                                     00002480
  87        XX1=OFF                                                    00002490
  88        WRITE(6,106) OP(K),VLIST(LL),VLIST(LL+1),VLIST(LL+2),C1(K),C2(K),  00002500
           1NDS(LL),NDS(LL+1),NDS(LL+2),XX1                            00002510
  89        CONTINUE                                                   00002520
            GO TO 5                                                    00002530
  90        WRITE (6,107)                                              00002540
            DO 91 J=1,NLIB                                             00002550
  91        WRITE (6,108) (AXX(J,I),I=1,10)                            00002560
            GO TO 5                                                    00002570
  93        DO 94 I=1,620                                              00002580
  94        JSAVE(I)=NWRD(I)                                           00002590
            NDFLG=6                                                    00002600
            CALL SEQNS                                                 00002610
            IF(NDFLG-2)12,1,92                                         00002620
  92        CALL DCSLN                                                 00002630
            IF(NDFLG-2)12,1,95                                         00002640
  95        CALL DYSLN                                                 00002650
            GOTO 1                                                     00002660
  C                                                                    00002670
  96      FORMAT (2H =)                                                00002680
  97      FORMAT (2(A2,A5),2A5)                                        00002690
  98      FORMAT (6H ERROR)                                            00002700
  99      FORMAT (11H MODEL NO.?)                                      00002710
 100      FORMAT (6H TYPE?)                                            00002720
 101      FORMAT (14H NO.,PARAMETER)                                   00002730
 102      FORMAT (7H NODES,,2A5)                                       00002740
 103      FORMAT (2X,22HDEVICE LIBRARY LISTING/)                       00002750
 104      FORMAT (1X,2A5,15H NOT IN LIBRARY)                           00002760
 105      FORMAT (9H NODES(8))                                         00002770
 106      FORMAT (1X,A2,A5,2(2X,2A5),1H=,3I3,1X,A5)                    00002780
 107      FORMAT (7X,1HR,11X,1HC,10X,3HPHI,10X,1HN,10X,2HIS,9X,5HTHETA13X  00002790
           1,5HALPHA,10X,1HK,10X,6HDEVICE/)                           00002800
 108      FORMAT (1X,2G13.6,2F11.3,G13.6,F11.2,10X,2G13.6,1X,2A5)      00002810
 109      FORMAT (14A5)                                               00002820
 112      FORMAT (1X,14A5/)                                           00002830
 113      FORMAT (1X,A2,A5,2I6,2X,G13.6,1X,2A5)                        00002840
            END                                                       00002850
            SUBROUTINE SEQNS                                          00002860
  C       SVEQNS GENERATES THE DEVICE EQUIVALENT CIRCUITS AND          00002870
  C       COMPUTES THE STATE-VARIABLE EQUATIONS                        00002880
            IMPLICIT REAL*8(A-H,O-Z)                                   00002890
            DIMENSION OP(7), IB(40,10), AX1(10)                        00002900
            COMMON NWRD(2),IPLOT,LDS,NTOT,NEL,NT2,NMF2,NJX,LPR          00002910
            COMMON NOP,NODES,NBR,NCHORD,LTEMP,NOE(1),NOCC,NOCK,NOCG,NOCL   00002920
            COMMON NOCI,NSV,NTBP,LEVH,LQGG,LEVI,LEVE,LCAP,LEVR,LZLL      00002930
            COMMON LZLC,LZCC,LZCL,LSLI,LSCI,LSLE,LSCE,LMTL,LMTC,LMTI     00002940
            COMMON LMTE,LSCR,LPBB,NSVL,NXC,NDEV,NDFLG,NLC,NSX,NSRCES      00002950
            COMMON ND(10),JNV(500),JNT(40),NTRSET(20),NDS(240),VLIST(240)  00002960
            COMMON WORD(1000),AX(20,8),X1(20),X2(20),IA(40,100)         00002970
            COMMON /SCRTCH/ BSCR(1),NSN(50),IPLUS(50),IMINUS(90),NJS,NJBR,TC,N00002980
           1JCH,NODE1,NODE2,NJXB,NJXC,NSS,NSP,NBPK,XMNEM,NBSK,LRPLUS,VALUE,LRM00002990
           2IN,NODEO                                                  00003000
            COMMON /DEVMOD/ TITLE(14),AXX(30,8),X11(30),X22(30),NLIB    00003010
            EQUIVALENCE (NOE(1),NOCE), (JNV(1),IB(1,1))                00003020
```

```
      EQUIVALENCE (AX1(1),AX(1,1))                            00003030
      DATA OP(1),OP(2),OP(3),OP(4),OP(5)/1HE,1HC,1HR,1HL,1HI/ 00003040
      DATA OP(6),OP(7)/2HMF,1HT/                              00003050
      DATA Z,S,XNODE/1HZ,1HS,4HNODE/                          00003060
      DATA XLINK,TREE,P,Q/4HLINK,4HTREE,1HP,1HQ/              00003070
      DATA BB,EMP,EMN,DC,PNP/1HB,3HEMP,3HEMN,1HD,3HPNP/       00003080
C         GENERATE DEVICE EQUIVALENT CIRCUITS                 00003090
      DO 17 J=1,40                                            00003100
      LLL=JNT(J)                                              00003110
      IF(LLL) 17,17,1                                         00003120
    1 NXC=NXC+1                                               00003130
      NODE0=NDS(LLL)                                          00003140
      NODE1=NDS(LLL+1)                                        00003150
      NODE2=NDS(LLL+2)                                        00003160
      DO 3 I=1,NLIB                                           00003170
      IF(X11(I)-VLIST(LLL+1)) 3,2,3                           00003180
    2 IF(X22(I)-VLIST(LLL+2)) 3,4,3                           00003190
    3 CONTINUE                                                00003200
    4 J1=(J-1)/10-1                                           00003210
      IF(J1) 5,6,15                                           00003220
    5 NMF2=NMF2+2                                             00003230
    6 NT2=NT2+2                                               00003240
      NTRSET(NXC)=NDS(LLL+3)                                  00003250
      K=I-1                                                   00003260
      DO 7 L=1,10                                             00003270
      AX(NXC,L)=AXX(K,L)                                      00003280
    7 AX(NXC+1,L)=AXX(I,L)                                    00003290
      X2(NXC)=VLIST(LLL)                                      00003300
      TC=OP(J1+7)                                             00003310
      X1(NXC)=TC                                              00003320
      XMNEM=X22(K)                                            00003330
      IF(XMNEM-PNP) 8,9,8                                     00003340
    8 IF(XMNEM-EMP) 10,9,10                                   00003350
    9 NJXC=NJX-1                                              00003360
      NJXB=NJXC-1                                             00003370
      NJX=NJXB                                                00003380
      ITEMP=NTOT+1                                            00003390
      GO TO 11                                                00003400
   10 NJXC=NTOT+1                                             00003410
      NJXB=NJXC+1                                             00003420
      ITEMP=NJXB+1                                            00003430
C         GENERATE EMITTER-BASE CAPACITOR AND SHUNT RESISTOR. 00003440
   11 CALL STORE (2,TC,NODE2,NJXB,1.)                         00003450
      CALL STORE (3,TC,NODE2,NJXB,1.E7)                       00003460
C         GENERATE BASE-COLLECTOR CAPACITOR AND SHUNT RESISTOR.00003470
      CALL STORE (2,TC,NJXB,NJXC,1.)                          00003480
      CALL STORE (3,TC,NJXB,NJXC,1.E7)                        00003490
C         GENERATE BASE COLLECTOR RESISTORS.                  00003500
      IF(XMNEM-EMN) 12,13,12                                  00003510
   12 IF(XMNEM-EMP) 14,13,14                                  00003520
   13 CALL STORE (1,OP(7),ITEMP,NJXB,AX(NXC,6))               00003530
      AX(NXC,6)=AX(NXC+1,6)                                   00003540
      NJXB=ITEMP                                              00003550
   14 CALL STORE (3,BB,NODE0,NJXB,AX1(NXC))                   00003560
      NXC=NXC+1                                               00003570
      CALL STORE (3,OP(2),NODE1,NJXC,AX1(NXC))                00003580
      GO TO 17                                                00003590
   15 DO 16 L=1,10                                            00003600
   16 AX(NXC,L)=AXX(I,L)                                      00003610
      X1(NXC)=DC                                              00003620
      X2(NXC)=VLIST(LLL)                                      00003630
      NJX=NJX-1                                               00003640
```

```
        VALUE=1000.*AX(NXC,7)                                                   00003650
C       GENERATE DIODE CAPACITOR AND SHUNT RESISTOR.                            00003660
        CALL STORE (2,DC,NJX,NODE1,1.)                                          00003670
        CALL STORE (3,OP(2),NODE1,NJX,VALUE)                                    00003680
C          GENERATE CATHODE RESISTOR                                           00003690
        CALL STORE (3,BB,NODEO,NJX,AX1(NXC))                                    00003700
17      CONTINUE                                                                00003710
        NJB=1-NJX                                                               00003720
        NODES=NTOT+NJB                                                          00003730
        NBR=NODES-1                                                             00003740
        NM1=NBR                                                                 00003750
        IF(NBR-40) 20,20,18                                                     00003760
18      WRITE (6,141)                                                           00003770
19      NDFLG=2                                                                 00003780
        GO TO 137                                                               00003790
20      NCHORD=NEL-NBR                                                          00003800
        DO 21 I=1,210                                                           00003810
21      NSN(I)=0                                                                00003820
C       GENERATE, ELEMENT BY ELEMENT, THE PROPER TREE AND INCIDENCE MATRIX     00003830
        DO 50 I=1,500                                                           00003840
        LEL=JNV(I)                                                              00003850
        IF(LEL) 50,50,22                                                        00003860
22      VALUE=VLIST(LEL+1)                                                      00003870
        IND=(I-1)/100+1                                                         00003880
        IF(IND-3) 27,23,27                                                      00003890
C          SORT RESISTORS TO FIND SMALLEST                                     00003900
23      J1=I                                                                    00003910
        DO 26 J=I,300                                                           00003920
        ITEMP=JNV(J)                                                            00003930
        IF(ITEMP) 26,26,24                                                      00003940
24      XMNEM=VLIST(ITEMP+1)                                                    00003950
        IF(VALUE-XMNEM) 26,26,25                                               00003960
25      VALUE=XMNEM                                                             00003970
        J1=J                                                                    00003980
26      CONTINUE                                                                00003990
        ITEMP=JNV(J1)                                                           00004000
        JNV(J1)=LEL                                                             00004010
        JNV(I)=ITEMP                                                            00004020
        LEL=ITEMP                                                               00004030
27      LRPLUS=NDS(LEL)                                                         00004040
        LRMIN=NDS(LEL+1)                                                        00004050
        NBPK=NJB+LRPLUS                                                         00004060
        NBSK=NJB+LRMIN                                                          00004070
        NSP=NSN(NBPK)                                                           00004080
        NSS=NSN(NBSK)                                                           00004090
        IF(NSP+NSS) 28,31,28                                                    00004100
28      IF(NSP) 29,32,29                                                        00004110
29      IF(NSS) 30,33,30                                                        00004120
30      IF(NSP-NSS) 34,40,34                                                    00004130
31      NJS=NJS+1                                                               00004140
        NSN(NBPK)=NJS                                                           00004150
        NSN(NBSK)=NJS                                                           00004160
        GO TO 37                                                                00004170
32      NSN(NBPK)=NSS                                                           00004180
        GO TO 37                                                                00004190
33      NSN(NBSK)=NSP                                                           00004200
        GO TO 37                                                                00004210
34      DO 36 J=1,NODES                                                         00004220
        IF(NSN(J)-NSS) 36,35,36                                                 00004230
35      NSN(J)=NSP                                                              00004240
36      CONTINUE                                                                00004250
37      NJBR=NJBR+1                                                             00004260
```

```
        MCOL=NJBR+NCHORD                                            00004270
        LIST=NCHORD+NJBR+LEVB-1                                     00004280
        XMNEM=TREE                                                  00004290
        IF(IND-3) 38,38,39                                          00004300
   38   NOE(IND)=NOE(IND)+1                                         00004310
        GO TO 42                                                    00004320
C           NO I OR L BRANCHES--NO V OR C CHORDS ALLOWED            00004330
   39   WRITE (6,138)                                               00004340
        GO TO 19                                                    00004350
   40   NJCH=NJCH+1                                                 00004360
        MCOL=NJCH                                                   00004370
        LIST=NJCH+LEVB-1                                            00004380
        XMNEM=XLINK                                                 00004390
        IF(IND-2) 39,39,41                                          00004400
   41   NOE(IND+1)=NOE(IND+1)+1                                     00004410
   42   WORD(LIST)=VALUE                                            00004420
        IF(LPR) 43,44,43                                            00004430
   43   WRITE (6,140) OP(IND),VLIST(LEL),LRPLUS,LRMIN,VALUE,XMNEM   00004440
   44   LEST=LIST+NEL                                               00004450
        WORD(LEST)=VLIST(LEL)                                       00004460
        IF(LRPLUS) 45,47,46                                         00004470
   45   LRPLUS=LRPLUS+1                                             00004480
   46   LRPLUS=LRPLUS-NJX                                           00004490
        IPLUS(MCOL)=LRPLUS                                          00004500
   47   IF(LRMIN) 48,50,49                                          00004510
   48   LRMIN=LRMIN+1                                               00004520
   49   LRMIN=LRMIN-NJX                                             00004530
        IMINUS(MCOL)=LRMIN                                          00004540
   50   CONTINUE                                                    00004550
C          ELIMINATE UNUSED NODES AND GENERATE NODE LIST           00004560
        K=NJB                                                       00004570
        DO 54 I=NJB,NM1                                             00004580
        ITEMP=0                                                     00004590
        DO 52 J=1,NEL                                               00004600
        IF(IPLUS(J)-I) 51,53,51                                     00004610
   51   IF(IMINUS(J)-I) 52,53,52                                    00004620
   52   CONTINUE                                                    00004630
        GO TO 54                                                    00004640
   53   NSN(K)=I+NJX                                                00004650
        K=K+1                                                       00004660
   54   CONTINUE                                                    00004670
        IF(NBR-K+1) 55,57,55                                        00004680
   55   NBR=K-1                                                     00004690
        DO 56 I=1,6                                                 00004700
   56   NOE(I)=0                                                    00004710
        GO TO 20                                                    00004720
   57   MCOL=NEL+NTBP                                               00004730
        DO 59 J=1,MCOL                                              00004740
        DO 58 I=1,NBR                                               00004750
   58   IA(I,J)=0                                                   00004760
        K=J                                                         00004770
        IF(J-NCHORD)585,585,581                                    00004780
  581   IF(J-NCHORD-NTBP)582,582,583                               00004790
  582   K=J-NCHORD                                                  00004800
        K=IABS(NO(K))                                              00004810
        DO 584 II=1,NODES                                          00004820
        IF(NSN(II)-K)584,586,584                                   00004830
  584   CONTINUE                                                    00004840
        GOTO 59                                                     00004850
  586   IA(II,J)=1                                                  00004860
        GOTO 59                                                     00004870
  583   K=J-NTBP                                                    00004880
```

```
585      LRPLUS=IPLUS(K)                                              00004890
         LRMIN=IMINUS(K)                                              00004900
         IA(LRPLUS,J)=1                                               00004910
         IA(LRMIN,J)=-1                                               00004920
59       CONTINUE                                                     00004930
         LOGG=LEVB+NOCG                                               00004940
         LEVI=LOGG+NOCL                                               00004950
         LEVE=LEVI+NOCI                                               00004960
         LCAP=LEVE+NOCE                                               00004970
         LEVR=LCAP+NOCC                                               00004980
         IF(LPR)504,506,506                                           00004990
504      DO 505 J=1,NBR                                               00005000
505      WRITE(6,1010)(IA(J,I),I=1,MCOL)                              00005010
506      DO 731 J=1,NBR                                               00005020
         LIST=NCHORD+J+NTBP                                           00005030
         DO 61 K=J,NBR                                                00005040
         IF(IA(K,LIST)) 60,61,62                                      00005050
60       INEG=K                                                       00005060
61       CONTINUE                                                     00005070
         ISIGN=-1                                                     00005080
         K=INEG                                                       00005090
         GO TO 63                                                     00005100
62       ISIGN=1                                                      00005110
63       DO 65 I=1,MCOL                                               00005120
         ITEMP=IA(J,I)                                                00005130
         IA(J,I)=ISIGN*IA(K,I)                                        00005140
         IF(J-K) 64,65,64                                             00005150
64       IA(K,I)=ITEMP                                                00005160
65       CONTINUE                                                     00005170
C        ELIMINATE -1 S BELOW DIAGONAL                                00005180
         J1=J+1                                                       00005190
         IF(J1-NBR)1000,1000,69                                       00005200
1000     DO 66 K=J1,NBR                                               00005210
         IF(IA(K,LIST)) 67,66,66                                      00005220
66       CONTINUE                                                     00005230
         GO TO 69                                                     00005240
67       DO 68 I=1,MCOL                                               00005250
68       IA(K,I)=IA(K,I)+IA(J,I)                                      00005260
C        ELIMINATE NONZEROS ABOVE THE DIAGONAL                        00005270
69       J1=J-1                                                       00005280
         IF(J1)731,731,1002                                           00005290
1002     DO 73 K=1,J1                                                 00005300
         ISIGN=1                                                      00005310
         IF(IA(K,LIST)) 71,73,70                                      00005320
70       ISIGN=-1                                                     00005330
71       DO 72 I=1,MCOL                                               00005340
72       IA(K,I)=IA(K,I)+ISIGN*IA(J,I)                                00005350
73       CONTINUE                                                     00005360
731      CONTINUE                                                     00005370
         IF(LPR)500,82,82                                             00005380
500      DO 501 J=1,NBR                                               00005390
501      WRITE(6,1010)(IA(J,I),I=1,MCOL)                              00005400
82       NSV=NOCL+NOCC                                                00005410
         NSRCES=NOCI+NOCE                                             00005420
         NSVL=NSV-NXC                                                 00005430
         LZLC=LZLL+NSV*NOCL                                           00005440
         LZCC=LZLC+NOCL                                               00005450
         LZCL=LZLL+NOCL                                               00005460
         LSLI=LZLL+NSV*NSV                                            00005470
         LSCI=LSLI+NOCL                                               00005480
         LSLE=LSLI+NSV*NOCI                                           00005490
         LSCE=LSLE+NOCL                                               00005500
```

```
        LMTL=LSLI+NSV*NSRCES                                  00005510
        LMTC=LMTL+NTBP*NOCL                                   00005520
        LMTI=LMTC+NTBP*NOCC                                   00005530
        LMTE=LMTI+NTBP*NOCI                                   00005540
        LPBB=LMTE+NTBP*NOCE                                   00005550
        LTEMP=LPBB+NOCG                                       00005560
        IF(LTEMP-3250) 86,86,85                               00005570
 85     WRITE (6,139)                                         00005580
        GO TO 19                                              00005590
C       ZERO OUT COEFFICIENT MATRICES.                       00005600
 86     DO 87 I=LZLL,LTEMP                                    00005610
 87     WORD(I)=0.0                                           00005620
        LAST=NOCG+NOCL                                        00005630
        LIST=NOCE+NOCC                                        00005640
        IF(NOCR) 100,100,88                                   00005650
 88     LT=NOCR*NOCR                                          00005660
        DO 150 I=1,LT                                         00005670
 150    WORD(I)=0.                                            00005680
        J=-NOCR                                               00005690
        DO 89 I=1,NOCR                                        00005700
        J=J+NOCR+1                                            00005710
        JJ=LEVR+I-1                                           00005720
 89     WORD(J)=1.0/WORD(JJ)                                  00005730
        IF(NOCG) 92,92,90                                     00005740
 90     DO 91 I=1,NOCG                                        00005750
        J=LPBB+I-1                                            00005760
        JJ=LEVR+I-1                                           00005770
 91     WORD(J)=1.0/WORD(JJ)                                  00005780
        CALL ZSEQNS (LPBB,1,33443,1321,LIST,0,LIST,0)        00005790
 92     CALL FBINV(NOCR,P)                                    00005800
C       CALCULATION OF ZLL                                   00005810
        IF(NOCL) 97,97,93                                     00005820
 93     CALL ZSEQNS (1,LZLL,75335,2112,LIST,NOCG,LIST,NOCG)  00005830
        IF(NTBP) 95,95,94                                     00005840
C       CALCULATION OF T L                                   00005850
 94     CALL ZSEQNS (1,LMTL,88335,2112,LIST,NCHORD,LIST,NOCG) 00005860
 95     IF(NOCI) 100,100,96                                   00005870
C       COMPUTE SLI                                           00005880
 96     CALL ZSEQNS (1,LSLI,75336,2112,LIST,NOCG,LIST,LAST)  00005890
 97     IF(NOCI) 100,100,98                                   00005900
 98     IF(NTBP) 100,100,99                                   00005910
C       COMPUTE T I                                           00005920
 99     CALL ZSEQNS (1,LMTI,88336,2112,LIST,NCHORD,LIST,LAST) 00005930
 100    IF(NOCG) 113,113,101                                  00005940
 101    LT=NOCG*NOCG                                          00005950
        DO 102 I=1,LT                                         00005960
 102    WORD(I)=0.                                            00005970
        J=-NOCG                                               00005980
        DO 103 I=1,NOCG                                       00005990
        J=J+NOCG+1                                            00006000
        JJ=LEVB+I-1                                           00006010
 103    WORD(J)=WORD(JJ)                                      00006020
        IF(NOCR) 105,105,104                                  00006030
 104    CALL ZSEQNS (LEVR,1,44334,2311,LIST,0,LIST,0)        00006040
 105    CALL FBINV(NOCG,Q)                                    00006050
        IF(NOCC) 108,108,106                                  00006060
C       IF THERE ARE CLASS C, COMPUTE ZCC                    00006070
 106    CALL ZSEQNS (1,LZCC,72442,1120,NOCE,0,NOCE,0)        00006080
        IF(NOCE) 108,108,107                                  00006090
C       IF THERE ARE CLASS C AND CLASS E, COMPUTE SCE        00006100
 107    CALL ZSEQNS (1,LSCE,72441,1120,NOCE,0,0,0)           00006110
 108    IF(NOCR) 113,113,109                                  00006120
```

```
C        IF THERE ARE CLASS G AND CLASS R, COMPUTE QGR.          00006130
  109    DO 112 I=1,NOCG                                          00006140
         DO 110 K=1,NOCR                                          00006150
         KEPS=LEVR+K-1                                            00006160
         BSCR(K)=0.0                                              00006170
         MY=LIST+K                                                00006180
          IJ=I                                                    00006190
         DO 110 J=1,NOCG                                          00006200
         IF(IA(MY,J))151,110,152                                  00006210
  151    BSCR(K)=BSCR(K)+WORD(KEPS)*WORD(IJ)                      00006220
         GOTO 110                                                 00006230
  152    BSCR(K)=BSCR(K)-WORD(KEPS)*WORD(IJ)                      00006240
  110    IJ=IJ+NOCG                                               00006250
         IK=I                                                     00006260
         DO 111 K=1,NOCR                                          00006270
         WORD(IK)=BSCR(K)                                         00006280
  111    IK=IK+NOCG                                               00006290
  112    CONTINUE                                                 00006300
  113    IF(NOCC) 117,117,114                                     00006310
  114    IF(NOCL) 117,117,115                                     00006320
C        IF THERE ARE CLASS L AND CLASS C, COMPUTE ZCL AND ZLC.  00006330
  115    CALL ZSEQNS (1,LZCL,72435,21111,NOCE,0,LIST,NOCG)       00006340
         LA1=LZCL-1                                               00006350
         LB1=LZLC-1                                               00006360
         DO 116 J=1,NOCL                                          00006370
         DO 116 I=1,NOCC                                          00006380
         JI=LB1+NSV*(I-1)+J                                       00006390
         IJ=LA1+NSV*(J-1)+I                                       00006400
  116    WORD(JI)=WORD(IJ)                                        00006410
  117    IF(NOCI) 119,119,118                                     00006420
C        IF THERE ARE CLASS L AND CLASS E, COMPUTE SLE.          00006430
  118    CALL ZSEQNS (1,LSLE,75341,12221,LIST,NOCG,0,0)          00006440
  119    IF(NOCI) 122,122,120                                     00006450
  120    IF(NOCC) 122,122,121                                     00006460
C        IF THERE ARE CLASS C AND CLASS I, COMPUTE SCI.          00006470
  121    CALL ZSEQNS (1,LSCI,72436,21111,NOCE,0,LIST,LAST)       00006480
  122    IF(NTBP) 127,127,123                                     00006490
C           COMPUTE T C AND T E                                  00006500
  123    IF(NOCC) 125,125,124                                     00006510
  124    CALL ZSEQNS (1,LMTC,88342,12221,LIST,NCHORD,NOCE,0)     00006520
  125    IF(NOCE) 127,127,126                                     00006530
  126    CALL ZSEQNS (1,LMTE,88341,12221,LIST,NCHORD,0,0)        00006540
C           MULTIPLY Z AND S BY 1/L AND -1/C                     00006550
  127    LEST=NSV+NOCI+NOCE                                       00006560
         DO 131 K=1,NSV                                           00006570
         IF(K-NOCL) 129,129,128                                   00006580
  128    LL=LCAP+K-1-NOCL                                         00006590
         AA=-1.0/WORD(LL)                                         00006600
         GO TO 130                                                00006610
  129    LL=LQGG+K-1                                              00006620
         AA=1.0/WORD(LL)                                          00006630
  130    DO 131 I=1,LEST                                          00006640
         LJ=LZLL+NSV*(I-1)+K-1                                    00006650
  131    WORD(LJ)=AA*WORD(LJ)                                     00006660
         IF(LPR) 132,133,132                                      00006670
  132    CALL MATOUT (LZLL,NSV,NSV,Z)                            00006680
         CALL MATOUT (LSLI,NSV,NSKCES,S)                          00006690
         CALL MATOUT (LMTL,NTBP,LEST,XNODE)                      00006700
         GO TO 19                                                 00006710
  133    IF(NOP-17) 136,135,134                                   00006720
  134    IF(NOP-21) 136,19,19                                     00006730
  135    NDFLG=11                                                 00006740
```

```
          GO TO 137                                            00006750
   136    NDFLG=8                                              00006760
   137    RETURN                                               00006770
  1010    FORMAT(2X,36I3)                                      00006780
   138    FORMAT (46H NO I OR L BRANCHES--NO V OR C CHORDS ALLOWED ) 00006790
   139    FORMAT (16H MEMORY OVERFLOW)                         00006800
   140    FORMAT (1X,A2,A5,2I5,1X,G13.6,1X,A5)                 00006810
   141    FORMAT (16H TOO MANY NODES.)                         00006820
          END                                                  00006830
          SUBROUTINE DCSLN                                     00006840
C         DCSLN PERFORMS THE NONLINEAR DC ANALYSIS             00006850
          IMPLICIT REAL*8(A-H,O-Z)                             00006860
          DIMENSION OP(17), C1(6), C2(6), SAVZ(400), XMARK(11), XPAR(11) 00006870
          DIMENSION X110(118),LROW(10),JROW(10)                00006880
          COMMON NWRD(2),IPLOT,LDS,NTOT,NEL,NT2,NMF2,NJX,LPR   00006890
          COMMON NOP,NODES,NBR,NCHORD,NJB,NOCE,NOCC,NOCR,NOCG,NOCL 00006900
          COMMON NOCI,NSV,NTBP,LEVB,LOGG,LEVI,LEVE,LEVC,LEVR,LZLL 00006910
          COMMON LZLC,LZCC,LZCL,LSLI,LSCI,LSLF,LSCE,LMTL,LMTC,LMTI 00006920
          COMMON LMTE,LSCR,LPBB,NSVL,NXC,NDEV,NDFLG,NPAR,NSX,NSRCES 00006930
          COMMON ND(10),JNV(540),NTRSET(20),NDS(240),VLIST(240),WORD(1000) 00006940
          COMMON R(40),PHI(40),XIS(20),THETA(20)               00006950
          COMMON ALPHA(40),X1(20),X2(20),ATR(20,20)            00006960
          COMMON PAR(1),VCE,VBE,VD,VDS,VGS,XC,XE,XB,XA,XS,XD,SPARE(8) 00006970
          COMMON ALZ(20,20),XLIST(1190),XIS1(20),THETA1(20)    00006980
          COMMON V(20),VO(20),CUR(20),XI(20)                   00006990
          COMMON /SCRTCH/ FLG(20),DELX(6),YLABEL(6),VV(8),A,B,C,D,XJ,XX2,VMA00007000
         1X,X,XMAX,XINC,XSAV,VMULT,VMULX,VMULV,EPS,SUM,JOFF     00007010
          COMMON /PLOT/ IPAR(10),IDEV(10),ND1(8),ND2(8),LPLOT,ICOND,ICUR,KNV00007020
          COMMON /DEVMOD/ TITLE(14)                            00007030
          EQUIVALENCE(ATR(1,1),X110(1)),(LROW,CUR),(XI,JROW)    00007040
          EQUIVALENCE(NWRD(1),IWIDTH)                          00007050
          DATA OP/1HE,1HI,2HTE,2HAL,2HIE,2HIC,1H ,2HMF,1HT,1HD,2HCL,2HIT, 00007060
         1 2HPL,2HDC,2HDY,2HTR,2HEN/                            00007070
          DATA C1/5HVOLTS,5HAMPER,5HDEGRE,5HALPHA,5HMICRO,5HMICRO/ 00007080
          DATA C2,BLANK/1H ,2HES,4HES C,1H ,4HAMPS,4HAMPS,1H /  00007090
          DATA PHAI,ALL,STAR,PLUS,ZLL/4HPHAI,2HAL,1H*,1H+,4HZ LL/ 00007100
          DATA XNAN,XMAI/1.E-9,2HMA/                            00007110
          DATA VOLT,XCUR,BLANK/1HV,1HI,1H /                     00007120
          DATA XPAR/2HCE,2HBE,2HAC,2HDS,2HGS,1HC,1HE,1HB,1HA,1HS,1HD/ 00007130
          DATA PLUS,XMARK/1H+,1H*,1HX,1H.,1HO,1HZ,1HI,1HV,1HN,1HT,1H-,1H / 00007140
          IF(IWIDTH-100)600,602,602                            00007150
   600    ICOORD=10                                            00007160
          WIDTH=50.                                            00007170
          GOTO 603                                             00007180
   602    ICOORD=20                                            00007190
          WIDTH=100.                                           00007200
   603    ITERA=70                                             00007210
          NPAR=0                                               00007220
          LSCR=101                                             00007230
          NSRCES=NOCE+NOCI                                     00007240
          NSVPS=NSV+NSRCES                                     00007250
          IVOLT=NTBP                                           00007260
          NSX=LSCR+NSVL                                        00007270
          NT11=LZLL+NSVL                                       00007280
          N1T1=LZLL+NSV*(NSVL)                                 00007290
          NTNT=N1T1+NSVL                                       00007300
          NN=NXC+1                                             00007310
          LA1=LZLL-NSV-1                                       00007320
          LZLL1=LZLL+(NSVL-1)*NSV+NSVL-1                       00007330
          DO 1 I=1,40                                          00007340
          IPAR(I)=0                                            00007350
          FLG(I)=BLANK                                         00007360
```

```
    1      XIS1(I)=XIS(I)                                                      00007370
           DO 4 I=1,NTBP                                                       00007380
           IF(ND(I)) 3,4,2                                                     00007390
    2      KNV=KNV+1                                                           00007400
           ND1(KNV)=ND(I)                                                      00007410
           ND2(KNV)=0                                                          00007420
           GO TO 4                                                             00007430
    3      ND2(KNV)=-ND(I)                                                     00007440
    4      CONTINUE                                                            00007450
           IF(NSVL) 249,249,5                                                  00007460
    C         SAVE Z MATRIX                                                    00007470
    5      NSV2=NSV*(NSV+NSRCES)                                               00007480
           DO 6 I=1,NSV2                                                       00007490
           J=LZLL+I-1                                                          00007500
    6      SAVZ(I)=WORD(J)                                                     00007510
           K=0                                                                 00007520
           DO 250 I=1,NSVL                                                     00007530
           KK=LZLL+(I-1)*NSV                                                   00007540
           DO 250 J=1,NSVL                                                     00007550
           K=K+1                                                               00007560
           WORD(K)=WORD(KK)                                                    00007570
  250      KK=KK+1                                                             00007580
           CALL FBINV(NSVL,ZLL)                                               00007590
           CALL TMPLY (NT11,1,N1T1,NTNT,NSV,NSV,NXC,NSVL,NXC)                  00007600
  249      IF(NOP-14)52,52,65                                                  00007610
    7      IF(IPLOT) 8,8,11                                                    00007620
    8      DO 10 I=1,NSRCES                                                    00007630
           J=1                                                                 00007640
           JJ=I+LEVI-1                                                         00007650
           KK=JJ+NEL                                                           00007660
           IF(I-NOCI) 9,9,10                                                   00007670
    9      J=2                                                                 00007680
   10      WRITE (6,207) OP(J),WORD(KK),WORD(JJ),C1(J),C2(J)                   00007690
   11      IPLOT=0                                                             00007700
   12      WRITE (6,198)                                                       00007710
           READ (5,199) TYPE,XX1,TYPEP,XX1P                                    00007720
           NNN=10                                                             00007730
           ICOND=0                                                            00007740
           DO 13 I=1,40                                                        00007750
           XIS(I)=XIS1(I)                                                      00007760
   13      FLG(I)=BLANK                                                        00007770
           DO 14 NOP=1,17                                                      00007780
           IF(OP(NOP)-TYPE) 14,16,14                                          00007790
   14      CONTINUE                                                            00007800
   15      WRITE (6,200)                                                       00007810
           GO TO 12                                                           00007820
   16      IF(NOP-2) 46,47,17                                                  00007830
   17      IF(NOP-7) 34,197,18                                                 00007840
   18      IF(NOP-11) 22,33,19                                                 00007850
   19      IF(NOP-13) 21,20,225                                               00007860
  225      IF(NOP-15)52,193,226                                               00007870
  226      IF(NOP-17)193,227,227                                              00007880
  227      NDFLG=1                                                            00007890
           GOTO 196                                                          00007900
   20      IPLOT=1                                                            00007910
           GO TO 12                                                           00007920
   21      WRITE (6,201)                                                       00007930
           READ (5,*) ITERA                                                    00007940
           GO TO 12                                                           00007950
   22      DO 24 I=1,NXC                                                       00007960
           IF(TYPE-X1(I)) 24,23,24                                            00007970
   23      IF(XX1-X2(I)) 24,25,24                                             00007980
```

```
  24      CONTINUE                                            00007990
          WRITE (6,202)                                       00008000
          GO TO 12                                            00008010
  25      IF(TYPEP-BLANK)252,251,252                          00008020
 251      WRITE (6,203)                                       00008030
          READ (5,199) TYPEP,XX1P                             00008040
 252      IF(TYPEP-VOLT) 26,27,26                             00008050
  26      ICUR=1                                              00008060
          GO TO 28                                            00008070
  27      IVOLT=1                                             00008080
  28      DO 29 J=1,11                                        00008090
          IF(XPAR(J)-XX1P) 29,30,29                           00008100
  29      CONTINUE                                            00008110
          GO TO 15                                            00008120
  30      NPAR=NPAR+1                                         00008130
          IF(NPAR+KNV-10) 32,32,31                            00008140
  31      WRITE (6,204)                                       00008150
          GO TO 12                                            00008160
  32      IPAR(NPAR)=J+1                                      00008170
          IDEV(NPAR)=I                                        00008180
          GO TO 12                                            00008190
  33      NPAR=0                                              00008200
          IVOLT=NTRP                                          00008210
          ICUR=0                                              00008220
          GO TO 12                                            00008230
  34      WRITE (6,205)                                       00008240
          READ (5,199) TYPE,XX2                               00008250
          XSAV=20.                                            00008260
          IF(TYPE-ALL) 35,51,35                               00008270
  35      IF(TYPE-BLANK) 36,51,36                             00008280
  36      DO 38 LEL=1,NXC                                     00008290
          IF(X1(LEL)-TYPE) 38,37,38                           00008300
  37      IF(X2(LEL)-XX2) 38,39,38                            00008310
  38      CONTINUE                                            00008320
          GO TO 12                                            00008330
  39      FLG(LEL)=PLUS                                       00008340
          IF(NOP-4) 40,42,45                                  00008350
  40      FLG(LEL)=STAR                                       00008360
          IF(LEL-NT2) 41,41,34                                00008370
  41      FLG(LEL+1)=STAR                                     00008380
          GO TO 34                                            00008390
  42      IF(XX1-PHAI) 44,43,44                               00008400
  43      LEL=LEL+1                                           00008410
  44      XSAV=ALPHA(LEL)                                     00008420
          GO TO 51                                            00008430
  45      LEL=LEL+NOP/6                                       00008440
          XSAV=XIS(LEL)/1.0E-6                                00008450
          GO TO 51                                            00008460
  46      JJ=LEVE                                             00008470
          KK=JJ+NOCE                                          00008480
          GO TO 48                                            00008490
  47      JJ=LEVI                                             00008500
          KK=JJ+NOCI                                          00008510
  48      DO 49 LEL=JJ,KK                                     00008520
          L=LEL+NEL                                           00008530
          IF(XX1-WORD(L)) 49,50,49                            00008540
  49      CONTINUE                                            00008550
          GO TO 12                                            00008560
  50      XSAV=WORD(LEL)                                      00008570
  51      WRITE (6,206) C1(NOP),C2(NOP)                       00008580
          READ (5,*) X,XMAX,XINC                             00008590
          IEND=-1                                             00008600
```

```
          IF(XINC) 52,57,52                                          00008610
   52     IEND=0                                                     00008620
          IF(IPLOT)539,539,53                                        00008630
  539      WRITE(6,217)TITLE                                         00008640
           GOTO 54                                                   00008650
   53     IF(NOP-14)531,12,655                                       00008660
  531      IF(IVOLT)533,533,532                                      00008670
  532      WRITE(6,212)C1(1)                                         00008680
           READ(5,*)YMIN,YMAX                                        00008690
  533     IF(ICUR)535,535,534                                        00008700
  534      WRITE(6,212)XMAI                                          00008710
           READ(5,*)ZMIN,ZMAX                                        00008720
  535      WRITE(6,209)TITLE                                         00008730
   54     IF(NOP-14) 57,65,65                                        00008740
  655      IPLOT=0                                                   00008750
           GOTO 65                                                   00008760
   55     X=X+XINC                                                   00008770
          IF(X-XMAX-.5*XINC) 57,57,56                                00008780
   56     X=XSAV                                                     00008790
          IEND=1                                                     00008800
   57     IF(NOP-3) 59,62,58                                         00008810
   58     IF(NOP-5) 60,61,61                                         00008820
   59     WORD(LEL)=X                                                00008830
          GO TO 62                                                   00008840
   60     ALPHA(LEL)=X                                               00008850
          GO TO 62                                                   00008860
   61     XIS(LEL)=X*1.0E-6                                          00008870
   62     IF(IEND) 12,63,7                                           00008880
   63     IF(IPLOT) 64,64,65                                         00008890
   64     WRITE (6,221) OP(NOP),XX1,X,C1(NOP),C2(NOP)                00008900
   65     JOFF=0                                                     00008910
          MODE2=0                                                    00008920
C         COMPUTE SOURCE CONTRIBUTIONS AT ZERO TIME.                 00008930
          II=LSCR                                                    00008940
          DO 67 I=1,NSV                                              00008950
          JJ=LSLI+I-1                                                00008960
          KK=LEVI                                                    00008970
          SUM=0.0                                                    00008980
          DO 66 J=1,NSRCES                                           00008990
          SUM=SUM+WORD(JJ)*WORD(KK)                                  00009000
          KK=KK+1                                                    00009010
   66     JJ=JJ+NSV                                                  00009020
          WORD(II)=SUM                                               00009030
   67     II=II+1                                                    00009040
          IF(NSVL) 69,69,68                                          00009050
C         COMPUTE -Z XL*Z LL INVERSE*S L AND ADD TO S X.            00009060
   68     CALL TMPLY (NT11,1,LSCR,NSX,NSVL,NXC,NXC,NSVL,1)           00009070
   69     ITERMN=2                                                   00009080
          NITER=ITERA/2                                              00009090
          VMULT=.5                                                   00009100
          VMULX=.5                                                   00009110
          VMULV=1.5                                                  00009120
          ITER=0                                                     00009130
          EPS=.05                                                    00009140
          IF(NOP-3)692,691,692                                       00009150
  691     XE=DEXP(.069*(X-20.))                                      00009160
          XC=293./(X+273.)                                           00009170
  692     DO 84 K=1,NXC                                              00009180
          IF(TYPE-ALL) 70,71,70                                      00009190
   70     IF(FLG(K)-STAR) 72,71,72                                   00009200
   71     THETA(K)=XC*THETA1(K)                                      00009210
          XIS(K)=XIS1(K)*XE                                          00009220
```

```
  72    IF(NT2-K) 79,83,73                                             00009230
  73    IF(K-2*(K/2)) 83,83,74                                         00009240
C          CHECK FOR OFF TRANSISTORS                                   00009250
  74    IF(NTRSET(K)) 75,76,75                                         00009260
  75    THETA(K)=.0001*THETA(K)                                        00009270
        THETA(K+1)=.0001*THETA(K+1)                                    00009280
        JOFF=1                                                         00009290
C          GENERATE ALPHA INVERSE                                      00009300
  76    A=1.0/(1.0-ALPHA(K)*ALPHA(K+1))                                00009310
        B=A*ALPHA(K+1)                                                 00009320
        C=A*ALPHA(K)                                                   00009330
C          COMPUTE ALPHA INVERSE Z  AND ALPHA INVERSE S.              00009340
        KK=K+NSX-1                                                     00009350
        D=WORD(KK+1)                                                   00009360
        CUR(K)=A*WORD(KK)+B*D                                          00009370
        CUR(K+1)=C*WORD(KK)+A*D                                        00009380
        DO 77 J=1,NXC                                                  00009390
        KJ=LZLL1+J*NSV+K                                               00009400
        D=WORD(KJ+1)                                                   00009410
        ALZ(K,J)=A*WORD(KJ)+B*D                                        00009420
  77    ALZ(K+1,J)=C*WORD(KJ)+A*D                                      00009430
        IF(ICOND) 78,78,81                                             00009440
  78    VO(K+1)=-.2                                                    00009450
        GO TO 81                                                       00009460
C          DIODES                                                      00009470
  79    KX=K+NSX-1                                                     00009480
        DO 80 J=1,NXC                                                  00009490
        KJ=LZLL1+J*NSV+K                                               00009500
  80    ALZ(K,J)=WORD(KJ)                                              00009510
        CUR(K)=WORD(KX)                                                00009520
  81    IF(ICOND) 82,82,83                                             00009530
  82    VO(K)=.5*PHI(K)                                                00009540
C          SET INITIAL VOLTAGES                                        00009550
  83    A=THETA(K)*VO(K)                                               00009560
        XI(K)=XIS(K)*DEXP(A)                                           00009570
        DELX(K)=THETA(K)*XI(K)                                         00009580
  84    CONTINUE                                                       00009590
  85    CONTINUE                                                       00009600
  86    DO 88 I=1,NXC                                                  00009610
        DO 87 J=1,NXC                                                  00009620
  87    ATR(I,J)=ALZ(I,J)                                              00009630
        ATR(I,NN)=XI(I)-XIS(I)-DELX(I)*VO(I)-CUR(I)                    00009640
  88    ATR(I,I)=ALZ(I,I)-DELX(I)                                      00009650
C        COMPUTE NEW VX VECTOR USING CROUT METHOD                      00009660
        VMAX=0.0                                                       00009670
        DO 93 L=1,NXC                                                  00009680
        IF(VMAX-DABS(VO(L))) 89,90,90                                  00009690
  89    VMAX=VO(L)                                                     00009700
  90    LLL=L-1                                                        00009710
        DO 93 I=L,NXC                                                  00009720
        II=I+1                                                         00009730
        A=ATR(L,II)                                                    00009740
        IF(LLL) 93,93,91                                               00009750
  91    B=ATR(I,L)                                                     00009760
        DO 92 K=1,LLL                                                  00009770
        B=B-ATR(I,K)*ATR(K,L)                                          00009780
  92    A=A-ATR(L,K)*ATR(K,II)                                         00009790
        ATR(I,L)=B                                                     00009800
  93    ATR(L,II)=A/ATR(L,L)                                           00009810
        DO 95 L=2,NXC                                                  00009820
        I=NN-L                                                         00009830
        VR=ATR(I,NN)                                                   00009840
```

```
        II=I+1                                                      00009850
        DO 94 K=II,NXC                                              00009860
94      VR=VR-ATR(I,K)*ATR(K,NN)                                    00009870
        ATR(I,NN)=VR                                               00009880
95      V(I)=VR                                                    00009890
        V(NXC)=ATR(NXC,NN)                                         00009900
        ITER=ITER+1                                                00009910
        ABSERR=7.E-3*EPS*VMAX                                      00009920
        DO 100 I=1,NXC                                             00009930
        IF(MODE2) 96,96,98                                         00009940
96      IF(V(I)) 97,98,98                                          00009950
97      IF(VO(I)) 100,98,98                                        00009960
98      IF(DABS(V(I)-VO(I))-EPS*DABS(V(I))-ABSERR) 100,100,99      00009970
99      IF(ITER-NITER) 111,107,107                                00009980
100     CONTINUE                                                   00009990
        IF(ITER-ITERMN) 111,111,101                               00010000
101     IF(MODE2) 102,102,103                                      00010010
102     MODE2=1                                                    00010020
        ITERMN=ITER+1                                              00010030
        NITER=ITERA                                                00010040
        EPS=.0005                                                  00010050
        VMULT=1.                                                   00010060
        VMULX=0.                                                   00010070
        VMULV=1.66                                                 00010080
        GO TO 111                                                  00010090
103     IF(JOFF) 104,129,104                                       00010100
C       IF ANY TRANSISTORS WERE SET *OFF*, RESET THEM.             00010110
104     JOFF=0                                                     00010120
        ITERMN=ITER+1                                              00010130
        DO 106 K=1,NT2                                             00010140
        IF(NTRSET(K)) 105,106,105                                  00010150
105     THETA(K)=THETA(K)*1.E4                                     00010160
        THETA(K+1)=THETA(K+1)*1.E4                                 00010170
C       IF ( V( K ) ) 340,340,380                                  00010180
C  340  IF ( V(K+1) ) 330,330,380                                  00010190
106     CONTINUE                                                   00010200
        GO TO 111                                                  00010210
107     IF(MODE2) 108,102,108                                      00010220
108     IF(ICOND)110,110,109                                       00010230
109     ICOND=0                                                    00010240
        GO TO 65                                                   00010250
110     WRITE (6,220)                                              00010260
        NCOUNT=NCOUNT+1                                            00010270
        IF(NOP-14) 55,7,55                                         00010280
111     DO 128 I=1,NXC                                             00010290
        VR=V(I)                                                    00010300
        B=THETA(I)                                                 00010310
        C=XIS(I)                                                   00010320
        VOI=VO(I)                                                  00010330
        XII=XI(I)                                                  00010340
C       MAKE VOLTAGE TEST.                                         00010350
        IF(VR) 116,112,112                                         00010360
112     XJ=XII+DFLX(I)*(VR-VOI)                                    00010370
        IF(XJ) 113,113,114                                         00010380
113     VOI=.8*VR                                                  00010390
        GO TO 119                                                  00010400
C          THE DIODE CURVE                                         00010410
114     XII=VMULT*XJ+VMULX*XII                                     00010420
        IF(I-NMF2) 117,117,115                                     00010430
C          EXPONENTIAL CURRENTS                                    00010440
115     A=XII/C                                                    00010450
        VOI=DLOG(A)/B                                              00010460
```

```
            GO TO 126                                                    00010470
     116    IF(MODE2) 117,117,118                                        00010480
     117    VOI=VR                                                       00010490
            GO TO 119                                                    00010500
     118    VOI=VMULT*VR+VMULX*VOI                                       00010510
     119    IF(I-NMF2) 120,120,123                                       00010520
C           MOSFETS                                                      00010530
     120    VOO=VOI                                                      00010540
            IF(VOO-B) 121,122,122                                        00010550
     121    VOO=B                                                        00010560
     122    XII=C*(.5*VOO*VOO-B*VOO+1.)                                  00010570
            DELX(I)=C*(VOO-B)                                            00010580
            GO TO 127                                                    00010590
     123    A=B*VOI                                                      00010600
C              EXPONENTIAL CURRENTS                                      00010610
            IF(A+30.) 124,124,125                                        00010620
     124    XII=0.0                                                      00010630
            GO TO 126                                                    00010640
     125    XII=1.E50                                                    00010650
            IF(A-115.)1251,1251,126                                      00010660
     1251   XII=C*DEXP(A)                                                00010670
C           COMPUTE THE DERIVATIVE OF THE DIODE CURVE.                   00010680
     126    DELX(I)=B*XII                                                00010690
     127    VO(I)=VOI                                                    00010700
     128    XI(I)=XII                                                    00010710
            VMULT=VMULV-VMULT                                            00010720
            VMULX=1.-VMULT                                               00010730
            GO TO 85                                                     00010740
     129    DO 135 J=1,NXC                                               00010750
            K=NXC-J+1                                                    00010760
            IF(IPLOT) 130,130,134                                        00010770
     130    KK=K+1                                                       00010780
            A=1000.*(XI(K)-XIS(K))                                       00010790
            IF(K-NT2) 131,131,133                                        00010800
     131    IF(K-2*(K/2)) 134,134,132                                    00010810
     132    B=1000.*(XI(KK)-XIS(KK))                                     00010820
            XE=B*ALPHA(KK)-A                                             00010830
            XC=A*ALPHA(K)-B                                              00010840
            XB=-1000.*(XE+XC)                                            00010850
            VCE=.001*R(KK)*XC-V(KK)+V(K)                                 00010860
            VR=.000001*R(K)*XB+V(K)                                      00010870
            WRITE (6,218) FLG(K),X1(K),X2(K),VCE,XC,VR,XB                00010880
            GO TO 134                                                    00010890
     133    VCE=V(K)+.001*R(K)*A                                         00010900
            WRITE (6,218) FLG(K),X1(K),X2(K),VCE,A                       00010910
     134    I=K+NSVL                                                     00010920
            VO(K)=V(K)                                                   00010930
     135    V(I)=V(K)                                                    00010940
            ICOND=1                                                      00010950
            IF(NSVL) 139,139,136                                         00010960
C           COMPUTE LINEAR VARIABLES.                                    00010970
C           COMPUTE V L=-Z LL INVERSE*S L.                              00010980
     136    DO 138 I=1,NSVL                                              00010990
            SUM=0.0                                                      00011000
            IJ=I                                                         00011010
            DO 137 J=LSCR,NSX                                            00011020
     137    SUM=SUM+WORD(IJ)*WORD(J)                                     00011030
            IJ=IJ+NSV                                                    00011040
     138    V(I)=-SUM                                                    00011050
C           COMPUTE V L=V L-Z LL INVERSE*Z LX*V X.                       00011060
            II=NSVL+3251                                                 00011070
            CALL TMPLY (1,N1T1,II,3251,NSV,NSV,NSVL,NXC,1)               00011080
```

```
  139     IF(NOP-14)1391,1391,193                                              00011090
 1391     IF(NTBP) 148,148,140                                                 00011100
  140     K=0                                                                  00011110
          DO 146 I=1,NTBP                                                      00011120
          SUM=0.                                                               00011130
          DO 143 J=1,NSVPS                                                     00011140
          XNS=V(J)                                                             00011150
          IF(J-NSV) 142,142,141                                               00011160
  141     KK=LEVI+J-1-NSV                                                      00011170
          XNS=WORD(KK)                                                        00011180
  142     L=LMTL+(J-1)*NTBP+I-1                                                00011190
  143     SUM=SUM+WORD(L)*XNS                                                  00011200
          IF(ND(I)) 144,145,145                                               00011210
  144     SUM=VV(K)-SUM                                                        00011220
          GO TO 146                                                            00011230
  145     K=K+1                                                                00011240
  146     VV(K)=SUM                                                            00011250
          IF(IPLOT) 147,147,151                                               00011260
  147     WRITE (6,208) (VOLT,ND1(I),ND2(I),VV(I),I=1,KNV)                     00011270
  148     IF(IPLOT) 149,149,151                                               00011280
  149     IF(NOP-14) 55,7,55                                                  00011290
  151     IF(NNN-10) 152,152,166                                              00011300
  152     NCOUNT=9                                                             00011310
          NLABL=0                                                              00011320
          NOUT=KNV+NPAR                                                        00011330
          IF(KNV) 154,154,153                                                 00011340
  153     WRITE (6,211) (XMARK(I),ND1(I),ND2(I),I=1,KNV)                      00011350
  154     IF(NPAR) 160,160,155                                                00011360
  155     DO 159 I=1,NPAR                                                      00011370
          L=IPAR(I)                                                           00011380
          IF(L-6) 156,156,157                                                 00011390
  156     A=VOLT                                                               00011400
          GO TO 158                                                            00011410
  157     A=XCUR                                                               00011420
  158     J=IDEV(I)                                                            00011430
          K=KNV+I                                                             00011440
  159     WRITE (6,210) XMARK(K),A,XPAR(L-1),X1(J),X2(J)                      00011450
  160     IF(ICUR) 1661,1661,164                                              00011460
  164     SCAL2=WIDTH/(ZMAX-ZMIN)                                             00011470
          DO 165 J=1,6                                                        00011480
  165     YLABEL(J)=ZMIN+FLOAT(J-1)*.2*WIDTH/SCAL2                            00011490
          IF(IWIDTH-100)605,606,606                                          00011500
  605     WRITE(6,222)XMAI,YLABEL                                            00011510
          GOTO 1661                                                          00011520
  606     WRITE (6,213) XMAI,YLABEL                                          00011530
 1661     IF(IVOLT)166,166,1662                                              00011540
 1662     SCALE=WIDTH/(YMAX-YMIN)                                            00011550
          DO 162 J=1,6                                                       00011560
  162     YLABEL(J)=YMIN+(FLOAT(J-1))*.2*WIDTH/SCALE                         00011570
          IF(IWIDTH-100)705,706,706                                          00011580
  705     WRITE(6,222)C1(1),YLABEL                                           00011590
          GOTO 166                                                           00011600
  706     WRITE(6,213)C1(1),YLABEL                                           00011610
  166     NCOUNT=NCOUNT+1                                                     00011620
          NLABL=NLABL+1                                                       00011630
          L=11                                                                00011640
          DO 169 J=1,IWIDTH                                                   00011650
          X110(J)=XMARK(NNN)                                                  00011660
          IF(NCOUNT-10) 169,167,169                                          00011670
  167     IF(J-L) 169,168,169                                                00011680
  168     X110(J)=PLUS                                                        00011690
          L=L+ICOORD                                                          00011700
```

```
169    CONTINUE                                                    00011710
       NNN=11                                                      00011720
       JJ=0                                                        00011730
       X110(11)=XMARK(10)                                          00011740
172    DO 185 J=1,NOUT                                             00011750
       JJ=JJ+1                                                     00011760
       IF(J-KNV) 173,173,175                                       00011770
173    YY=VV(J)                                                    00011780
174    L=(YY-YMIN)*SCALE+11.5                                      00011790
       GO TO 182                                                   00011800
175    KK=J-KNV                                                    00011810
       L=IDEV(KK)                                                  00011820
       K=L                                                         00011830
       IF(KNV) 177,177,176                                         00011840
176    K=L+NSVL                                                    00011850
177    A=1000.*(XI(L)-XIS(L))                                      00011860
       IF(L-NT2) 178,178,179                                       00011870
178    B=1000.*(XI(L+1)-XIS(L+1))                                  00011880
       XE=B*ALPHA(L+1)-A                                           00011890
       XC=A*ALPHA(L)-B                                             00011900
       XB=-XC-XE                                                   00011910
       XS=XE                                                       00011920
       XD=XC                                                       00011930
       VCE=.001*R(L+1)*XC-V(K+1)+V(K)                              00011940
       VBE=.001*R(L)*XB+V(K)                                       00011950
       VDS=VCE                                                     00011960
       VGS=VBE                                                     00011970
       GO TO 180                                                   00011980
179    XA=A+V(K)/ALPHA(L)                                          00011990
       VD=V(K)+.001*R(L)*XA                                        00012000
180    K=IPAR(KK)                                                  00012010
       YY=PAR(K)                                                   00012020
       IF(K-6) 174,174,181                                         00012030
181    L=(YY-ZMIN)*SCAL2+11.5                                      00012040
182    IF(L) 185,185,183                                           00012050
183    IF(L-IWIDTH) 184,184,185                                    00012060
184    X110(L)=XMARK(JJ)                                           00012070
185    CONTINUE                                                    00012080
       LAST=0                                                      00012090
       DO 610 I=1,IWIDTH                                           00012100
       IF(X110(I)-BLANK)609,610,609                                00012110
609    LAST=I                                                      00012120
610    CONTINUE                                                    00012130
       IF(NLABL-5) 189,192,189                                     00012140
189    IF(NCOUNT-10) 190,191,191                                   00012150
190    WRITE (6,214) (X110(I),I=1,LAST)                            00012160
       GO TO 55                                                    00012170
191    NCOUNT=NCOUNT-10                                            00012180
       WRITE (6,215) X,(X110(I),I=11,LAST)                         00012190
       GO TO 55                                                    00012200
192    WRITE (6,216) OP(NOP),XX1,(X110(I),I=11,LAST)               00012210
       GO TO 55                                                    00012220
193    IF(NSVL) 196,196,194                                        00012230
194    DO 195 I=1,NSV2                                             00012240
       J=LZLL+I-1                                                  00012250
195    WORD(J)=SAVZ(I)                                             00012260
       GOTO 196                                                    00012270
197    NOFLG=2                                                     00012280
196    RETURN                                                      00012290
C                                                                  00012300
198    FORMAT (2H +)                                               00012310
199    FORMAT (A2,A5,2X,A2,A5)                                     00012320
```

```
200    FORMAT (6H ERROR)                                              00012330
201    FORMAT (18H NO. OF ITERATIONS)                                 00012340
202    FORMAT (23H DEVICE NOT IN CIRCUIT.)                            00012350
203    FORMAT (10H PARAMETER)                                         00012360
204    FORMAT (18H ONLY 10 WAVEFORMS)                                 00012370
205    FORMAT (8H DEVICE?)                                            00012380
206    FORMAT (22H MIN,MAX,INCREMENT IN ,2A5)                         00012390
207    FORMAT (2X,A2,A5,1H=,G13.5,1X,2A5)                             00012400
208    FORMAT (2(5X,A1,I2,4H - V,I2,3H = ,G13.5))                     00012410
209    FORMAT (2X,14A5/)                                              00012420
210    FORMAT (10X,A5,2A2,2X,A2,A5)                                   00012430
211    FORMAT (10X,A5,1HV,I2,4H - V,I2)                               00012440
212    FORMAT (13H MIN AND MAX ,A5)                                   00012450
213    FORMAT (/1X,A5,1X,G11.4,5(9X,G11.4))                          00012460
214    FORMAT (2X,118A1)                                              00012470
215    FORMAT (1X,G11.4,118A1)                                        00012480
216    FORMAT (3X,A2,A5,2X,118A1)                                     00012490
217    FORMAT (1X,14A5 /7H DEVICE,3X,13HVCE,VD OR VDS,                00012500
      13X,8HIC OR ID,4X,10HVBE OR VGS,3X,8HIB OR IG/12X,7H(VOLTS)     00012510
      2,9X,4H(MA),7X, 7H(VOLTS),3X,13H(MICROAMPERE)/)                 00012520
218    FORMAT (1X,A1,A2,A5,2X,5(1X,G12.5))                           00012530
220    FORMAT (15H NO CONVERGENCE)                                    00012540
221     FORMAT(/2X,A2,A5,1H=,G13.5,1X,2A5)                            00012550
222     FORMAT(/1X,A5,2X,6G10.3)                                      00012560
       END                                                           00012570
       SUBROUTINE DYSLN                                              00012580
C         TRANSISTOR DYNAMICS SOLUTION                               00012590
       IMPLICIT REAL*8(A-H,O-Z)                                      00012600
       DIMENSION XMARK(11), OP(12), FISAV(2), XPAR(15), RK5(30)      00012610
       COMMON NWRD(2),IPLOT,LDS,NTOT,NEL,NT2,NMF2,NJX,LPR            00012620
       COMMON NOP,NODES,NBR,NCHORD,LTEMP,NOCE,NOCC,NOCR,NOCG,NOCL    00012630
       COMMON NOCI,NSV,NTBP,LEVB,LOGG,LEVI,LEVE,LEVC,LEVR,LZLL       00012640
       COMMON LZLC,LZCC,LZCL,LSLI,LSCI,LSLE,LSCE,LMTL,LMTC,LMTI      00012650
       COMMON LMTE,LSCR,LPBB,NSVL,NXC,NDEV,NDFLG,NLC,NSX,NSRCES      00012660
       COMMON ND(10),JNV(560),NDS(240),VLIST(240),WORD(1000)         00012670
       COMMON R(20),C(20),PHI(20),XN(20),XIS(20),THETA(20)           00012680
       COMMON ALPHA(20),XK(20),X1(20),X2(20)                         00012690
       COMMON PAR(1),VCE,VBE,VD,VDS,VGS,XC,XE,XB,XA,XJC,XJE,XJB       00012700
       COMMON A,XS,XD                                                00012710
       COMMON T1(30),T2(30),TREP(30),T11(30),T22(30),VSMAX(30)       00012720
       COMMON VSMIN(30),CCUR(59),HBY3                                00012730
       COMMON VS(30),XNN(30),VO(30),RK1(30)                          00012740
       COMMON RK2(30),RK3(30),RK4(60),LV(30),IPAR(15),IDEV(15)       00012750
       COMMON YLABEL(6),X110(110),V(11),XLIST(1367),XNS(60),XI(20)   00012760
       COMMON /DEVMOD/ TITLE(14)                                     00012770
       EQUIVALENCE (RK3,RK5),(NWRD(1),IWIDTH)                        00012780
       DATA OP/2HTI,2HPL,2HCO,1H ,1HE,1HI,2HMF,1HT,1HD,2HCL,2HER,2HEN/00012790
       DATA VV,XCUR,PLUS,OFF,XNAN,XMAI/1HV,1HI,1H+,1HO,1.E-9,2HMA/    00012800
       DATA XPAR/2HCE,2HBE,2HAC,2HDS,2HGS,1HC,1HE,1HB,1HA,2HC*,2HE*,2HB*/00012810
       DATA XMARK/1H*,1HX,1H.,1HO,1HZ,1HI,1HV,1HN,1HT,1H-,1H /        00012820
       IF(IWIDTH-100)600,602,602                                     00012830
600    ICOORD=10                                                     00012840
       WIDTH=50.                                                     00012850
       NWIDTH=60                                                     00012860
       GOTO 603                                                      00012870
602    ICOORD=20                                                     00012880
       WIDTH=100.                                                    00012890
       NWIDTH=108                                                    00012900
603    NSVPS=NSV+NSRCES                                              00012910
       NT3=NT2+NSVL                                                  00012920
       EPS1=.005                                                     00012930
       NDFLG=6                                                       00012940
```

```
            ICUR=0                                                    00012950
            IVOLT=NT8P                                                00012960
            NPAR=0                                                    00012970
            DO 1 I=1,30                                               00012980
            VO(I)=XNS(I)                                              00012990
            IPAR(I)=0                                                 00013000
       1    LV(I)=0                                                   00013010
       2    WRITE (6,107)                                             00013020
            READ (5,108) TYPE,XX1,TYPEP,XX1P                          00013030
            DO 3 NOP=1,12                                             00013040
            IF(TYPE-OP(NOP)) 3,5,3                                    00013050
       3    CONTINUE                                                  00013060
       4    WRITE (6,109)                                             00013070
            GO TO 2                                                   00013080
       5    IF(NOP-2) 6,30,7                                          00013090
       6    READ (5,110) TITLE                                        00013100
            GO TO 2                                                   00013110
       7    IF(NOP-4) 24,8,9                                          00013120
       8    NDFLG=2                                                   00013130
     138    RETURN                                                    00013140
       9    IF(NOP-6) 25,26,10                                        00013150
      10    IF(NOP-10) 11,22,139                                      00013160
     139    IF(NOP-12)23,140,140                                      00013170
     140    NDFLG=1                                                   00013180
            GOTO 138                                                  00013190
      11    DO 13 I=1,NXC                                             00013200
            IF(TYPE-X1(I)) 13,12,13                                   00013210
      12    IF(XX1-X2(I)) 13,14,13                                    00013220
      13    CONTINUE                                                  00013230
            WRITE (6,111)                                             00013240
            GO TO 2                                                   00013250
      14    IF(TYPEP-XMARK(11))142,141,142                            00013260
     141    WRITE (6,112)                                             00013270
            READ (5,108) TYPEP,XX1P                                   00013280
     142    IF(TYPEP-VV) 15,16,15                                     00013290
      15    ICUR=1                                                    00013300
            GO TO 17                                                  00013310
      16    IVOLT=1                                                   00013320
      17    DO 18 J=1,15                                              00013330
            IF(XPAR(J)-XX1P) 18,19,18                                 00013340
      18    CONTINUE                                                  00013350
            GO TO 4                                                   00013360
      19    NPAR=NPAR+1                                               00013370
            IF(NPAR+NTBP-10) 21,21,20                                 00013380
      20    WRITE (6,113)                                             00013390
            GO TO 2                                                   00013400
      21    IPAR(NPAR)=J+1                                            00013410
            IDEV(NPAR)=I                                              00013420
            GO TO 2                                                   00013430
      22    NPAR=0                                                    00013440
            GO TO 2                                                   00013450
      23    WRITE (6,114)                                             00013460
            READ (5,*) EPS1                                           00013470
            GO TO 2                                                   00013480
   C          CONTINUED SOLUTION                                      00013490
      24    WRITE (6,115)                                             00013500
            READ (5,*) TFINL                                          00013510
            TFINL=TFINL*XNAN                                          00013520
            GO TO 32                                                  00013530
      25    JJ=LEVE                                                   00013540
            KK=JJ+NOCE                                                00013550
            GO TO 27                                                  00013560
```

```
   26    JJ=LEVI
         KK=JJ+NOCI                                                       00013580
   27    DO 28 LEL=JJ,KK                                                  00013590
         L=LEL+NEL                                                        00013600
         IF(XX1-WORD(L)) 28,29,28                                         00013610
   28    CONTINUE                                                         00013620
         GO TO 4                                                          00013630
   29    WRITE (6,116)                                                    00013640
         READ (5,*) T111,T222,TREPP,VSMX,VSMN                            00013650
         T111=T111*XNAN                                                   00013660
         T222=T222*XNAN                                                   00013670
         TREPP=TREPP*XNAN                                                 00013680
         NVS=LEL-LEVI+1                                                   00013690
         LV(NVS)=1                                                        00013700
         TREP(NVS)=TREPP                                                  00013710
         T1(NVS)=T111                                                     00013720
         T11(NVS)=T111                                                    00013730
         T2(NVS)=T111+T222                                                00013740
         T22(NVS)=T111+T222                                               00013750
         VSMAX(NVS)=VSMX                                                  00013760
         VSMIN(NVS)=VSMN                                                  00013770
         GO TO 2                                                          00013780
   C        NEW SOLUTION                                                  00013790
   30    DO 31 I=1,30                                                     00013800
         XNS(I)=V0(I)                                                     00013810
         T22(I)=T2(I)                                                     00013820
   31    T11(I)=T1(I)                                                     00013830
   C        TRANSIENT SOLUTION                                            00013840
         WRITE(6,117)                                                     00013850
         READ (5,*) NPTS,TFINL                                           00013860
         IF(IVOLT)151,151,150                                            00013870
   150   WRITE(6,121)                                                     00013880
         READ(5,*)YMIN,YMAX                                              00013890
   151   IF(ICUR)154,154,152                                             00013900
   152   WRITE(6,123)                                                     00013910
         READ(5,*)ZMIN,ZMAX                                              00013920
   154   TFINL=TFINL*XNAN                                                 00013930
         PERIOD=TFINL/FLOAT(NPTS)                                         00013940
         TBGIN=0.                                                         00013950
         H=PERIOD/2.                                                      00013960
         HBY3=H/3.                                                        00013970
         NSTEP=2                                                          00013980
         EPS15=5.*EPS1                                                    00013990
         EPS2=EPS15/64.                                                   00014000
         OUTPUT=0.                                                        00014010
         NN=10                                                           00014020
   32    CONTINUE                                                         00014030
   C        RUNGE-KUTTA MERSON METHOD                                     00014040
   C        SOURCE CONTRIBUTION                                           00014050
   33    DO 41 J=1,NSRCES                                                 00014060
         JJ=J+NSV                                                        00014070
         IF(LV(J)) 40,40,34                                              00014080
   34    L=J                                                             00014090
         IF(TBGIN-T11(L)-TREP(L)) 36,35,35                              00014100
   35    T11(L)=T11(L)+TREP(L)                                           00014110
         T22(L)=T22(L)+TREP(L)                                           00014120
   36    IF(TBGIN-T11(L)) 37,38,38                                       00014130
   37    VSJ=VSMIN(L)                                                     00014140
         GO TO 41                                                        00014150
   38    IF(TBGIN-T22(L)) 39,37,37                                       00014160
   39    VSJ=VSMAX(L)                                                     00014170
         GO TO 41                                                        00014180
```

```
    40    KK=LEVI+J-1                                                         00014190
          VSJ=WORD(KK)                                                        00014200
    41    XNS(JJ)=VSJ                                                         00014210
          IF(NN-10) 42,63,42                                                  00014220
    42    DO 44 I=1,NSV                                                       00014230
          SUM=0.                                                             00014240
          DO 43 J=1,NSRCES                                                    00014250
          JJ=J+NSV                                                            00014260
          VSJ=XNS(JJ)                                                         00014270
          JJ=LSLI+(J-1)*NSV+I-1                                               00014280
    43    SUM=SUM+VSJ*WORD(JJ)                                                00014290
    44    VS(I)=SUM                                                           00014300
          OUTPUT=TBGIN+PERIOD                                                 00014310
          ISTEP=NSTEP                                                         00014320
    45    DO 46 I=1,NSV                                                       00014330
    46    XNN(I)=XNS(I)                                                       00014340
          GO TO 49                                                            00014350
    47    DO 48 I=1,NSV                                                       00014360
    48    XNS(I)=XNN(I)                                                       00014370
    49    CALL DER (RK1)                                                      00014380
          DO 50 I=1,NSV                                                       00014390
    50    XNS(I)=XNN(I)+RK1(I)                                                00014400
          CALL DER (RK2)                                                      00014410
          DO 51 I=1,NSV                                                       00014420
    51    XNS(I)=XNN(I)+0.5*(RK1(I)+RK2(I))                                   00014430
          CALL DER (RK3)                                                      00014440
          DO 52 I=1,NSV                                                       00014450
    52    XNS(I)=XNN(I)+0.375*(RK1(I)+3.0*RK3(I))                             00014460
          CALL DER (RK4)                                                      00014470
          DO 53 I=1,NSV                                                       00014480
          XNS(I)=XNN(I)+1.5*(RK1(I)-3.0*RK3(I))+6.0*RK4(I)                    00014490
    53    RK2(I)=XNS(I)                                                       00014500
    54    CALL DER (RK5)                                                      00014510
          L=0                                                                00014520
          DO 58 I=1,NSV                                                       00014530
          XNS(I)=XNN(I)+0.5*(RK1(I)+RK5(I))+2.0*RK4(I)                        00014540
          WRD=RK2(I)                                                          00014550
          DIFF=DABS(WRD-XNS(I))                                               00014560
          WRD=DABS(WRD)                                                       00014570
          IF(DIFF-WRD*EPS15) 56,55,55                                         00014580
    55    HBY3=0.5*HBY3                                                       00014590
          ISTEP=ISTEP+ISTEP                                                   00014600
          NSTEP=NSTEP+NSTEP                                                   00014610
          GO TO 47                                                            00014620
    56    IF(DIFF-WRD*EPS2) 58,58,57                                          00014630
    57    L=1                                                                00014640
    58    CONTINUE                                                            00014650
          ISTEP=ISTEP-1                                                       00014660
          IF(ISTEP) 80,80,59                                                  00014670
    59    IF(L) 60,60,45                                                      00014680
    C         CAN DOUBLE H                                                    00014690
    60    IF(NSTEP-2) 45,45,61                                                00014700
    61    IF(ISTEP-2*(ISTEP/2)) 62,62,45                                      00014710
    62    HBY3=HBY3+HBY3                                                      00014720
          NSTEP=NSTEP/2                                                       00014730
          ISTEP=ISTEP/2                                                       00014740
          GO TO 45                                                            00014750
    63    WRITE (6,118) TITLE                                                 00014760
          CALL DER (RK2)                                                      00014770
          NCOUNT=9                                                            00014780
          NOUT=NTBP+NPAR                                                      00014790
          IF(NPAR) 69,69,64                                                   00014800
```

```
  64      DO 68 I=1,NPAR                                                00014810
          L=IPAR(I)                                                     00014820
          IF(L-6) 65,65,66                                              00014830
  65      WRD=VV                                                        00014840
          GO TO 67                                                      00014850
  66      WRD=XCUR                                                      00014860
  67      J=IDEV(I)                                                     00014870
          K=NTBP+I                                                      00014880
  68      WRITE (6,119) XMARK(K),WRD,XPAR(L-1),X1(J),X2(J)              00014890
  69      IF(NTBP) 74,74,70                                             00014900
  70      I=1                                                           00014910
 701      K=I                                                           00014920
          J=ND(I+1)                                                     00014930
          IF(J) 72,71,71                                                00014940
  71      J=0                                                           00014950
          GO TO 73                                                      00014960
  72      I=I+1                                                         00014970
          J=-J                                                          00014980
  73      WRITE (6,120) XMARK(K),ND(K),J                                00014990
          I=I+1                                                         00015000
          IF(I-NTBP)701,701,74                                          00015010
  74      IF(ICUR) 791,791,78                                           00015020
  78      SCAL2=WIDTH/(ZMAX-ZMIN)                                       00015030
          DO 79 J=1,6                                                   00015040
  79      YLABEL(J)=ZMIN+FLOAT(J-1)*.2*WIDTH/SCAL2                      00015050
          IF(NWIDTH-100)605,606,606                                     00015060
 605       WRITE(6,222)XMAI,YLABEL                                      00015070
          GOTO 791                                                      00015080
 606      WRITE (6,122) XMAI,YLABEL                                     00015090
 791      IF(IVOLT)80,80,792                                            00015100
 792      SCALE=WIDTH/(YMAX-YMIN)                                       00015110
          DO 76 J=1,6                                                   00015120
  76      YLABEL(J)=YMIN+(FLOAT(J-1))*.2*WIDTH/SCALE                    00015130
          IF(NWIDTH-100)705,706,706                                     00015140
 705       WRITE(6,222)VV,YLABEL                                        00015150
          GOTO 80                                                       00015160
 706      WRITE(6,122)VV,YLABEL                                         00015170
  80      DO 81 J=1,NWIDTH                                              00015180
  81      X110(J)=XMARK(NN)                                             00015190
          NN=11                                                         00015200
          JJ=0                                                          00015210
          X110(1)=XMARK(10)                                            00015220
  C          COMPUTE NODE VOLTAGES                                      00015230
          IF(NTBP) 85,85,82                                             00015240
  82      DO 84 I=1,NTBP                                                00015250
          SUM=0.                                                        00015260
          DO 83 J=1,NSVPS                                               00015270
          L=LMTL+(J-1)*NTBP+I-1                                         00015280
  83      SUM=SUM+WORD(L)*XNS(J)                                        00015290
  84      V(I)=SUM                                                      00015300
  85      J=0                                                           00015310
 851      J=J+1                                                         00015320
          JJ=JJ+1                                                       00015330
          IF(J-NTBP) 86,86,89                                          00015340
  86      YY=V(J)                                                       00015350
          IF(ND(J+1)) 87,88,88                                          00015360
  87      J=J+1                                                         00015370
          YY=YY-V(J)                                                    00015380
  88      L=(YY-YMIN)*SCALE+1.5                                         00015390
          GO TO 94                                                      00015400
  89      KK=J-NTBP                                                     00015410
          L=IDEV(KK)                                                    00015420
```

```
          K=L+NSVL                                                        00015430
          A=1000.*(XI(L)-XIS(L))                                          00015440
          IF(K-NT3) 90,90,91                                              00015450
    90    B=1000.*(XI(L+1)-XIS(L+1))                                      00015460
          XJE=B*ALPHA(L+1)-A                                              00015470
          XJC=A*ALPHA(L)-B                                                00015480
          XJB=-XJC-XJE                                                    00015490
          XE=XJE-1000.*CCUR(L)                                            00015500
          XC=XJC-1000.*CCUR(L+1)                                          00015510
          XS=XE                                                           00015520
          XD=XC                                                           00015530
          XB=-XC-XE                                                       00015540
          VCE=.001*R(L+1)*XC-XNS(K+1)+XNS(K)                              00015550
          VBE=.001*R(L)*XB+XNS(K)                                         00015560
          VDS=VCE                                                         00015570
          VGS=VBE                                                         00015580
          GO TO 92                                                        00015590
    91    XA=A+1000.*CCUR(L)+XNS(K)/ALPHA(L)                              00015600
          VD=XNS(K)+.001*R(L)*XA                                          00015610
    92    K=IPAR(KK)                                                      00015620
          YY=PAR(K)                                                       00015630
          IF(K-6) 88,88,93                                                00015640
    93    L=(YY-ZMIN)*SCAL2+1.5                                           00015650
    94    IF(L) 95,95,96                                                  00015660
    95    X110(1)=OFF                                                     00015670
          GO TO 99                                                        00015680
    96    IF(L-NWIDTH) 97,97,99                                           00015690
    97    X110(L)=XMARK(JJ)                                               00015700
    99    IF(J-NOUT)851,98,98                                             00015710
    98    NCOUNT=NCOUNT+1                                                 00015720
          LAST=0                                                          00015730
          DO 610 I=1,NWIDTH                                               00015740
          IF(X110(I)-BLANK)609,610,609                                    00015750
   609    LAST=I                                                          00015760
   610    CONTINUE                                                        00015770
          IF(NCOUNT-10) 102,103,102                                       00015780
   102    WRITE (6,124) (X110(I),I=1,LAST)                                00015790
          GO TO 105                                                       00015800
   103    DO 104 JJ=1,NWIDTH,ICOORD                                       00015810
   104    X110(JJ)=PLUS                                                   00015820
          NCOUNT=0                                                        00015830
          XTIM=OUTPUT/XNAN                                                00015840
          WRITE (6,125) XTIM,(X110(I),I=1,NWIDTH)                         00015850
   105    CONTINUE                                                        00015860
   106    TBGIN=OUTPUT                                                    00015870
          OUTPUT=TBGIN+PERIOD                                             00015880
          IF(TFINL-TBGIN)2,33,33                                          00015890
  C                                                                       00015900
   107    FORMAT (3H ..)                                                  00015910
   108    FORMAT (A2,A5,2X,A2,A5)                                         00015920
   109    FORMAT (6H ERROR)                                               00015930
   110    FORMAT (14A5)                                                   00015940
   111    FORMAT (23H DEVICE NOT IN CIRCUIT.)                             00015950
   112    FORMAT (11H PARAMETER?)                                         00015960
   113    FORMAT (18H ONLY 10 WAVEFORMS)                                  00015970
   114    FORMAT (6H ERROR)                                               00015980
   115    FORMAT (11H TFINAL(NS))                                         00015990
   116    FORMAT (43H TDELAY,TWIDTH,TREP,MAX,MIN (NS,VOLTS OR A))         00016000
   117    FORMAT (28H NO. OF INTERVALS,TFINAL(NS))                        00016010
   118    FORMAT (5X,14A5/)                                               00016020
   119    FORMAT (10X,A5,2A2,2X,A2,A5)                                    00016030
   120    FORMAT (10X,A5,4HNODE,I3,1H-,I3)                                00016040
```

```
   121    FORMAT (18H MIN AND MAX VOLTS)                                  00016050
   122    FORMAT (/3X,A4,G11.4,5(9X,G11.4))                              00016060
   123    FORMAT (15H MIN AND MAX MA)                                     00016070
   124    FORMAT (10X,110A1)                                             00016080
   125    FORMAT (G10.3,110A1)                                           00016090
   222    FORMAT(/1X,A4,1X,6G10.3)                                        00016100
         END                                                             00016110
         SUBROUTINE DER (DV)                                             00016120
         IMPLICIT REAL*8(A-H,O-Z)                                        00016130
         DIMENSION DV(20)                                                00016140
         COMMON NWRD(2),IPLOT,LDS,NTOT,NEL,NT2,NMF2,NJX,LPR              00016150
         COMMON NOP,NODES,NBR,NCHORD,LTEMP,NOCE,NOCC,NOCK,NOCG,NOCL      00016160
         COMMON NOCI,NSV,NTBP,LEVB,LOGG,LEVI,LEVE,LEVC,LEVR,LZLL         00016170
         COMMON LZLC,LZCC,LZCL,LSLI,LSCI,LSLE,LSCE,LMTL,LMTC,LMTI        00016180
          COMMON LMTE,LSCR,LPBB,NSVL,NXC,NDEV,NDFLG,NLC,NSX,NSRCES       00016190
         COMMON ND(810),VLIST(240),WORD(1000)                            00016200
         COMMON R(20),C(20),PHI(20),XN(20),XIS(20),THETA(20)             00016210
         COMMON ALPHA(20),XK(20),X1(20),X2(20),PAR(16)                   00016220
         COMMON T1(30),T2(30),TREP(30),T11(30),T22(30),VSMAX(30)         00016230
         COMMON VSMIN(30),CCUR(59),HBY3,VS(30)                           00016240
         COMMON XLIST(1734),V(60),XI(20)                                 00016250
         DO 27 I=1,NSV                                                    00016260
         SUM=VS(I)                                                        00016270
         JJ=LZLL+I-1                                                      00016280
         DO 1 J=1,NSV                                                     00016290
         SUM=SUM+V(J)*WORD(JJ)                                           00016300
    1    JJ=JJ+NSV                                                        00016310
         II=I-NSVL                                                        00016320
         IF(II) 26,26,2                                                   00016330
    2    IF(II-NT2) 3,3,17                                                00016340
    3    IF(II-2*(II/2)) 20,20,4                                          00016350
C        TRANSISTORS                                                      00016360
    4    IF(II-NMF2) 5,5,10                                               00016370
C        MOSFETS                                                          00016380
    5  · A=V(I)                                                           00016390
         B=THETA(II)                                                      00016400
         IF(A-B) 6,7,7                                                    00016410
    6    A=B                                                              00016420
    7    AA=XIS(II)*(.5*A*A-B*A+1.)                                       00016430
         A=V(I+1)                                                         00016440
         B=THETA(II+1)                                                    00016450
         IF(A-B) 8,9,9                                                    00016460
    8    A=B                                                              00016470
    9    BB=XIS(II+1)*(.5*A*A-B*A+1.)                                     00016480
         GO TO 16                                                         00016490
   10    A=THETA(II)*V(I)                                                 00016500
         B=THETA(II+1)*V(I+1)                                             00016510
         AA=XIS(II)                                                       00016520
         BB=XIS(II+1)                                                     00016530
         IF(A+40.) 11,11,12                                               00016540
   11    AA=-AA                                                           00016550
         XII=0.                                                           00016560
         GO TO 13                                                         00016570
   12    IF(A-115.)121,121,120                                           00016580
  120      XII=AA*1.E50                                                   00016590
         GOTO 122                                                         00016600
  121      XII=AA*DEXP(A)                                                 00016610
  122      AA=XII-AA                                                      00016620
   13    IF(B+40.) 14,14,15                                               00016630
   14    BB=-BB                                                           00016640
         XIII=0.                                                          00016650
         GO TO 16                                                         00016660
```

```
   15      IF(B-115.)151,151,150                                              00016670
   150       XIII=BB*1.0E50                                                   00016680
           GOTO 152                                                          00016690
   151     XIII=BB*DEXP(B)                                                    00016700
   152     BB=XIII-BB                                                         00016710
   16      XEC=AA-ALPHA(II+1)*BB                                             00016720
           XEC1=BB-ALPHA(II)*AA                                             00016730
           GO TO 21                                                          00016740
C          DIODES                                                            00016750
   17    A=THETA(II)*V(I)                                                    00016760
           IF(A+40.) 18,18,19                                               00016770
   18      XII=0.                                                           00016780
           XEC=-XIS(II)                                                     00016790
           GO TO 21                                                          00016800
   19      IF(A-115.)191,191,190                                            00016810
   190     XII=XIS(II)*1.0E50                                               00016820
           GOTO 192                                                         00016830
   191     XII=XIS(II)*DEXP(A)                                              00016840
   192     XEC=XII-XIS(II)                                                  00016850
           GO TO 21                                                          00016860
C          COLLECTOR JUNCTIONS OF TRANSISTORS                                00016870
   20      XEC=XEC1                                                         00016880
           XII=XIII                                                         00016890
   21      XI(II)=XII                                                       00016900
           CAP=C(II)                                                        00016910
           IF(II-NMF2) 25,25,22                                            00016920
   22      PHIV=PHI(II)-V(I)                                                00016930
           IF(PHIV) 24,24,23                                               00016940
   23      CAP=C(II)/(PHIV**XN(II))                                         00016950
   24      CAP=CAP+XK(II)*XII                                               00016960
   25      SUM=(SUM-XEC)/CAP                                                00016970
           CCUR(II)=CAP*SUM                                                 00016980
   26      DV(I)=SUM*HBY3                                                    00016990
   27      CONTINUE                                                          00017000
           RETURN                                                            00017010
           END                                                              00017020
           SUBROUTINE STORE (I,X,NODE1,NODE2,VAL)                            00017030
           IMPLICIT REAL*8(A-H,O-Z)                                         00017040
           COMMON NWRD(2),IPLOT,LDS,NTOT,NEL,NSR(54),JNV(560),NDS(240)       00017050
           COMMON VLIST(240),WORD(1000)                                     00017060
           N1=NODE1                                                         00017070
           N2=NODE2                                                         00017080
           IF(N2-N1) 1,2,2                                                  00017090
   1       L=N1                                                             00017100
           N1=N2                                                            00017110
           N2=L                                                             00017120
   2       IF(NTOT-N2) 3,4,4                                                00017130
   3       NTOT=N2                                                          00017140
   4       DO 9 J=1,100                                                     00017150
           K=100*I-J+1                                                      00017160
           L=JNV(K)                                                         00017170
           IF(L) 5,5,9                                                      00017180
   5       KSAV=K                                                           00017190
           LSAV=L                                                           00017200
   9       CONTINUE                                                         00017210
           NEL=NEL+1                                                        00017220
           IF(LSAV) 10,11,11                                                00017230
   10    L=-LSAV                                                            00017240
           GO TO 14                                                         00017250
   11      L=LDS                                                            00017260
           LDS=LDS+2                                                        00017270
   14      JNV(KSAV)=L                                                      00017280
```

```
                NDS(L)=N2                                                      00017290
                NDS(L+1)=N1                                                    00017300
                VLIST(L)=X                                                     00017310
       15       VLIST(L+1)=VAL                                                 00017320
                RETURN                                                         00017330
                END                                                           00017340
                SUBROUTINE ZSEQNS (LB,LD,LOC,NTYP,LYA,LZA,LYC,LZC)             00017350
       C        ZSEQNS MULTIPLIES MATRIX A(INTEGER) TIMES B(NORMAL OR DIAGONAL) 00017360
       C        TIMES C(INTEGER) AND STORES THE RESULT IN D                    00017370
                IMPLICIT REAL*8(A-H,O-Z)                                       00017380
                COMMON NWRD(2),NSCR(10),NBR,LLL,LTEMP,NLOC(605)                00017390
                COMMON NDS(240),VLIST(240),WORD(1200),IA(40,100)               00017400
                COMMON /SCRTCH/ KK(1),NZC,NYC,NZA,NYA,NYS,K2(1),ISIGN,NTYPC,NTYPB, 00017410
               1NTYPA,NTYPS,JA(70),TEMP(10)                                    00017420
                NPAR=LOC                                                       00017430
                LAST=NTYP                                                      00017440
                I=5                                                           00017450
                NDEN=10000                                                     00017460
       1        NN=NPAR/NDEN                                                   00017470
                NT=LAST/NDEN                                                   00017480
                KK(I+1)=NLOC(NN)                                               00017490
                K2(I+1)=NT-1                                                   00017500
                NPAR=NPAR-NN*NDEN                                              00017510
                LAST=LAST-NT*NDEN                                              00017520
                I=I-1                                                         00017530
                NDEN=NDEN/10                                                   00017540
                IF(I) 2,2,1                                                    00017550
       2        NTYPB=NTYPB-1                                                  00017560
                IF(NTYPB) 3,4,3                                                00017570
       3        LAST=NZA                                                       00017580
                GO TO 5                                                        00017590
       4        LAST=NYC                                                       00017600
       5        LB1=LB-LAST-1                                                  00017610
                IF(NTYPB) 7,6,8                                                00017620
       6        NASTOP=LAST                                                    00017630
                GO TO 9                                                        00017640
       7        NASTOP=NYC                                                     00017650
                GO TO 9                                                        00017660
       8        NASTOP=LAST                                                    00017670
       9        LD1=LD-1                                                       00017680
                IF(NZC)43,43,91                                                00017690
       91       DO 42 I=1,NZC                                                  00017700
                L1=LYC+I                                                       00017710
                L2=LZC+I                                                       00017720
                IF(NASTOP)93,93,92                                             00017730
       92       DO 12 K=1,NASTOP                                               00017740
                IF(NTYPC) 11,11,10                                             00017750
       10       L2=LZC+K                                                       00017760
                GO TO 12                                                       00017770
       11       L1=LYC+K                                                       00017780
       12       JA(K)=IA(L1,L2)                                                00017790
       93       IF(NZA)97,97,94                                                00017800
       94       DO 25 J=1,NZA                                                  00017810
                SIJ=0.0                                                        00017820
                IF(NTYPB) 16,20,13                                             00017830
       13       LJ=LB-1+J                                                      00017840
                IF(JA(J)) 14,24,15                                             00017850
       14       SIJ=-WORD(LJ)                                                  00017860
                GO TO 24                                                       00017870
       15       SIJ=WORD(LJ)                                                   00017880
                GO TO 24                                                       00017890
       16       JK=J+LB1                                                       00017900
```

```
          IF(NASTOP)24,24,95                                        00017910
   95     DO 19 K=1,NASTOP                                          00017920
          JK=JK+LAST                                               00017930
          IF(JA(K)) 17,19,18                                       00017940
   17     SIJ=SIJ-WORD(JK)                                         00017950
          GO TO 19                                                 00017960
   18     SIJ=SIJ+WORD(JK)                                         00017970
   19     CONTINUE                                                 00017980
          GO TO 24                                                 00017990
   20     KJ=J*LAST+LB1                                            00018000
          IF(NASTOP)24,24,96                                       00018010
   96     DO 23 K=1,NASTOP                                         00018020
          KJ=KJ+1                                                  00018030
          IF(JA(K)) 21,23,22                                       00018040
   21     SIJ=SIJ-WORD(KJ)                                         00018050
          GO TO 23                                                 00018060
   22     SIJ=SIJ+WORD(KJ)                                         00018070
   23     CONTINUE                                                 00018080
   24     TEMP(J)=SIJ                                              00018090
   25     CONTINUE                                                 00018100
   97     IF(NYA)42,42,98                                          00018110
   98     DO 41 J=1,NYA                                            00018120
          L1=LYA+J                                                 00018130
          L2=LZA+J                                                 00018140
          SIJ=0.                                                  00018150
          IF(NZA) 33,33,26                                         00018160
   26     DO 32 K=1,NZA                                            00018170
          IF(NTYPA) 27,27,28                                       00018180
   27     L2=LZA+K                                                 00018190
          GO TO 29                                                 00018200
   28     L1=LYA+K                                                 00018210
   29     IF(IA(L1,L2)) 30,32,31                                   00018220
   30     SIJ=SIJ-TEMP(K)                                          00018230
          GO TO 32                                                 00018240
   31     SIJ=SIJ+TEMP(K)                                          00018250
   32     CONTINUE                                                 00018260
   33     IJ=J+LD1                                                 00018270
          IF(ISIGN) 40,34,39                                       00018280
   34     IF(NTYPS) 38,35,36                                       00018290
   35     L1=I+LYC                                                 00018300
          L2=J+LZA                                                 00018310
          GO TO 37                                                 00018320
   36     L1=J+LYA                                                 00018330
          L2=I+LZC                                                 00018340
   37     WORD(IJ)=FLOAT(IA(L1,L2))+SIJ                            00018350
          GO TO 41                                                 00018360
   38     WORD(IJ)=WORD(IJ)+SIJ                                    00018370
          GO TO 41                                                 00018380
   39     SIJ=-SIJ                                                 00018390
   40     WORD(IJ)=SIJ                                             00018400
   41     CONTINUE                                                 00018410
   42     LD1=LD1+NYS                                              00018420
   43     RETURN                                                  00018430
          END                                                     00018440
          SUBROUTINE MATOUT (L,NR,NC,XNAME)                       00018450
          IMPLICIT REAL*8(A-H,O-Z)                                00018460
          COMMON NWRD(620),NDS(240),VLIST(240),WORD(1000)         00018470
          IF(NR) 4,4,5                                            00018480
    5     WRITE (6,2) XNAME                                       00018490
          NRC=NR*NC+L-1                                           00018500
          DO 1 I=1,NR                                             00018510
          LL=L+I-1                                                00018520
```

```
 1      WRITE (6,3) (WORD(J),J=LL,NRC,NR)                                    00018530
 4      RETURN                                                               00018540
 2      FORMAT (/5X,A5,6HMATRIX/)                                            00018550
 3      FORMAT (2X,8G13.5/(18X,7G13.5))                                      00018560
        END                                                                  00018570
        SUBROUTINE TMPLY (LCA,LCB,LCC,LCD,NYC,NYS,LA,LC,MC)                  00018580
        IMPLICIT REAL*8(A-H,O-Z)                                             00018590
        COMMON NWRD(21),NSV,II(21),NSVL,NCR(576),NDS(240),VLIST(240)         00018600
        COMMON WORD(1000)                                                    00018610
        LA1=LCA-NSV-1                                                        00018620
        LB1=LCB-NSV-1                                                        00018630
        LC1=LCC-NYC-1                                                        00018640
        LD1=LCD-NYS-1                                                        00018650
        DO 3 J=1,MC                                                          00018660
        LC1J=LC1+J*NYC                                                       00018670
        LD1J=LD1+NYS*J                                                       00018680
        DO 3 I=1,LA                                                          00018690
        IJ=LD1J+I                                                            00018700
        SIJ=0.                                                               00018710
        DO 2 K=1,NSVL                                                        00018720
        IK=LA1+K*NSV+I                                                       00018730
        SKJ=0.                                                               00018740
        DO 1 L=1,LC                                                          00018750
        KL=LB1+L*NSV+K                                                       00018760
        LJ=LC1J+L                                                            00018770
 1      SKJ=SKJ+WORD(KL)*WORD(LJ)                                            00018780
 2      SIJ=SIJ+WORD(IK)*SKJ                                                 00018790
 3      WORD(IJ)=WORD(IJ)-SIJ                                                00018800
        RETURN                                                               00018810
        END                                                                  00018820
        BLOCK DATA                                                           00018830
        IMPLICIT REAL*8(A-H,O-Z)                                             00018840
        COMMON /DEVMOD/ TITLE(14),AXX(30,8),X11(30),X22(30),NLIB            00018850
        DATA NLIB/13/                                                        00018860
        DATA TITLE/14*1H /                                                   00018870
        DATA AXX/.001,120.,4.4,30.,7.4,62.,20.,30.,5.,30.,5.,               00018880
     1 55.,16.,17*0.,                                                       00018890
     2 2.3E-11,1.3E-11,2.5E-11,3.7E-11,2.5E-11,4.9E-12,                    00018900
     2 8.2E-12,4*2.E-12,3.7E-12,3.3E-12,17*0.,                             00018910
     3 1.,4.,4.,9.,8,6*1.,1.,1.,17*0.,                                     00018920
     4 .48,.49,.49,.41,.1,.29,.23,4*.5,.34,.1,17*0.,                       00018930
     5 4.2E-9,3.E-9,3.51E-9,3.24E-14,1.3E-13,5.11E-15,2.9E-14,             00018940
     5 1.85E-4,1.E-4,7.E-4,7.E-4,3.55E-15,2.03E-12,17*0.,                  00018950
     6 22.1,39.1,39.1,40.1,38.1,39.61,32.2,2.,-.1,-2.4,-.1,               00018960
     6 40.1,29.4,17*0.,                                                    00018970
     7 2.E7,.99491,.881,.9881,.7651,.977,.5,1.,.9999,1.,.9999,            00018980
     7 .9859,.2391,17*0.,                                                  00018990
     8 1.5E-4,2.91E-7,3.4E-5,3.043E-8,7.3E-5,1.16E-8,                     00019000
     8 7.245E-7,4.7E-9,246.1E-9,8*0./                                      00019010
        DATA X11/5H1N400,1H ,5H2N404,1H ,5H2N718,1H ,5H2N236,             00019020
     1 1H ,4HNMOS,1H ,4HPMOS,1H ,5H2N236/                                 00019030
        DATA X22/1H3,3HPNP,1H ,3HNPN,1H ,3HNPN,1H8,3HEMN,1H ,3HEMP,1H ,   00019040
     1 3HNPN,1H9/                                                         00019050
        END                                                                  00019060
        SUBROUTINE FBINV(N,X)                                                00019070
        IMPLICIT REAL*8(A-H,O-Z)                                             00019080
        DIMENSION L(40),M(50)                                                00019090
        COMMON NWRD(860),VLIST(240),WORD(1200)                               00019100
        NK=-N                                                                00019110
        DO 80 K=1,N                                                          00019120
        NK=NK+N                                                              00019130
        L(K)=K                                                               00019140
```

```
          M(K)=K                                                    00019150
          KK=NK+K                                                   00019160
          BIGA=WORD(KK)                                             00019170
          DO 20 J=K,N                                               00019180
          IZ=N*(J-1)                                                00019190
          DO 20 I=K,N                                               00019200
          IJ=IZ+I                                                   00019210
  10      IF(DABS(BIGA)-DABS(WORD(IJ)))15,20,20                     00019220
  15      BIGA=WORD(IJ)                                             00019230
          L(K)=I                                                    00019240
          M(K)=J                                                    00019250
  20      CONTINUE                                                  00019260
          J=L(K)                                                    00019270
          IF(J-K)35,35,25                                           00019280
  25      KI=K-N                                                    00019290
          DO 30 I=1,N                                               00019300
          KI=KI+N                                                   00019310
          HOLD=-WORD(KI)                                            00019320
          JI=KI-K+J                                                 00019330
          WORD(KI)=WORD(JI)                                         00019340
  30      WORD(JI)=HOLD                                             00019350
  35      I=M(K)                                                    00019360
          IF(I-K)45,45,38                                           00019370
  38      JP=N*(I-1)                                                00019380
          DO 40 J=1,N                                               00019390
          JK=NK+J                                                   00019400
          JI=JP+J                                                   00019410
          HOLD=-WORD(JK)                                            00019420
          WORD(JK)=WORD(JI)                                         00019430
  40      WORD(JI)=HOLD                                             00019440
  45      IF(BIGA)48,46,48                                          00019450
  48      DO 55 I=1,N                                               00019460
          IF(I-K)50,55,50                                           00019470
  50      IK=NK+I                                                   00019480
          WORD(IK)=WORD(IK)/(-BIGA)                                 00019490
  55      CONTINUE                                                  00019500
          DO 65 I=1,N                                               00019510
          IK=NK+I                                                   00019520
          HOLD=WORD(IK)                                             00019530
          IJ=I-N                                                    00019540
          DO 65 J=1,N                                               00019550
          IJ=IJ+N                                                   00019560
          IF(I-K)60,65,60                                           00019570
  60      IF(J-K)62,65,62                                           00019580
  62      KJ=IJ-I+K                                                 00019590
          WORD(IJ)=HOLD*WORD(KJ)+WORD(IJ)                           00019600
  65      CONTINUE                                                  00019610
          KJ=K-N                                                    00019620
          DO 75 J=1,N                                               00019630
          KJ=KJ+N                                                   00019640
          IF(J-K)70,75,70                                           00019650
  70      WORD(KJ)=WORD(KJ)/BIGA                                    00019660
  75      CONTINUE                                                  00019670
          WORD(KK)=1./BIGA                                          00019680
  80      CONTINUE                                                  00019690
          K=N                                                       00019700
  100     K=K-1                                                     00019710
          IF(K)150,150,105                                          00019720
  105     I=L(K)                                                    00019730
          IF(I-K)120,120,108                                        00019740
  108     JQ=N*(K-1)                                                00019750
          JR=N*(I-1)                                                00019760
```

```
          DO 110 J=1,N                                              00019770
          JK=JQ+J                                                   00019780
          HOLD=WORD(JK)                                             00019790
          JI=JR+J                                                   00019800
          WORD(JK)=-WORD(JI)                                        00019810
   110    WORD(JI)=HOLD                                             00019820
   120    J=M(K)                                                    00019830
          IF(J-K)100,100,125                                        00019840
   125    KI=K-N                                                    00019850
          DO 130 I=1,N                                              00019860
          KI=KI+N                                                   00019870
          HOLD=WORD(KI)                                             00019880
          JI=KI-K+J                                                 00019890
          WORD(KI)=-WORD(JI)                                        00019900
   130    WORD(JI)=HOLD                                             00019910
          GOTO 100                                                  00019920
    46    WRITE(6,1000)X                                            00019930
  1000    FORMAT(12H THE MATRIX ,A5,12H IS SINGULAR )               00019940
   150    RETURN                                                    00019950
          END                                                       00019960
```

(*Courtesy The Instruction and Research Computer Center, The Ohio State University, Columbus, Ohio.*

Index